Cultivation for Climate Change Resilience

Volume 2

Temperate Fruit Trees

Editors

Adel A. Abul-Soad

Tropical Fruit Research Department
Horticulture Research Institute, Agricultural Research Center
Giza, Egypt

Jameel M. Al-Khayri

Department of Agricultural Biotechnology
College of Agriculture and Food Sciences
King Faisal University
Saudi Arabia

CRC Press is an imprint of the
Taylor & Francis Group, an **informa** business
A SCIENCE PUBLISHERS BOOK

First edition published 2023
by CRC Press
6000 Broken Sound Parkway NW, Suite 300, Boca Raton, FL 33487-2742

and by CRC Press
4 Park Square, Milton Park, Abingdon, Oxon, OX14 4RN

CRC Press is an imprint of Taylor & Francis Group, LLC

Library of Congress Cataloging-in-Publication Data (applied for)

ISBN: 978-1-032-39736-8 (hbk)
ISBN: 978-1-032-39739-9 (pbk)
ISBN: 978-1-003-35115-3 (ebk)

DOI: 10.1201/9781003351153

Typeset in Palatino
by Radiant Productions

Preface

Awareness of the adverse impact associated with the global climate change on the future of agriculture, researchers are devoting efforts for finding solutions to mitigate undesired effects based on intelligent predictions and improved utilization of available genetic resources. This book highlights the contemporary knowledge of the impacts of abiotic and biotic stresses inflected by climate changes on the production, horticultural practices and physiological processes of various fruit tree species. Moreover, it describes the adaptation of innovative approaches to mitigate the climatic adverse effects and enhance resilience characteristics of fruit crops.

This book consists of 2-volume set. This Volume 2 subtitled Temperate Fruit Trees contains 11 chapters grouped in 3 parts. Part I consists of 2 chapters addressing conceptual framework emphasizing climate change impact on agriculture and food security and mitigation strategies. Part II describes the impact of climate change on some of warm temperate fruits species including citrus, common fig, cactus pear, date palm, grape, olive and pomegranate. Part III covers cool temperate fruit crops including apricot and plum.

Each chapter addresses one fruit crop and updates available information in relation to the various concepts. Topics discussed include a general introduction on climatic requirements of a fruit crop, significant symptoms of climate change impacts, crop management under changed climate conditions, natural adaptation of genetic resources, mitigation strategies against biotic and abiotic stresses, remote sensing and environmental certification, and concludes with future prospects and literature.

This book is essential for researchers and students concerned with improving the productivity and quality of fruit crops to achieve sustainable fruit cultivation and conservation of this important nutritional food source for future generations. This book is geared toward a variety of general readers, fruit farmers, research scholars and scientists interested in learning about the impact of climate changes on fruit cultivation. It is a valuable source of information for students to increase their awareness of fruit cultivation under climate change conditions.

Our sincere gratitude is presented to the contributing authors for their generous cooperation and much appreciation to the CRC Press for the opportunity to publish this book.

Adel A. Abul-Soad
Giza, Egypt

Jameel M. Al-Khayri
Al-Ahsa, Saudi Arabia

Contents

Part III: Cool Temperate Fruits

Part I

Conceptual Framework

1

Implications of Climate Change on Agriculture and Food Security

Adam Ahmed,[1,2,*] *Azharia Elbushra*[1,3] and *Mutasim Elrasheed*[1]

1. Introduction

1.1 Importance of Agriculture and Fruit Cultivation

Agricultural sector plays an important role in livelihood, food security and economic development for the majority of the world. This contribution is well documented in the literature. According to the Food and Agriculture Organization of the United Nation (FAO, 2008), about 36% of the total world labor force depends primarily for its livelihood on the state agricultural sector. However, contribution of workforce varies according to region and population density. For instance, it constitutes about 4.2% in Europe (EUROSTAT, 2018); 20% in Arab countries (World Bank, 2020); more than one-third in Southeast Asia (Zhai and Zhuang, 2009); 40–50% in highly populated countries of Asia and the Pacific (FAO, 2008); 48% for the total African population (NEPAD, 2013) and up to 66% in sub-Saharan Africa (FAO, 2008). Furthermore, a number of studies have addressed the

[1] Department of Agribusiness and Consumer Sciences, College of Agricultural and Food Sciences, King Faisal University, P.O. Box 400, Al-Ahsa 31982, Saudi Arabia.
[2] Department of Agricultural Economics, Faculty of Agriculture, University of Khartoum, P.O. Box 32, Postal Code 13314, Khartoum North, Sudan.
[3] Department of Agricultural Economics, College of Agriculture, University of Bahri, 12327, Postal Code 111111, Khartoum North, Sudan.
Emails: aaali@kfu.edu.sa; melrasheed@kfu.edu.sa
* Corresponding author: ayassin@kfu.edu.sa; adamelhag2002@yahoo.com

contribution of agriculture to labor force in different countries. According to EUROSTAT (2018) statistics, the contribution of agricultural sector to the total labor employment accounts for 23% in Romania, 17.5% in Bulgaria, 10.7% in Greece and 10.1% in Poland, and about 1% in Canada and Germany (World Bank, 2020). While the share of agriculture in the total employment in the African countries showed a wide spectrum, it is only one-fourth in Egypt, about half in Kenya, two-third in Ethiopia, two-fifth in Sudan and about three-quarter in Uganda. Similarly, it constitutes about 25, 29 and 37% in China, Indonesia and Pakistan, respectively (World Bank, 2020).

A report produced by the Food and Agricultural Organization (FAO) of the United Nations estimated that 2.5 billion out the 3 billion, who lived in the rural areas, depend on agriculture as the main source of livelihood (FAO, 2013). It is worth mentioning that the demand for agricultural products has increased three times from 1960 to date to meet the needs of the growing world population. Edame et al. (2011) stated that agricultural production needs to be increased by more than two-third to satisfy the food demands of the growing population by 2050.

Evidence-based research revealed that the percentage share of the agriculture sector to the gross domestic product (GDP) of the higher income countries is very small, indicating an inverse relation between national income and the percentage share of agriculture in GDP (FAO, 2019). As depicted in Table 1, the average agricultural contribution to GDP at the global level represented 3.4% in 2017 whereas it contributed about one-third in the developing countries (FAO, 2013).

The contribution of the agricultural sector to GDP is stable at around 4% since 2000 at the global level (FAO, 2021) and varies considerably among different countries and regions in the world. It amounted to 5.4% in USA (USDA, 2019); 15% in Africa, where it ranged from 3% in Botswana and South Africa to 50% in Chad (OECD-FAO, 2016); more than

Table 1. Percentage contribution of agricultural sector value added to the total Gross Domestic Product at global and regional level (2017–2018).

Regions	Year	
	2017	**2018**
Arab world	5.4	5.2
Europe and Central Asia	2.0	1.9
European Union	1.7	1.7
Middle East and North Africa	5.1	4.0
Sub-Saharan Africa	15.8	15.7
World	3.4	-

Source: World Bank, 2020

10% in Southeast Asia (Zhai and Zhuang, 2009); and less than 2% in Gulf countries (Woertz, 2017).

The main objective of this chapter is to describe the implications of climate change on agriculture and food security. It particularly reviews the effect of greenhouse gas (GHG) emissions, temperature, precipitation and the associated weather phenomenon, such as drought, dry and wet spells, floods and heat waves on food systems and food security pillars. Moreover, fruits' contributions to kilocalories, protein and fat supply as well as fruits' footprint of land use, fresh water withdrawal, greenhouse gases and eutrophying emissions are covered. In this context, Ritchie and Roser (2020) define eutrophying emissions as the 'runoff of excess nutrients into the surrounding environment and waterways, which affect and pollute ecosystems with nutrient imbalances and they are measured in kilograms of phosphate equivalents.'

This chapter is divided into five sections: Section one is concerned with the introduction, which highlights the importance of the agricultural sector and the basic concepts of food security and climate change issues; Section two covers interlinks among climate change, food systems and food security; the third section is devoted to fruits, climate change and food security, while climate change adaptation and mitigation measures are presented in Section four. Lastly, Section five is devoted to the conclusions drawn.

1.2 *Concept and Causes of Climate Change*

The concept of climate change has been addressed by many researchers and concerned agencies. According to a study conducted by FAO (2008), the numbers of climate change definitions have been stated. For instance, the World Metrological Organization (WMO) defines climate change as 'long-term changes in average weather conditions', while Global Climate Observing System (GCOS) refers to it as 'all changes in the climate system, including the drivers of change, the changes themselves and their effects' and the United Nation Framework Convention on Climate Change (UNFCCC) defines it as 'only human-induced changes in the climate system'. Contrarily to UNFCCC, the Intergovernmental Panel on Climate Change (IPCC) refers to climate change as not only changes induced by human activity but also by natural variability (IPCC, 2007a). It is worth noting that most of the experts stress the importance of differentiation between climate change and climate variability based on the time horizon. Accordingly, climate change is defined as a significant long-term variation in weather patterns, including global temperature, precipitation, wind, rainfall and snowfall in a certain place or a planet as whole (FAO, 2012; UCDAVIS, 2019; Takepart, 2020), while climate variability refers to short-term variation (daily, seasonal, inter-annual, several years) in the weather patterns (FAO, 2012).

On the one hand, climate change can be attributed to natural factors and human activities. The natural causes of climate change include among others, volcanic eruptions, changes in solar energy (EPA, 2016a), earth orbit and CO_2 level in the atmosphere (FAO, 2012). The United States Environmental Protection Agency (EPA) summarizes the main causes of climate change as the energy balance (between energy entering and leaving the earth); greenhouse effects, changes in the sun's energy and reflectivity (EPA, 2016a). On the other hand, human activities, including GHG emissions from burning fossil fuel, usually result in global warming (FAO, 2012). Relevant researches in the area indicate that burning of oil, gas and coal results in release of CO_2 into the air and hence, the rising planet temperature (Met. Office, 2020). For example, human activities increased the global greenhouse gas emissions by 70% during the period from 1970 to 2004 (IPCC, 2007c).

1.3 Concepts of Food Security and Food Security Pillars

The FAO (1996) defines food security as a 'situation that exists when all people, at all times, have physical, social, and economic access to sufficient, safe and nutritious food that meets their dietary needs and food preferences for an active and healthy life.' Whereas IPCC (2013) defines food security as 'a state that prevails when people have secure access to sufficient amounts of safe and nutritious food for normal growth, development, and an active and healthy life.' Additionally, FAO (2003) stated that 'food security depends more on socio-economic conditions than on agroclimatic ones, and on access to food rather than the production or physical availability of food.' It also stated that to evaluate the potential impacts of climate change on food security, 'it is not enough to assess the impacts on domestic production in food-insecure countries. One also needs to (a) assess climate change impacts on foreign exchange earnings; (b) determine the ability of food surplus countries to increase their commercial exports or food aid; and (c) analyze how the incomes of the poor will be affected by climate change.'

Based on the available literature, food security is decomposed into four pillars, namely: availability, accessibility, utilization and stability.

Food availability refers to the existence of food in a particular place at a particular time (Brown et al., 2015), which means that each individual in the country has a sufficient amount of food in terms of quantity and quality from the country's own production, or via imports, food aid and donation (Stamoulis and Zezza, 2003; Scialabba, 2011). It also indicates that there is adequate food for everyone at the household, community, regional, state and international levels (Clay, 2003). Food availability reflects the supply side of food security (Brown et al., 2015). The components of food security supply side include food production, distribution, exchange (Armstrong-Mensah, 2017) as well as stored and processed food (FAO, 2008).

Food accessibility refers to the individual's economical and physical ability to access appropriate and nutritious food (Schmidhuber and Tubiello, 2007; Brown et al., 2015). Furthermore, Gericke (2003) decomposed food accessibility into four dimensions: physical, economic, social and technological access. The physical access dimension is concerned with infrastructure, such as market, transportation and distribution facilities, while economic access refers to the ability of individuals to produce their own food and/or have enough money to purchase it. Furthermore, the culturally accepted food by the community reflects the social access dimensions, while technological access refers to the technological resources owned by the households for food preparation and/or preservation. The many factors that affect food accessibility and stability are: food price, market infrastructure, distance from market, income, consumer's preferences, self-sufficiency, transportation costs, socio-economic condition and technological facilities. Moreover, according to Brown and others (2015), food accessibility and stability are highly affected by changing climate variables.

Food utilization refers to how efficiently available and accessible food is utilized in order to obtain the adequate nutritional value required for individual needs. Food utilization is influenced by many factors, such as food safety, quality, consumption patterns and trends (Brown et al., 2015), food processing, storage and preservation facilities. In this context, some researchers emphasize the importance of incorporating non-food inputs, such as health when dealing with the concept of food utilization (Stamoulis and Zezza, 2003). Lastly, food stability refers to the sustainability of all or any one of the other food security dimensions (availability, accessibility and utilization) over time (Schmidhuber and Tubiello, 2007).

1.4 Concepts of Food Systems

Researchers have defined food systems as complex and connected activities including production, processing, transportation, consumption and food waste (FAO, 2008; Oxford University Press, 2017; FAO, 2018). According to Grubinger et al. (2010), the food system could be defined as 'an interconnected web of activities, resources and people that extends across all domains involved in providing human nourishment and sustaining health, including production, processing, packaging, distribution, marketing, consumption and disposal of food. The organization of food systems reflects and responds to social, cultural, political, economic, health and environmental conditions and can be identified at multiple scales, from a household kitchen to a city, county, state or nation.'

Each activity of the food system is further disaggregated into assets (FAO, 2008) and subsystems (FAO, 2018), which interact with each other to affect the food system activities and hence food security pillars. Further, breakdown of the food system activities, elements of food security pillars as well as outcomes and their interaction are well captured in the Global Environmental Change and Food Systems' (GECAFS) definition. They state that 'food systems encompass (i) activities related to the production, processing, distribution, preparation and consumption of food; and (ii) the outcomes of these activities contributing to food security (food availability, with elements related to production, distribution and exchange; food access, with elements related to affordability, allocation and preference; and food use, with elements related to nutritional value, social value and food safety). The outcomes also contribute to environmental and other securities (e.g., income). Interactions between and within biogeophysical and human environments influence both the activities and the outcomes' (GECAFS, 1994).

The FAO (2008) differentiates between food systems and food chain. The former tackles the issue from a holistic approach consisting of simultaneously interacting food activities, while, the latter refers to linear sequential activities that happen for people to get their food.

To fulfil the purpose of this chapter, climate change and food security, a framework document is used to highlight the impact of climatic change factors (environmental) on food systems and their effect on food security pillars.

2. Interlinks Among Climate Change, Food Systems and Food Security

Sustainable agriculture and food systems are important in achieving food security. However, agriculture and food systems are threatened by climate change. Accordingly, mitigation and adaptation strategies that are used to address the impact of the climate change (MDGs 13) are essential for achieving the Millennium Development Goals (MDGS): no poverty (MDGs 1), zero hunger (MDGs 2) and clean water and sanitation (MDGS 6). Available evidence declared that 924 million peoples were severe food insecure in 2020 (FAO, IFAD, UNICEF, WFP and WHO, 2021), while during 2012–14, approximately 805 million people were food insecure at the globe, where 98% of them lived in the developing countries (FAO, IFAD and WFP, 2014), and at least 2 billion live with inadequate nutritious food (Pinstrup-Andersen, 2009). Overweight and obesity reached 2.1 billion in 2013 compared to only 857 thousand in 1980 (Ng et al., 2014).

Agriculture is the main sector for providing food to the growing population in the world. Agricultural production is generally governed by the availability of resources, such as arable land, water, production inputs, technologies, human resources and the environment. Interlink between agriculture production and climate change is widely acknowledged in the literature. Climate change risk factors, such as extreme changes in temperature and precipitation, wind and other extreme weather events and their associated phenomenon; negatively influence agricultural production and hence food availability. Moreover, these environmental factors exert influence on food accessibility, utilization and their stability over time. These weather events-associated phenomena include among others, heat waves, dry spells, shifting seasons, biodiversity, drought, deforestation, desertification, soil fertility, erosion, flood, rising sea level, melting glacier, water scarcity, pollution and land scarcity.

The impact of climate change on food production, processing, distribution, consumption and waste disposable could be summarized in a wide range of effects at the local, national, regional and global levels. Its impact on production activity is manifested in agricultural yield (crops, livestock, forestry, poultry and fisheries) and labor productivity fluctuation; changing cropping patterns and irrigation water requirements; and soil types. It is worth stating that most agricultural products are characterized by seasonality in production. This necessitates performing storage and processing functions in order to sustain food security all through the year. The effect of climate change on storage and processing activity is illustrated in terms of needs for energy and chemical requirement as well as the availability of sufficient and hygienic water. A report released by FAO (2008) declared that the impact of climate change on global food systems could be reflected not only via influencing market price of food processing, storing, transporting and production, but also through food price.

The distribution activity requires the presence of facilities in terms of vehicles, paved roads, railways, cargo airplane, ships to deliver agricultural products from production to consumption centers. Any disruption in the distribution activity channels as a consequence of climate change might result in increase of agricultural products' marketing costs and hence negatively influence the food security situation. In this regard, McGuirk and others (2009) stated that efficiency and effectiveness of land, water and air transportation facilities are prone to climate change adverse events and are ultimately reflected in higher transportation costs.

The effect of changing temperature, precipitation, as well as other extreme weather events affect food consumption activity in different ways. For instance, high temperature may result in reducing the shelf-life of perishable agricultural products, increasing water consumption and energy requirements for storage facilities. Moreover, rainfall fluctuation might hamper agricultural production and in turn reduce the availability of diversified food to satisfy the consumer preference for acquiring healthy and nutritious food.

Different argumentations might be initiated with reference to food systems and food security determinants other than climate change factors. Some researchers summarized these factors into technological and structural changes in food system's activities; population growth, socioeconomic and demographic changes; changes in food consumption patterns; disasters and changes in both energy availability and usage (Brown et al., 2015); conflict and economic downturn (FAO, IFAD, UNICEF, WFP and WHO, 2021) as well as the incidence of pandemics, such as Coronavirus.

2.1 *Food Systems and Greenhouse Gas Emissions*

The contribution of the agriculture sector to the global emissions from human activities, during the period from 2007 to 2016, were estimated to be 13, 44 and 82% of carbon dioxide, methane and nitrous oxide, respectively. These contributions collectively accounted for 23% of total net anthropogenic GHG emissions (IPCC, 2019). In a study focused on trend in global GHG emissions conducted by Olivier and his colleagues (2005), it was found that agricultural contribution to the global man-made methane emissions constitutes about 43%, out of which, animal fermentation; rice farming and animal waste and Savanah burning contributed about 25, 12 and 6%, respectively. Moreover, they confirm that agriculture is the main source of origin of global anthropogenic nitrous oxide emissions. It contributes about 38% from animal waste, grazing and manure; 13% from crop cultivation; 12% fertilizers and 19% from indirect agricultural sources.

Based on FAO report that focused on climate change, agriculture and food security, changing land use systems and land degradation contributed significantly to the total GHG emissions (FAO, 2016). One example of it is that about 5 billion metric tons of CO_2eq to the atmosphere was produced because of deforestation of peat-lands during the periods 2005–2017. In this context, only 20 countries accounted for more than 66% of the total agricultural GHG emissions, out of which, only four countries (China, India, Brazil and the United States of America) shared more than 50% of the total agricultural GHG emissions, with Asia being the biggest contributor (FAO, 2019).

Worldwide food systems accounted for about more than one-third (34%) of the GHG emissions (Crippa et al., 2021). According to IPCC, the contribution of the food systems to GHG emissions in term of CO_2 and non-CO_2 gases was from different agriculture sources. These sources are:

within the farm gate (10–12%); land use and land use change dynamics (8–10%); and food supply chains activities beyond farm gate (5–10%), ending up with a total food system share of 25–30% (IPCC, 2019; Poore and Nemecek, 2018). Moreover, scientists stated that food systems, including crop and livestock production, are responsible for 19–29% of the global anthropogenic GHG emissions, out of which the contribution of agricultural production might reach up to 86% (Sonja et al., 2012).

In this context, Table 2 summarizes the global agricultural sector emissions of CO_2eq (CO_2eq), from CH_4 and (CO_2eq) from NO_2 (in million Giga grams) during the period 1961–2017. Asia contributes the highest

Table 2. Global agricultural emissions of CO_2eq, CH_4 and NO_2 in million Giga grams (1961–2017).

Regions	**Years**											
	1961–70		**1971–80**		**1981–90**		**1991–2000**		**2001–2010**		**2011–2017**	
	value	**%**	**Value**	**%**	**value**	**%**	**value**	**%**	**value**	**%**	**value**	**%**
Emissions (CO_2eq)												
Africa	2.6	8.6	3.1	8.7	3.8	9.4	6.2	13.5	7.6	15.5	6.2	16.8
Americas	7.7	25.7	9.5	26.4	10.4	25.5	11.4	24.8	12.6	25.8	9.2	25.1
Asia	10.9	36.1	12.8	35.6	15.3	37.5	19.0	41.3	20.9	42.6	16.0	43.4
Caribbean	0.2	0.7	0.2	0.6	0.2	0.6	0.2	0.5	0.2	0.5	0.2	0.5
Europe	7.6	25.2	9.0	25.0	9.7	23.9	7.3	15.9	5.9	12.1	4.0	10.9
Oceania	1.3	4.4	1.5	4.3	1.5	3.7	2.1	4.5	2.0	4.1	1.4	3.9
Emissions (CO_2eq) from CH_4												
Africa	1.5	7.4	1.8	7.8	2.1	8.6	3.2	12.0	3.9	14.3	3.3	16.1
Americas	5.0	24.6	5.9	25.5	6.3	25.4	6.7	25.4	7.4	26.8	5.3	25.8
Asia	0.1	0.7	0.1	0.6	0.1	0.6	0.1	0.6	0.1	0.5	0.1	0.5
Caribbean	8.3	40.5	9.2	39.8	10.1	40.5	11.5	43.4	12.2	44.1	9.1	44.8
Europe	4.8	23.5	5.3	22.9	5.5	22.0	4.0	15.0	3.0	11.0	2.0	9.7
Oceania	0.8	4.0	0.9	4.0	0.9	3.5	1.1	4.1	1.1	3.8	0.7	3.6
Emissions (CO_2eq) from NO_2												
Africa	1.1	11.1	1.3	10.3	1.7	10.6	3.0	15.4	3.6	17.0	2.9	17.7
Americas	2.7	27.8	3.6	27.8	4.0	25.4	4.7	23.9	5.2	24.4	4.0	24.1
Asia	0.1	0.7	0.1	0.6	0.1	0.6	0.1	0.4	0.1	0.4	0.1	0.4
Caribbean	2.6	26.6	3.6	27.8	5.2	32.7	7.5	38.3	8.7	40.4	6.8	41.4
Europe	2.8	28.4	3.7	28.7	4.3	26.8	3.3	17.0	2.9	13.4	2.0	12.3
Oceania	0.5	5.4	0.6	4.7	0.6	3.9	1.0	5.0	0.9	4.4	0.7	4.2

Source: Calculated from FAOSTAT, 2020

share with an increasing trend of CO_2 emissions in the world as it produced 36% during the sixties and reaching up to 43.4% during 2011–2017. Similarly, Africa's share of the global CO_2 emissions show an increasing trend as it was almost double during the period 1961–2017. Americas are ranked second in terms of CO_2 emissions; however, its contribution is almost constant during the entire period, ranging from 24.8 to 26.4%. In contrast, Europe's share to the global CO_2 emissions show a declining trend as the emissions decreased from 25% in the sixties to less than 11% in 2011–2017.

On the other hand, although Caribbean shows the lowest share of CO_2eq emissions in the world, it had the highest share of CO_2eq emissions from methane (CH_4), ranging from 39.8 to 44.8% for the period 1961–2017. Americas ranked second in CO_2eq emissions from methane followed by Europe. However, a constant trend of CO_2eq emissions from methane is noticed for Americas and a decreasing one for Europe, where it decreased from 23.5% in 1961–70 to only 9.7% in 2011–17 of the global emissions.

Pertaining to CO_2eq emissions of NO_2, Europe, Americas and Caribbean collectively represented more than three-quarters of the global emissions. It was noticed that Americas and Europe showed a decreasing trend in the percentage share of global CO_2eq emissions from NO_2 over time (1961–2017). However, Caribbean showed an increasing trend during the same period (Table 2).

On the other hand, the negative impacts of the elevating CO_2 level and its consequences in global warming can be manifested in increasing the potential effect of insects, weeds and pests in damaging agricultural production (U.S. Global Change Research Program, 2009a) and reducing crops absorption of some essential nutrients such as zinc and iron (Myers et al., 2014). In contrast, the positive impact can be depicted in flourishing biomass and improving crops yields (FAO, 2008; Cho, 2018).

2.2 *Food Systems and Temperature*

The impact of rising temperature on climate change is well documented in the literature. Globally, rise in temperature will result in significant economic losses (Gurdeep et al., 2021). The rise or decline in the mean temperature influenced agricultural production, food processing and consumption (FAO, 2008) and consequently food security. Its impact on production can be witnessed in the decline of plants' and animals' productivities; change in crop and livestock pattern; lower labor productivity and increase in pest and diseases infestation thresholds. These impacts are widely tackled by researchers, for instance, EPA (2016a) declared that temperature plays a crucial role in the determination of plant and animal types that survive in certain locations on the planet. In this sense, FAO (2016) reported that rising temperatures and frequent occurrences of extremely dry and wet years in sub-Saharan Africa were expected to have a negative effect on crop and livestock production.

On the same lines, some researchers mentioned that the occurrence of heat waves adversely affected livestock fertility; dairy cows' productivity (Cho, 2018); and increase in production and capital expenditure (U.S. Global Change Research Program, 2009b). Moreover, livestock become prone to parasite infestations and diseases (Cho, 2018).

Temperature could have direct and indirect impacts on agricultural production. The direct effect is noticed in yield reduction as response to temperature changes beyond the optimal level. For instance, IPCC (2007b) modelling studies indicated that slight increase in mean temperature ranging between 1–3°C accompanied with an increase in the atmospheric carbon dioxide (CO_2) and changing rainfall are expected to improve crop productivity in the temperate zones. Furthermore, in low-latitude areas, moderate increases in the mean temperature between 1–2°Centigrade are likely to cause decline in the major cereal crops, where warming of more than 3°Centigrade negatively affects all regions (IPCC, 2007b). According to the National Research Council (2011), any global warming by 1°Celsius may result in 5–10% changes in both precipitation across many regions and stream flow of rivers; 5–15% decrease in crop productivities; and increase in the severity of summers and ocean acidity.

Labor plays an important factor in the agricultural sector, particularly in developing countries, where agriculture is a labor-intensive activity. Agricultural labor productivity is expected to be influenced by economic, social and environmental factors. The economic factors include all factors that affect agricultural labor efficiency, such as wages, labor availability, skills and assets ownerships. On the other hand, social factors include demographic characteristics and dependency ratio. The environmental factors influencing agricultural labor are weather conditions, such as temperature, precipitation and other extreme events (wind patterns, snow, floods). The impact of extreme changes in temperature is expected to seriously decrease labor productivity. Thus, the consequences of reduction in labor productivity might have a severe effect on production costs, food availability and soaring food priced, ending up with reduction in access to food.

Regarding food processing and storage, any increase in temperature may cause an additional cost in food storage, distribution and transportation. This could be attributed to the additional need of energy requirements for cool transportation and storage. Absence of appropriate storage facilities and high temperature-related effects might not only result in drastic increase in postharvest losses due to storage affected by pests' attacks, but also lowering in food quality. Ultimately, due to the seasonality of many agricultural products, particularly vegetables and fruits, well-equipped and proper processing and storage facilities might improve food quality preservation, food safety, longer shelf-life and lower food disposal, thus playing a significant role in food security pillars.

One of the main challenges of achieving global food and nutrition security is the provision of enough supply of nutritive, healthy and accessible food. Food processing and associated marketing functions play an important role in achieving food security. Food processing is the process of transformation of raw agricultural products into food ready for consumption (Lund, 2003). Agro-processed food improves foods' nutritional ingredients and quality. Moreover, agro-processing reduces post-harvest losses and thus conserves healthy food for the consumer.

Based on the available literature, a number of studies addressed the impact of rising temperatures and changing weather events on food systems. McGuirk and others (2009) state one example—increasing temperature and the occurrence of extreme weather events usually influence food distribution systems at different levels. Others state that the delivery of safe and quality food necessitates the use of proper food processing, storing, packaging and transporting facilities. They also added that any obvious increase in temperature as a consequence of climate change would lead to food poisoning and spoilage, which require improved cold storage facilities (James and James, 2010).

In discussing the concept of food system, food consumption is defined as the processes that relates to food utilization and consumption. It is broadly concerned with food processing; household's decision-making regarding food distribution practices and choices; as well as individual access to health care, sanitation and knowledge (FAO, 1997). The exposure to extreme temperature alters an individual's food consumption and increases incidences of heat-related diseases, such as heat stroke and dehydration particularly for agricultural outdoor labor, thereby resulting in lower labor productivity.

2.3 *Food Systems and Altered Precipitation Patterns*

Precipitation means any liquid or frozen water that constituted in the atmosphere and falls back to the Earth in different shape such as rain, snow and sleet (Graham et al., 2010). Rainfall is one of the main environmental factors that influence agricultural production, particularly rain-fed crops, where about 60 and 90% of the stable food are produced under rain-fed globally and in sub-Saharan Africa, respectively (Savenije, 2001). Furthermore, globally, rain-fed agriculture constitute about 83% of all cultivated land and three fifth of all food (FAO, 2002). Shortage and/or extreme rainfall affect agriculture and livestock production through different phenomenon such as the occurrence of droughts, dry and wet spells, heavy precipitations and floods. The frequency, severity and duration of these phenomena determine the expected impact level on food security and the livelihoods of the population. Drought and floods adversely affect access to clean water for drinking and agricultural uses in different ways. For instance, drought might generate water stress problem due to water scarcity, while flood might lead to water contamination.

Drought could be defined as the situation in which the lack of precipitation causes a prolonged period of dry weather, which result in water shortages (EPA, 2016b). While the United Nations Convention to Combat Desertification (UNCCD, 1994) define it as 'drought means the naturally occurring phenomenon that exists when precipitation has been significantly below normal recorded levels, causing serious hydrological imbalances that adversely affect land resource production systems.'

The risk of drought varies across the countries depending on the degree of severity (moderate, severe, extreme and exceptional drought), thus most countries are susceptible to different levels of damages and losses in crops, pastures, and water sources. On the other hand, dry spell is defined as the deficit of precipitation for a short period, while, the occurrence of intensive periods of precipitation is known as wet spell (EPA, 2016b). It is worth mentioning that, the definition of dry spell required the determination of different threshold values for the successive dry days relative to location, crop types and researcher's aims. Whereas, heavy precipitation is the situation in which the amount of precipitation in a certain place considerably exceeds the normal level (EPA, 2016b). Lastly, flood can be defined as a situation in which dry land is changed into wetland or open water resulting from climate change in form of heavy rains and/or increasing sea water level (EPA, 2016b).

The expected impact of drought, dry & wet spells, heavy precipitations and flood could be depicted in soil moisture contents, soil fertility, cropping pattern, length of the growing season, pasture, biodiversity, and pest & disease infestations. Consequently, this might significantly influence food production, storing, processing and consumption activities and eventually food security.

On the production side, the occurrence of precipitation phenomenon such as heavy precipitation, wet and dry spells and drought might change cropping pattern, interrupt crops growth and increase uncertainty of crops and livestock production. However, it is worth mentioning that, shortages of rainfall might necessitate the need for supplementary irrigation as water become scarcer to fulfil crops water requirements.

Drought and dry spell might negatively affect food availability through affecting crops' yields and quality, soil fertility, and might even lead to complete crop loss. Moreover, their impact on livestock is manifested in overgrazing, water stress, decrease in livestock yield and may even cause animal death. Some researchers argued that incidences of drought and dry spell cause changes in forests and other ecosystems, resulting in deforestation and desertification (Olagunju, 2015). In this context, a study investigating the impact of drought on crop production losses, during the period 1983–2009, found that about 75% of the harvested areas from maize, rice, soybeans, and wheat worldwide are subject to yield loss amounting to about USD 166 billion due to drought (Kim et al., 2019). Moreover, a study focused on climate change and grazing herds in Ethiopia concluded that the intensity and durations of drought not only affect animal stocking rates, but also forage production (Godde et al., 2019). Drought also usually affects rural farmers' income generated from agricultural production as they resort to felling of trees for charcoal and wood as a source of income, thereby aggravating the severity of desertification.

The impact of climate change events, such as heavy precipitation, drought and dry and wet spells, might be extended to not only affecting agricultural and livestock production activities but also disrupting international food trade and increasing the number of households to face food insecurity and malnourished; and hence increase the need for food aid.

On the other side, flooding, which results in the conversion of arable lands that is used for food production and as livestock pasture into wet land or open water, will negatively impact agricultural activities, food prices and hence, food security. The situation is more obvious in countries located near the oceans or in locations facing heavy precipitation. For instance, South Asian countries, such as Bangladesh, India and Pakistan are more prone to severe, extreme and frequent floods due to climate changes, such as rise in temperatures and sea levels (Mirza, 2010). Moreover, the pasture carrying capacity of non-flooded grazing land gets seriously affected by overgrazing. This situation of overgrazing is usually accompanied with widespread livestock diseases and pest infestations which in turn lower the quality and quantity of livestock products, thereby affecting the contribution of animal-origin products to the total dietary energy supply (DES) of the majority of the population in the affected areas. Moreover, overgrazing accompanied with the occurrence of drought results in a high livestock mortality rate in Mongolia (Nandintsetseg, 2018).

Another related adverse effect of drought and dry spells is obvious as it causes agricultural labor to migrate to urban centers, resulting in skilled agricultural labor resorting to small trading and marginal activities. This is in line with the findings of Nawrotzki et al. (2017) who reported that the probability of rural-urban migration would increase by 3.6 per cent because of a monthly increase of drought.

Drought, dry and wet spells and precipitation might impair farmers' assets as well as public infrastructure concerned with food production, storage, possessing and distribution facilities, thereby exacerbating the problem of food availability and accessibility. Moreover, in such circumstances, consumers may shift their food consumption patterns into less preferable and non-diversified food as a coping mechanism to climate change and ultimately influence food utilization. A study conducted by Carpena (2019) to assess the effect of drought on food spending and macronutrient consumption in rural India revealed a shift of household consumption pattern from highly nutritive food into lower ones. Further, he concluded that household incomes generated during drought is the main cause of consuming lower-quality food. Another study concerned with droughts and famine in Kenya attributed the drought-triggered food insecurity factors to not only rainfall but also assets and labor constraints; poor policy implementation and poverty (Speranza, 2008).

2.4 *Food System and Extreme Weather Events*

Climate change is sometimes associated with the occurrence of extreme weather events, such as storms, high winds and frequent floods. The impact of these events varies across locations according to their degree of frequency and intensity. Extreme weather events influence households' human, social, financial, physical and natural assets. For instance, high winds and storms may destroy the physical productive assets, such as crops, livestock, buildings, stores and equipment. The implication of these events would be reflected in the reduction of food availability; diversity and quality; rise in food prices; increase in water pollution; increase in pest infestations and diseases; disruption in food supply chains; destruction of infrastructure facilities; decline in soil fertility; deterioration of human health and nutritional status; changing land-use pattern; threat to household's income sources and hence negatively affect food availability, accessibility and utilization (FAO, 2008).

In this regard, agriculture, rangelands and livestock in the Near East are expected to be seriously affected by climate change in terms of rising temperatures and the associated events, such as drought, floods and soil degradation. The impact varies across countries; for instance, the national production of rice and soya beans in Egypt will decrease by 11% and 28%, by the year 2050, respectively, relative to the current situation (FAO, 2014). In the same vein, the occurrence of severe heat waves in Russia 2010–11 resulted in huge grain crop loss, amounting to 30% of the total production. Likewise, the severe floods that happened in the 1980's led to a considerable reduction in the cultivated land area (14%) (Yadav, 2019).

3. Fruits, Climate Change and Food Security

3.1 *Role of Fruits in Enhancing Food Security*

Fruits and vegetables is one of the major five food groups that plays a crucial role in food security. Their consumption is considered as an important source of vitamins, minerals and fibers, which may reduce the risk of heart disease and certain types of cancer (Pérez, 2002), lower occurrences of cardiovascular diseases and obesity (Slavin and Lloyd, 2012). Moreover, the collective contribution of fruits, nuts and vegetables as sources of vitamins in the U.S. diet amounts to approximately more than 90% of vitamin C, about half of vitamin A, more than one-quarter of vitamin B6, and less than one-fifth each of thiamine and niacin (Kader, 2001).

Worldwide, the average fruit supply amounted to 75 kilograms per capita per year during the period 2014–2017. However, it varied across regions, ranging from 64 kg/capita/year in Africa to 166 kg/capita/year in Caribbean (Table 3). The recommended amount of fruits and vegetables consumption per capita/year is estimated to be 146 kg/year (WHO/FAO, 2003). Table 3 revealed that fruit consumption in Africa was less than the world average, while Asia, Oceania and Europe consumes an amount that was almost similar to the world average. However, Caribbean and Americas consumed 2.2 and 1.3-fold of world average, respectively. Similarly, the percentage share of fruit to food supply (kcal/capita/day), protein and fat supply (gram/capita/day) showed slight variability among all regions except Caribbean, where fruit contributed about 8% of food supply, 3.5 and 3.8% of protein and fat supply, respectively.

Table 3. Fruit Food Supply (FS) (kg/capita/yr) and macronutrient contribution at global and regional levels (2014–2017).

	FS(kg/capita/yr)	FS (kcal/c/d)		Protein (g/c/d)		Fat (g/c/d)	
		Value	%	Value	%	Value	%
World	75	99	3.4	1.2	1.4	0.7	0.8
Africa	64	105	4.0	1.2	1.8	0.5	1.0
Americas	100	129	3.9	1.5	1.6	1.2	1.0
Caribbean	166	229	8.2	2.4	3.5	2.8	3.8
Asia	71	89	3.2	1.0	1.3	0.6	0.8
Europe	78	108	3.2	1.3	1.2	0.8	0.6
Oceania	72	105	3.2	1.1	1.1	1.1	0.8

Source: FAO, 2020

3.2 *Impact of Climate Change on Fruit Crops*

A recent study that addresses the environmental impact of food production stated that agriculture represents about half of the global habitable land, out of which, more than three-quarters (77%) is used for meat and dairy production, while crops occupy less than one-quarter. Although livestock used more than three-quarters of the agricultural land, it contributed less than one-fifth (18%) and lesser than two-fifth (37%) of the global calorie and protein supply, respectively. Moreover, agriculture used more than two-third (70%) of fresh water and produced more than three-quarters of the global ocean and fresh-water pollutants (Ritchie and Roser, 2020).

Table 4 illustrated the footprints of land use, freshwater withdrawal, GHG and eutrophying emissions for select crops, mainly fruits, cereal and nuts. For comparison purposes, the footprints were measured in kilograms of food products, 100 grams of protein and 1000 Kilocalories for each of the selected crops.

With reference to the land-use footprint, all fruits used less land per square meters to produce 1 kg of food products as compared to nuts and cereals crops. Contrarily, in general, fruits used more area for supplying both 100 grams of proteins and 1000 Kcal compared to cereals and nuts.

On the other hand, fresh water withdrawal footprint refers to the amount of fresh water in liters required to produce one kg of food products or for supplying 100 grams of proteins and 1000 Kcal. It is worth mentioning that fruits, such as citrus, banana, apple, berries and grapes withdraw less amounts of water to produce one kg of food products compared to nuts, rice and wheat. The highest amount of water required for supplying 100 grams of proteins was reported for banana, followed by berries and grapes, with least water withdrawal for wheat. On the other hand, berries and grapes ranked first with reference to the amount of fresh water withdrawal for supplying 1000 Kcal which is relatively similar to nuts and rice.

In the same vein, GHG emission footprint refers to the amount of GHG emissions in Kg of CO_2eq required to produce 1 kg of product, 100 grams of proteins and 1000 Kcal. Table 4 shows that GHG emissions from producing 1 kg of rice outnumbered all the selected crops. In other words, GHG emissions generated from producing 1 kg of rice is equivalent to that produced from 5.7 kg banana, 10 kg apple and 13.3 kg citrus fruits.

According to Ritchie and Roser (2020), eutrophying emissions is defined as 'runoff of excess nutrients into the surrounding environment and waterways, which affect and pollute ecosystems. They are measured in grams of phosphate equivalents (PO_4eq).' The ecosystem pollutants in grams of PO_4eq generated from the production of one kg of rice corresponds to the pollutants resulting from the production of 24 kg of citrus, 16 kg of apple, 11 kg of banana and 6 kg of berries and grapes. In fact, fruits are perishable products, which require proper handling at pre- and post-harvest processes. Accordingly, they are highly susceptible to climate change. For instance, high temperature is associated with additional costs in transportation and storages facilities. According to Moretti and others

Table 4. Footprint for land use, freshwater withdrawal, and GHG and eutrophying emissions of fruit species compared to field crops in 2018.

Crop	Land use (m²)			Fresh water (Liter)			GHG emissions (Kg Co_2eq)			Eutrophying emissions (g)		
	Kg product	100 g protein	1000 kcal	kg of product	100 protein	1000 kcal	Kg product	100 protein	1000 kcal	KG of product	100 g protein	1000 kcal
Citrus	0.9	14.3	2.7	83	1378	258	0.3	6.5	1.2	2.2	37.3	7.0
Banana	1.9	21.4	3.2	115	1272	191	0.7	9.6	1.4	3.3	36.6	5.5
Nuts	13.0	7.9	2.1	4134	2531	672	0.3	0.3	0.1	19.2	11.7	3.1
Apple	0.7	21	1.3	180	6003	375	0.4	14.3	0.9	1.5	48.3	3.0
Berries and grapes	2.4	24.1	4.2	420	4196	736	-	15.3	2.7	6.1	61.2	10.7
Wheat and rye	3.9	3.2	1.4	648	531	242	1.4	1.3	0.6	7.2	5.9	2.7
Rice	2.8	3.9	0.8	2248	3167	610	4.0	6.3	1.2	35.1	49.4	9.5
Maize	2.9	3.1	0.7	216	93	48	1.0	1.8	0.4	4.0	4.2	0.9

Source: Ritchie and Roser (2020)

(2010), high levels of temperature, CO_2 and O_3 have direct and indirect impact on both quantity and quality of fruits and vegetables.

4. Adaptation and Mitigation to Climate Changes

Scientists stress the importance of adaptation and mitigation and differentiate between them as strategies for addressing climate change. They define any strategy concerned with the reduction of sources and amount of GHG emissions as mitigation, whereas adaptation is referred to any 'adjustment in natural or human systems in response to actual or expected climatic stimuli or their effects, which moderates harm or exploits beneficial opportunities' (IPCC, 2001). Moreover, Tol (2005) distinguished between them according to spatial (global vs. district and region) and temporal (long-term vs. current and short-terms) scales and economic sectors (energy, transportation vs. urban planning, water, agriculture). Researchers differentiate between them according to their causal association; hence, mitigation tackles the root causes of climate change, whereas adaptation addresses the effects of climate change (CIFOR, 2011; Duguma et al., 2014).

Climate change effect is widely noticed at global, regional, country, sector, local and household levels. Accordingly, different measures for adaptation and mitigation are necessary. These measures vary according to the frequency, intensity, duration and severity of the climatic change events. In this sense, different adaptation and mitigation strategies and policies are used to mitigate the impact of climate change on agriculture and food security. In general, these might include, among others, (a) food supply, demand and any disruption in food chain; (b) access to services, such as credit, insurance and extension services; (c) soil and water management strategies; and (d) livelihood and agricultural and non-agricultural income activities.

Some of the climate change mitigation and adaptation strategies are usually used to improve the efficiency of agricultural products through use of different measures. These include adoption of improved and high-yielding varieties that are tolerant to climate change events; use of modern technologies; use of clean energy sources, such as solar energy; activation of agricultural extension agents; and encouragement to agricultural research and development. Moreover, provision of agricultural services in term of credit and insurance would increase and sustain agricultural production and lighten the adverse effect of climate change. Additionally, soil management strategies might be applied to reduce the climate change's negative effect on soil fertility, soil moisture contents, soil erosion and waterlogged land.

The adaptation and mitigation strategies not only cover the agricultural production side but also extend to cover other food system activities (processing, storing, distributions, consumption and disposal) to elevate the adverse effects of climate change and hence realize sustainable food security (Brown et al., 2015). Moreover, different countries adopted different adaptation and mitigation strategies to modify food demand and these might include loss and waste reduction and recycling; taxes and subsidies; awareness campaign.

In addition to this, additional adaptation and mitigation measures for achieving sustainable fruits production could be highlighted, based on the fruit's long life cycle which makes them more prone to environmental changes relative to short-duration crops. These measures include among others, diversification of cropping pattern; selection of crops varieties that suit the environment (Sarkar et al., 2021); and implementing precise strategies for water management (Mukhopadhyay and Mandal, 2021).

Fruits' long life cycles make them more prone to environmental changes relative to short-duration crops. This necessitates additional adaptation and mitigation measures for achieving sustainable fruit production. The measures include, among others, diversification of cropping pattern; selection of crops varieties that suit the environment (Sarkar et al., 2021); and implementation of precise strategies for water management (Mukhopadhyay and Mandal, 2021).

5. Conclusion and Prospects

Climate change risk factors, such as extreme changes in GHG emissions, temperature, precipitation, wind and other extreme weather events and their associated phenomena negatively influence the agricultural sector, food systems and hence, food security. The agricultural sector contributes significantly to global warming as it generates about one-fourth of the total net anthropogenic GHG emissions. Analysis of the global agricultural emissions of CO_2eq during the period 1961–2017 revealed:

- Asia has the highest share (36–43%) with increasing trends.
- Americas ranked second with almost constant trends (24.8–26.4%).
- Africa displays an increasing trend (8.6–16.8%).
- Europe shows a declining trend (25–11%).

Rising temperatures in the tropical zone negatively influence the food systems. The impact of changing temperatures on agricultural production can be manifested in the decline of plants' and animals' productivities; change in crop and livestock patterns; lower labor productivity and increase in pest infestations and diseases. According to the National Research Council (2011), any global warming by 1°Celsius may result in 5–10% changes in both precipitation across many regions and stream flow of rivers; 5–15% decrease in crop productivities; and increase in the severity of summers and ocean acidity. On the other hand, food systems are also affected by precipitation events, such as heavy precipitation, drought, flood and dry and wet spells. These effects could lead to disruption in international food trade, impairment of farmers' assets, changes in food consumption patterns, and eventually rendering the number of households insecure and malnourished by way of food.

With reference to land use, fresh water withdrawal and GHG emissions footprints:

- All fruits used less land in square meters to produce 1 kg of food products compared to nuts and cereals crops. Contrarily, they used more area for supplying both 100 grams of proteins and 1000 Kcal compared to cereals and nuts.
- Some fruits, such as citrus, banana, apple, berries and grapes withdraw less amounts of water to produce one kg of food products as compared to nuts, rice and wheat.
- GHG emissions generated from producing 1 kg of rice are equivalent to that produced from 5.7 kg banana, 10 kg apple and 13.3 kg citrus fruits.

Based on the aforementioned conclusion, adaptation and mitigation measures should be implemented to reduce the adverse implications of climate change on food security.

References

Armstrong-Mensah, E.A. (2017). *Global Health: Issues, Challenges, and Global Action.* John Wiley & Sons. https://www.wiley.com/enus/Global+Health%3A+Issues%2C+Challenges%2C+and+Global+Action-p-9781119110217.

Brown, M.E., Antle, J.M., Backlund, P., Carr, E.R., Easterling, W.E., Walsh, M.K., Ammann, C., Attavanich, W., Barrett, C.B., Bellemare, M.F., Dancheck, Funk, V.C., Grace, K., Ingram, J.S.I., Jiang, H., Maletta, H., Mata, T., Murray, A., Ngugi, M., Ojima, D., O'Neill, B. and Tebaldi, C. (2015). *Climate Change, Global Food Security and the U.S. Food System.* 146 pp. http://www.usda.gov/oce/climate_change/FoodSecurity2015Assessment/FullAssessment.pdf.

Carpena, F. (2019). How do droughts impact household food consumption and nutritional intake? A study of rural India. *World Development*, 122: 349–369. https://doi.org/10.1016/j.worlddev.2019.06.005.

Cho, R. (2018). *Agriculture and Climate: How Climate Change will Alter Our Food, State of the Planet.* Earth Institute, Colombia University. https://blogs.ei.columbia.edu/2018/07/25/climate-change-food-agriculture/.

Clay, E. (2003). Chapter 2. *Food Security: Concepts and Measurement in Food Reforms and Food Security: Conceptualizing the Linkages Expert Consultation.* FAO. http://www.fao.org/3/y4671e/y4671e06.htm#bm06.

Crippa, M., Solazzo, E., Guizzardi, D. et al. (2021). Food systems are responsible for a third of global anthropogenic GHG emissions. *Nat. Food*, 2: 198–209. https://doi.org/10.1038/s43016-021-00225-9.

Duguma, L.A., Wambugu, S.W., Minang, P.A. and Noordwijk, M.V. (2014). A systematic analysis of enabling conditions for synergy between climate change mitigation and adaptation measures in developing countries. *Environmental Science & Policy*, 42: 138–148. http://dx.doi.org/10.1016/j.envsci.2014.06.003.

Edame, G.E., Anam, B.E., Fonta, W.M. and Duru, E. (2011). Climate change, food security and agricultural productivity in Africa: Issues and policy directions. *International Journal of Humanities and Social Science*, 1(21): 205–223. https://www.ijhssnet.com/journals/Vol_1_No_21_Special_Issue_December_2011/21.pdf.

EPA. (2016a). *Causes of Climate Change*. United States Environmental Protection Agency. https://19january2017snapshot.epa.gov/climate-change-science/causes-climate-change_.html.

EPA. (2016b). *Climate Change Indicators in the United States: Drought*. United States Environmental Protection Agency. https://www.epa.gov/sites/production/files/2016-08/documents/print_drought-2016.pdf.

EUROSTAT. (2018). *Agriculture, Forestry and Fishery Statistics*. Luxembourg: Publications Office of the European Union. https://ec.europa.eu/eurostat/documents/3217494/9455154/KS-FK-18-001-EN-N.pdf/a9ddd7db-c40c-48c9-8ed5-a8a90f4faa3f.

FAO. (1996). *Technical Background Document Executive Summary*. World Food Summit, 13–17 Nov. 1996, Rome, Italy. Available at: http://www.fao.org/3/w2612e/w2612e00.htm.

FAO. (2002). Crops and Drops: Making the Best Use of Water for Agriculture. Rome, Italy. *In*: Brown, M.E., Antle, J.M., Backlund, P., Carr, E.R., Easterling, W.E., Walsh, M.K., Ammann, C., Attavanich, W., Barrett, C.B., Bellemare, M.F., Dancheck, V., Funk, C., Grace, K., Ingram, J.S.I., Jiang, H., Maletta, Mata, H.,T., Murray, A., Ngugi, M., Ojima, D., O'Neill, B. and Tebaldi, C. (2015). *Climate Change, Global Food Security, and the U.S. Food System*. Available online at http://www.usda.gov/oce/climate_change/FoodSecurity2015Assessment/FullAssessment.pdf.

FAO. (2003). World agriculture: Toward 2015/2030, Chapter 13, Rome, *Earthscan*. http://www.fao.org/3/a-y4252e.pdf.

FAO. (2008). *Climate Change and Food Security: A Framework Document*. http://www.fao.org/docrep/010/k2595e/k2595e00.htm.

FAO. (2012). *Climate Change and Food Security, Food and Agricultural Organization of the United Nations e-Learning Courses*. This course is funded by the European Union's Food Security Thematic Programme and implemented by the Food and Agriculture Organization of the United Nations. Available online: https://elearning.fao.org/course/view.php?id=143.

FAO. (2013). *FAO Statistical Yearbook*. World Food and Agriculture Organization. http://www.fao.org/3/i3107e/i3107e.pdf.

FAO. (2014). *World Food and Agriculture – Statistical Year Book 2014*. Rome. http://www.fao.org/3/i3591e/i3591e.pdf.

FAO. (2016). *The State of Food and Agriculture, Climate Change, Agriculture and Food Security Report*. Available at: http://www.fao.org/3/a-i6030e.pdf; accessed 2 June 2020.

FAO. (2018). *A Sustainable Food System Conceptual and Framework*. Food and Agriculture Organization, Rome. http://www.fao.org/3/ca2079en/CA2079EN.pdf.

FAO. (2019). *Statistical Pocketbook 2019*. World Food and Agriculture Organization, Rome. http://www.fao.org/3/ca6463en/CA6463EN.pdf.

FAO. (2021). *Statistical Yearbook 2021*. Food and Agriculture Organization of the United Nations, Rome. https://www.fao.org/3/cb4477en/online/cb4477en.html#chapter-1.

FAO, IFAD and WFP. (2014). *The State of Food Insecurity in the World: Strengthening the Enabling Environment for Food Security and Nutrition*. Rome, Italy. Retrieved from http://www.fao. org/3/a-i4030e.pdf.

FAO, IFAD, UNICEF, WFP and WHO. (2021). *The State of Food Security and Nutrition in the World: Transforming Food Systems for Food Security, Improved Nutrition and Affordable Healthy Diets for All*. Rome, FAO. https://doi.org/10.4060/cb4474en.

FAOSTAT. (2020). http://www.fao.org/faostat/en/#data/FBS. Accessed 2 June 2020.

GECAFS. (1994). About GECAFS. *In: FAO 2008: Climate Change and Food Security: A Framework Document*. http://www.fao.org/docrep/010/k2595e/k2595e00.htm.

Gericke, G.J. (2003). Better eating for better health: principles and practices of planning a healthful diet. Schönfeldt, H.C. (ed.). *Fundamentals of Nutrition Security in Rural Development: Graduate Readings*, vol. 3, University of Pretoria, Pretoria, 7 pp.

Godde, C., Dizyee, K., Ash, A., Thornton, P., Sloat, L., Roura, E., Henderson, B. and Herrero, M. (2019). Climate change and variability impacts on grazing herds: Insights from a system dynamics approach for semi-arid Australian rangelands. *Glob. Change Biol.*, 25: 3091–3109. Doi: 10.1111/gcb.14669.

Graham, S., Parkinson, C. and Chahine, M. (2010). *The Water Cycle*. NASA Earth Observatory. Accessed on 19 June 2020. https://earthobservatory.nasa.gov/features/Water.

Grubinger, V., Berlin, L., Berman, E., Fukagawa, N., Kolodinsky, J., Neher. D., Parsons, B., Trubek, A. and Wallin, K. (2010). University of Vermont Transdisciplinary Research Initiative Spire of Excellence Proposal: Food Systems. Proposal, Burlington, University of Vermont. *In:* Chase, L.P and Grubinger, V. (2014). *Introduction to Food Systems*. In the book, Chase, L. (ed.). *Food, Farms and Community: Exploring Food Systems*. University Press of New England. https://muse.jhu.edu/book/36007; https://www.cifor.org/fileadmin/fileupload/cobam/ENGLISH-Definitions%26ConceptualFramework.pdf.

Gurdeep, S., Manpreet, M. and Prashant K. (2021). Impact of climate change on agriculture and its mitigation strategies: a review. *Sustainability*, 13(3): 1318. https://doi.org/10.3390/su13031318.

IPCC. (2001). Climate Change 2001, Synthesis Report, Cambridge University Press. *In*: CIFOR (2011). *Climate Change and Forests in the Congo Basin Synergies between Adaptation and Mitigation in a Nutshell*. Center for International Forestry Research. Retrieved in 30 May 2020, 19:30 GMT.

IPCC. (2007a). *Climate Change 2007 - The Physical Science Basis Contribution of Working Group I to the Fourth Assessment Report of the IPCC*. ISBN 978 0521 88009-1 Hardback; 978 0521 70596-7 Paperback. https://www.ipcc.ch/site/assets/uploads/2018/05/ar4_wg1_full_report-1.pdf.

IPCC. (2007b). *Climate Change 2007: Impacts, Adaptation and Vulnerability*. Contribution of Working Group II to the Fourth Assessment Report of the IPCC. https://www.ipcc.ch/site/assets/uploads/2018/03/ar4_wg2_full_report.pdf.

IPCC. (2007c). *Climate Change 2007 – Mitigation of Climate Change*. Contribution of Working Group III to the Fourth Assessment Report of the IPCC. https://www.ipcc.ch/site/assets/uploads/2018/03/ar4_wg3_full_report-1.pdf.

IPCC. (2013). Glossary. *In: Climate Change 2014: Impacts, Adaptation, and Vulnerability*. Annex II, Contribution of Working Group II to the Fifth Assessment Report of the Intergovernmental Panel on Climate Change, Cambridge University Press, Cambridge and New York, NY, p.13. Available at www.ipcc.ch/report/ar5/wg2/.

IPCC. (2019). *IPCC Special Report on Climate Change, Desertification, Land Degradation, Sustainable Land Management, Food Security, and Greenhouse Gas Fluxes in Terrestrial Ecosystems*. IPCC Intergovernmental Panel for Climate Change. https://www150.statcan.gc.ca/n1/en/pub/82-003-x/2001003/article/6103-eng.pdf?st=qASktFSv.

James, S.J. and James, C. (2010). The food cold-chain and climate change. *Food Res. Int.*, 43: 1944–1956. https://www.sciencedirect.com/science/article/abs/pii/S0963996910000566. doi:10.1016/j.foodres.2010.02.001.

Kader, A. (2001). Importance of fruits, nuts, and vegetables in human nutrition and health, Department of Pomology, UC Davis, No. 106. *Perishables Handling Quarterly*. https://ucanr.edu/datastoreFiles/234-104.pdf.

Kim, W., Iizumi, T. and Nishimori, M. (2019). Global patterns of crop production losses associated with droughts from 1983 to 2009. *Journal of Applied Meteorology and Climatology*, 58: 1233–2144. Doi: 10.1175/JAMC-D-18-0174.1.

Lund, D. (2003). Predicting the impact of food processing on food constituents. *Journal of Food Engineering*, 56: 113–117. https://dokumen.tips/download/link/predicting-the-impact-of-food-processing-on-food-constituents.

McGuirk, M., Shuford, S., Peterson, T.C. and Pisano, P. (2009). Weather and climate change implications for surface transportation in the USA. *WMO Bulletin*, 58(2): 84–93. https://public.wmo.int/en/bulletin/weather-and-climate-change-implications-surface-transportation-usa.

Met. Office. (2020). *Weather and Environment, Causes of Climate Change*. https://www.metoffice.gov.uk/weather/climate-change/causes-of-climate-change; accessed 19 May 2020.

Mirza, M.M.Q. (2011). Climate change, flooding in South Asia and implications. *Reg. Environ Change*, 11: 95–107. https://doi.org/10.1007/s10113-010-0184-7.

Moretti, C.L., Mattos, L.M., Calbo, A.G. and Sargent, S.A. (2010). Climate changes and potential impacts on postharvest quality of fruit and vegetable crops: A review. *Food Research International*, 43: 1824–1832. Doi:10.1016/j.foodres.2009.10.013.

Mukhopadhyay, S. and Mandal, A.K. (2021). Impact of climate change on groundwater resource of India: A geographical appraisal. pp. 125–154. *In*: Md. N. Islam and Amstel, A. (eds.). *India: Climate Change Impacts, Mitigation and Adaptation in Developing Countries*. Springer Nature, Switzerland AG, 2021. https://doi.org/10.1007/978-3-030-67865-4.

Myers, S.S., Zanobetti, A., Kloog, I. et al. (2014). Increasing CO_2 threatens human nutrition. Nature, 510: 139–142. *In*: John Hopkins Center of a Livable Future (undated), *Food System Premier – Food and Climate Change*. Retrieved on 30 May 2020; 16:30 GMT. http://www.foodsystemprimer.org/food-production/food-and-climate-change/.

Nandintsetseg, B., Shinoda, M. and Erdenetsetseg, B. (2018). Contributions of multiple climate hazards and overgrazing to the 2009/2010 winter disaster in Mongolia. *Nat. Hazards*, 92: 109–126. https://doi.org/10.1007/s11069-017-2954-8.

National Research Council. (2011). *Climate Stabilization Targets: Emissions, Concentrations, and Impacts over Decades to Millennia*. Washington, DC: The National Academies Press. https://doi.org/10.17226/12877.

Nawrotzki, R.J., DeWaard, J., Bakhtsiyarava, M. and Ha, J.T. (2017). Climate shocks and rural-urban migration in Mexico: Exploring nonlinearities and thresholds. *Clim. Change*, 140(2): 243–258. Doi:10.1007/s10584-016-1849-0.

NEPAD. (2013). *Agriculture in Africa, Transformation and Outlook*. NEBAD transforming Africa. https://www.un.org/en/africa/osaa/pdf/pubs/2013africanagricultures.pdf.

Ng, M., Fleming, T., Robinson, M., Thomson, B., Graetz, N., Margono, C., and Gakidou, E. (2014). Global, regional, and national prevalence of overweight and obesity in children and adults during 1980–2013: A systematic analysis for the Global Burden of Disease Study 2013. *Lancet*, 384(9945): 766–781. https://www.thelancet.com/action/showPdf?pii=S0140-6736%2814%2960460-8.

OECD/FAO. (2016). *OECD-FAO Agricultural Outlook 2016–2025*. OECD Publishing, Paris. http://dx.doi.org/10.1787/agr_outlook-2016-en. http://www.fao.org/3/a-i5778e.pdf.

Olagunju, T.E. (2015). Drought, desertification and the Nigerian environment: A review. *Journal of Ecology and the Natural Environment*, 7(7): 196–209. Doi: 10.5897/JENE2015.0523.

Olivier, J.G.J., Aardenne, J.A.V., Dentener, F.J., Pagliari, V., Ganzeveld, L.N. and Peters, J.A.H.W. (2005). Recent trends in global greenhouse gas emissions: Regional trends

1970–2000 and spatial distribution of key sources in 2000. *Environmental Sciences*, 2: 2–3, 81–99. DOI: 10.1080/15693430500400345.

Oxford University Press. (2015). *What is the Food System?* Oxford Martin Programme on the future of food. Available on line at https://www.futureoffood.ox.ac.uk/what-food-system; last access 15 June 2020.

Pérez, C.E. (2002). Fruit and vegetable consumption. *Health Reports*, Statistics Canada, Catalogue No. 82-003, 13(3). https://www150.statcan.gc.ca/n1/en/pub/82-003-x/2001003/article/6103-eng.pdf?st=qASktFSv.

Pinstrup-Andersen, P. (2009). Food security: Definition and measurement. *Food Security*, 1: 5–7. https://link.springer.com/content/pdf/10.1007/s12571-008-0002-y.pdf.

Poore, J. and Nemecek, T. (2018). Reducing food's environmental impacts through producers and consumers. *Science*, 360(6392): 987--992. *In*: Ritchie, H. and Roser, M. (2020). *Environmental Impacts of Food Production*. published online at OurWorldInData.org. https://ourworldindata.org/environmental-impacts-of-food.

Ritchie, H. and Roser, M. (2020). *Environmental Impacts of Food Production*. published online at Our World in Data.org. Retrieved from: https://ourworldindata.org/environmental-impacts-of-food.

Sarkar, T., Roy, A., Choudhary, S.M. and Sarkar, S.K. (2021). Impact of climate change and adaptation strategies for fruit crops. pp. 79–98. *In*: Md. N. Islam and Amstel, A. (eds.). *India: Climate Change Impacts, Mitigation and Adaptation in Developing Countries*. Springer Nature, Switzerland AG, 2021. https://doi.org/10.1007/978-3-030-67865-4.

Savenije, H.H.G. (2001). The Role of Green Water in Food Production in sub-Saharan Africa. FAO, the Netherlands. *In*: Alam, M., Toriman, M., Siwar, C. and Talib, B. (2010). Rainfall variation and changing pattern of agricultural cycle. *American Journal of Environmental Sciences*, 7(1): 82–89. https://www.researchgate.net/publication/279407992_Rainfall_Variation_and_Changing_Pattern_of_Agricultural_Cycle.

Schmidhuber, J. and Tubiello, F.N. (2007). Global food security under climate change. *Proceedings of the National Academy of Sciences (PNAS) of the United States of America*, PNAS, December 11, 2007, 104(50): 19703–19708. https://doi.org/10.1073/pnas.0701976104.

Scialabba, N.E. (2011). *Food Availability and Natural Resource Use FAO/OECD*. Expert Meeting on Greening the Economy with Agriculture Paris, 5–7 September Natural Resources Management and Environment Department, FAO. http://www.fao.org/fileadmin/user_upload/suistainability/Presentations/Availability.pdf.

Slavin, J.L. and Lloyd, B. (2012). Health benefits of fruits and vegetables. *Advances in Nutrition*, 3(4): 506–516. https://doi.org/10.3945/an.112.002154.

Sonja, J.V., Campbell, B.M. and Ingram, J.S.I. (2012). Climate change and food systems. *Annu. Rev. Environ. Resour.*, 37: 195–222. Doi:10.1146/annurev-environ-020411-130608.

Speranza, C., Kiteme, B. and Wiesmann, U. (2008). Droughts and famines: The underlying factors and the causal links among agro-pastoral households in semi-arid Makueni district, Kenya. *Global Environmental Change*, 18(1): 220–233. https://doi.org/10.1016/j.gloenvcha.2007.05.001.

Stamoulis, K. and Zezza, A. (2003). A Conceptual Framework for National Agricultural, Rural Development and Food Security Strategies and Policies, *ESA Working Paper*, No. 03–17 November 2003, FAO. Available at http://www.fao.org/3/ae050e/ae050e00.pdf. accessed 2 June 2020.

Takepart. (2020). *What is Climate Change?* http://www.takepart.com/flashcards/what-is-climate-change/ accessed 19 May 2020.

Tol, R.S.J. (2005). Adaptation and mitigation: Trade-offs in substance and methods. *Environmental Science and Policy*, 8(6): 572–578. *In*: CIFOR (2011). *Climate Change and Forests in the Congo Basin Synergies between Adaptation and Mitigation in a Nutshell*. Center for International Forestry Research. Retrieved in 30 May 2020, 19:30 GMT. https://

www.cifor.org/fileadmin/fileupload/cobam/ENGLISH-Definitions%26ConceptualFramework.pdf.

UCDAVIS. (2019). *Science and Climate*. Available online: https://climatechange.ucdavis.edu/science/climate-change-definitions/.

UNCCD. (1994). *United Nations Convention to Combat Desertification in Countries Experiencing Serious Drought and/or Desertification, particularly in Africa (UNCCD)*. https://www.jus.uio.no/english/services/library/treaties/06/6-02/combat-desertification.xml#:~:text=%22drought%22%20means%20the%20naturally%20occurring,d.

U.S. Global Change Research Program. (2009a). Global Climate Change Impacts in the United States, Cambridge, New York, Melbourne, Madrid, Cape Town, Singapore, São Paulo, Delhi: Cambridge University Press. *In*: John Hopkins Center of a Livable Future (2020). *Food System Premier – Food and Climate Change*. Retrieved on 30 May 2020; 16:30 GMT. http://www.foodsystemprimer.org/food-production/food-and-climate-change/.

U.S. Global Change Research Program. (2009b). *Global Climate Change Impacts in the United States*, Karl, T.R., Melillo, J.M. and Peterson, T.C. (eds.). Cambridge University Press, 2009. http://www.iooc.us/wp-content/uploads/2010/09/Global-Climate-Change-Impacts-in-the-United-States.pdf.

WHO/FAO. (2003). *Expert Report on Diet, Nutrition and the Prevention of Chronic Diseases*, Technical Report Series 916, https://apps.who.int/iris/bitstream/handle/10665/42665/WHO_TRS_916.pdf;jsessionid=D2551BA86C8E19B8A0A03AB71FDAD003?sequence=1.

Woertz, E. (2017). Agriculture and development in the wake of the arab spring. *International Development Policy, Revue Internationale de Politique de développement* [Online], 8.1 | 2017, Online since 12 February 2017, connection on 23 February 2017. URL: http://poldev.revues.org/2274. Doi: 10.4000/poldev.2274.

World Bank. (2020). *Agriculture, Forestry, and Fishing, Value Added (% of GDP)*. https://data.worldbank.org/indicator/NV.AGR.TOTL.ZS. Accessed 2 June 2020.

Yadav, S.S., Hegde, V.S., Habibi, A.B., Dia, M. and Verma, S. (2019). Climate change, agriculture and food security. pp 1–24. *In*: Yadav, S.S., Redden, R.J., Hatfield, J.L., Ebert, A.W. and Hunter, D. (eds.). *Food Security and Climate Change*. first edition, 2019, John Wiley & Sons Ltd., published 2019 by John Wiley & Sons Ltd. https://www.researchgate.net/publication/329870678_Climate_Change_Agriculture_and_Food_Security.

Zhai, F. and Zhuang, J. (2009). *Agricultural Impact of Climate Change: A General Equilibrium Analysis with Special Reference to Southeast Asia*, ADBI Working Paper 131, Tokyo: Asian Development Bank Institute. Available: https://www.adb.org/sites/default/files/publication/155986/adbi-wp131.pdf.

2

Instigating Adaptation and Mitigation Strategies to Combat Impact of Global Climate Change in Fruit Crops

Rinny Swain,[1,3] *Jannela Praveena*,[2] *Mamata Behera*[3] and *Gyana Ranjan Rout*[1,*]

1. Introduction

Climate change is the chief concern and an inevitable threat to sustainable development of agriculture worldwide. The surge in surface mean temperature leading to melting of glaciers and the rise in sea level are the allusion of worldwide global warming and the newest challenge for human community in the 21st century (Ayala et al., 2020). As per the Intergovernmental Panel on Climate Change (IPCC, 2019) report, till 2015 the mean land surface air temperature had increased by 1.53°C more than the global mean surface (land and ocean) temperature (GMST), which is

[1] Department of Agricultural Biotechnology, College of Agriculture, Odisha University of Agriculture & Technology, Bhubaneswar- 751003, Odisha, India; presently working at Crop Improvement Division, School of Agriculture, GIET University, Gunupur-765022, Rayagada, Odisha, India.

[2] Department of Fruit Science and Horticulture Technology, College of Agriculture, Odisha University of Agriculture & Technology, Bhubaneswar-751003, Odisha, India.

[3] Crop Improvement Division, School of Agriculture, GIET University, Gunupur-765022, Rayagada, Odisha, India.

Emails: swain.rinny12@gmail.com; jannela.praveena17@gmail.com

* Corresponding author: grrout@rediffmail.com

0.87°C. In 2015, after the Paris Agreement convention, all governments worldwide agreed to limit global warming below 2°C by putting in efforts to limit GMST to 1.5°C, thereby enhancing adaptation to the adverse impacts of climate change. According to the projected Intergovernmental Panel on Climate Change report (IPCC), there is a high probability of 10–40% loss in crop production by 2080–2100 due to global warming in India. Many of the commercial tropical, sub-tropical and temperate fruits will perform in a poor and unpredictable manner due to climate change. With an increase in average global temperature, several commercial fruit varieties have changed or shifted to new latitudinal belts with favorable climatic zones. Therefore, the crops that are used to be productive in one particular area may no longer be so, or the other way round. Climate change is believed to be a variation of climate over a long period of time and is credited directly or indirectly to human activities that have altered the total composition of the global atmosphere. Climate change continues to create challenges to life and livelihoods worldwide, causing changes in water quantity and quality, and shifts in geographical areas, seasonal activities, migration patterns, changes in species abundance and interactions between many terrestrial as well as freshwater or marine species with more negative than positive impacts on the yields of most crops (IPPC, 2021). Climatic variability has occurred mostly due to enormous emissions of greenhouse gases rapidly since the 1900s. The change in the atmosphere has severely affected agriculture by causing changes in crop physiology, biochemistry, and floral biology, biotic and abiotic stresses. This has eventually resulted in the reduction of yield and quality of crop plants. The climate change has led to loss of vigor, early or late flowering, low fruit-bearing capability, reduction in fruit size, reduction in juice content, less color, reduced shelf-life, and escalating pest attacks, consequentially resulting in low production and poor quality of fruit crops. An economic literature was also published recently, exploring the impacts of climate change on farm yields in several parts of the world (Chandio et al., 2020). Mitigation strategies are essential measures to reduce the devastating consequences of climate change. Also, the negative impacts of climate change can be controlled through implementation of adaptation strategies that are relevant, robust and effortlessly operated by scientists and farmers. In view of the rising concern at worldwide climate change, this chapter focuses on the imminent impact of changing climate on the horticultural crops, especially fruit crops and strategies to lessen the risk of crop failure and minimize monetary loss involved in crop production.

1.1 Global Climate Change

After a vivid analysis of parameters like ice cores, glacier lengths, ocean sediments, tree rings, pollen remains and studding the change in earth's orbit movement around the sun, scientist have listed out prime explanations of climate change (Wuebbles et al., 2017). Natural causes (like continental drift, earth's tectonic tilts, intensity of solar radiation and reflectivity, volcanic activities, ocean current and variation in CO_2 concentration), anthropogenic activities (leading to emission of Green House Gases, GHGs and reflectivity of sun) and land use changes (like deforestation and urbanization) are the contributing factors.

Natural causes: Even been considered as the contributing factors of climate change, they are not assumed to be the reason of the rise in mean temperature or global warming now.

Continental drift was hypothesized to be one of the foremost factors disturbing climate in past. It indicates the changes in the earth's orbit and its axis of rotation which showed that continents in earth is gradually moving through geological time over a deep-seated viscous zone. It was observed that the coldest part of the last glacial period (or ice age) was about 11°F colder and the peak of the last interglacial period was at most 2°F warmer than it is today respectively (Wuebbles et al., 2017).

Earth's tectonic tilts, revolution and rotation are other factors that influence the climate change. Earth currently has an axial tilt of about 23.44°, which affects the climate change and causes changes in the seasons.

Intensity of solar radiation and reflectivity is another factor that affect the mean surface temperature. Earth's climate hinge on the subtle balance amid incoming solar radiation, outgoing thermal radiation as well as the composition of Earth's atmosphere. The changes in the sun's energy output affects the intensity of the sunlight that can reach the earth's surface and influence the earth's climate. Studies showed that there is no net surge in the sun's output, even after the global mean surface temperatures have risen (NAS, 2020). Also, among the total light received 70% is absorbed by land, ocean, and atmosphere, while around 30% of the solar energy that are reflected back into space (Fahey et al., 2017).

Volcanic activities arises with a crack in the Earth crust and allows the gushing hot volcanic ash as well as gases to escape from magma below the surface of Earth crust. In case of explosive or violent volcanic eruptions large quantity of CO_2 is released into environment and SO_2 into the upper atmosphere, which can reflect abundant sunlight back to the space to cool the surface of the planet for quite long years space (Fahey et al., 2017). Hence, it affects the global climate by reducing the volume of solar radiation reaching the Earth's surface, sinking temperatures in the troposphere, emitting large quantity of GHGs and altering atmospheric air circulation patterns.

Ocean current are caused by continuous flow of ocean water, surface winds, partly by the temperature and salinity gradients, and earth's rotation. They act as a conveyor line indulge in transporting warm water and precipitation from the equator toward the poles and cold water from the poles back to the tropics. Thus, they regulate global climate, helping to counteract the effect of uneven distribution of solar radiation reaching Earth's surface

Variation in Carbon Dioxide Concentrations is a continuous process over the past numerous years and it changes with glacial cycles. Through the warm interglacial periods, CO_2 levels were high, whereas during cool glacial periods its levels were lower. The alternate heating or cooling of the earth's surface and oceans disrupt the balance of natural source and sink of gases, and enhances GHGs concentrations in the atmosphere (EPA, 2022a).

Anthropogenic activities: After the Industrial Revolution, the concentrations of greenhouse gases have drastically augmented due to human actions. Gas concentrations like Carbon dioxide (CO_2), methane (CH_4), nitrous oxide (NO_2), and water vapor had become more abundant in the earth's atmosphere than any other period in the last 800,000 years (NAS, 2020). Such gases create a layer over the atmosphere, causing trapping of sunlight and hence increases temperature.

Carbon dioxide released in environment absorbs infrared radiation emitted from earth's surface, thus trapping heat and instigating surface mean temperature. Recently, it was estimated that the human activities accounts over 30 billion tons of CO_2 into the atmosphere every year (Hayhoe et al., 2018).

Methane primarily contributors to the formation of ground-level ozone layer, a hazardous air toxin pollutant and which enhance 1 million premature deaths every year globally (EPA, 2022a).

The rise in global mean surface temperature (GMST) is mostly attributed to the alarming increase in the concentration of greenhouse gases (GHGs), like carbon dioxide (CO_2), methane (CH_4), nitrous oxide (N_2O), sulfur dioxide (SO_2) and chlorofluorocarbons (CFC) primarily due to accelerated growth of industrialization worldwide (Singh, 2010). The increase in GHGs has occurred as a result of manmade interventions, like coolants in industrial enterprises, excessive automobiles, production of energy-intensive agro-chemicals including fertilizers that has led to ozone depletion. The global GHG emissions reached 31.2% and above in 2016 when compared with the GHG level, at an average annual increase of 0.9% since 2010 (UNFCCC, 2019). As per the United Nations Framework

Convention on Climate Change (UNFCC), sectors that contributed a major share in GHG emissions during 2016 are energy (34%), industry (22%) and transport (14%). In addition, about 23% of net emissions of GHG was from anthropogenic activities like agriculture, forestry and other land use accounting to 13% of CO_2, 44% of CH_4 and 82% of N_2O emissions globally during 2007–2016. After fossil fuel, land-use change, forest deforestation and degradation are the biggest emitters of CO_2 (Baumert et al., 2009). The CO_2 emission is primarily accountable for 77% of global warming over the period of past 100-years and hence is the most significant aspect of GHG emissions (Climate Analysis Indicators Tool, 2011). Khan et al. (2021) reported that energy use, CO_2 emissions, labor force, harvested land and fruit crop growth have a long-term co-integrated relationship. The long-term CO_2 emission and rural population have negative influence on fruit crop production. They suggested that 1% rise in rural population and CO_2 emissions will reduce fruit crop production by –0.59% and –1.97% in the long term.

The International Kyoto Protocol, set out and adopted in 1997 for binding emission reduction commitments to industrialized countries, came into force in 2005. In 2012, a second commitment to the Doha Amendment was adopted and which pertained to a new period up to 2020. Finally, in December 2015, the UNFCC adopted the Paris Agreement that came into force on November 2016 with an outcome to develop a legal instrument applicable to all parties involved in reducing the GHG emissions and which would be implemented from 2020. It was decided to build climate resilience and foster climate finance flows consistent with a pathway towards low emission of GHGs. The increase in GMT in relation to industrial development involved the threat of desertification, water scarcity and land degradation. All these events can ultimately cause instability between crop yield and food supply, leading to food insecurities. Such an adverse climate change will also aggravate land degradation processes through increase in rainfall, flooding, frequent droughts, heat stress, dry spells, strong wind, rise in sea level and high tides, permafrost thaw with gravely devastating outcomes. The enhanced land surface air temperature, evapotranspiration and decreased rainfall in interaction with climate variability and anthropogenic activities are mainly accountable for increased desertification in some dry land areas, like Sub-Saharan Africa, parts of East and Central Asia, and Australia (UNFCCC, 2019). To combat desertification, activities like climate change adaptation with mitigation as well as halting biodiversity loss through sustainable development, are being carried out. Avoidance and reduction of desertification would enhance soil fertility, increase carbon storage in soil and biomass, and provide benefit to agricultural productivity and food security. It is preferable to attempt to restore degraded land for better

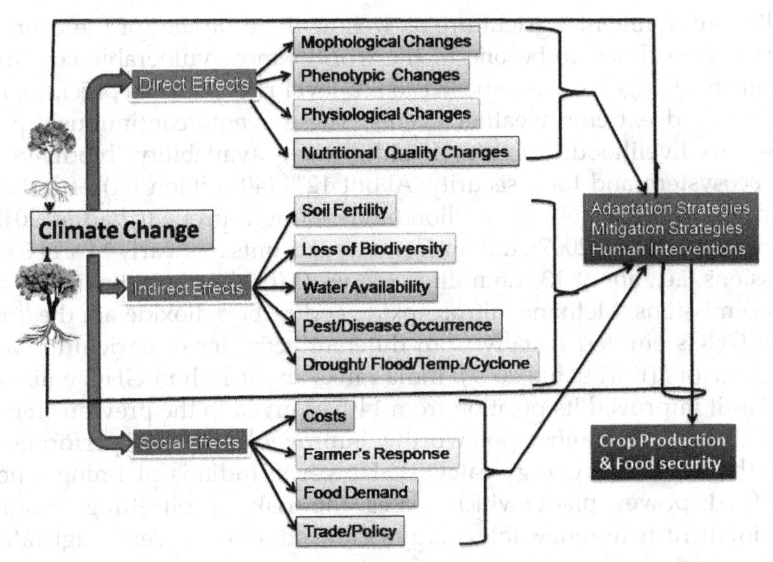

Fig. 1. Diagrammatic representation of climate change impact on agriculture.

utilization and to reduce pressure on land. On the other hand, IPCC has estimated that the sea level in 2100 will be 40 cm higher than what it is today and nearly 80 million residents in coastal areas of Asia will be flooded. Most flooded areas will lie in South Asia, particularly Bangladesh and India. Climate change has had three deleterious effects on agriculture and crop production (Fig. 1). Farmers have to adopt different strategies (adaptation, mitigation and human interventions) to manage the climate change scenario and food security.

1.2 Climate Change Indian Contest

India is the second most densely populated developing country in the world after China, that supports 17% of the world population, approximately. India among the total land resource available in world possesses only about 2.16%. The population density in India is also high, approximately 382 persons/km^2 higher than majority of the developing and developed countries in world (Srivastava, 2021). The exponential population has caused overexploitation and mismanagement of natural resources that increased concern in the society influencing the world communities. Thus, humans' negligence in sustaining population, excessive use of fossils, inadvertent agricultural and industrialization activities had instigated unusual imbalance in atmospheric gases and enhancing uncontrolled emission of greenhouse gases causing global warming and further climate change.

The changes in Indian climatic conditions are consistent with global trends over the last 100 years and thus represent a serious risk to human health, environment, agriculture as well as the economy of the country. India is considered to be one of the world's most vulnerable countries to climate change as it is exposed to sea-level rise, shifts in precipitation patterns and extreme weather events. These events continuously pose a risk to livelihoods, human health, water availability, biodiversity, and ecosystem and food security. About 12% (40 million ha) of India is flood-prone, while 16% (51 million ha) is drought-prone (Chadha, 2015). According to IPCC (2007), Indian economy accounts for nearly 14% of GHG emissions, i.e., about 1331.6 million tonnes (Mt) of carbon dioxide in the GHG emissions. Methane, nitrous oxide and carbon dioxide are the three main GHGs emitted usually from different activities of agriculture and allied sectors (Puri et al., 2017). India ranks about 11th in GHG emission level as it improved its position from 14th as given in the previous report (CCPI, 2019). The most noteworthy improvement in its performance is in the renewable energy category. However, India is planning a new coal-fired power plant which poses the risk of offsetting positive development in the renewable energy sector. India has an overall high rating in the emission category because of comparatively low levels of per capita GHG emissions and a relatively determined mitigation target plan for 2030 (IPCC, 2009). In 2019, it was reported that states like Assam, Mizoram, and Jammu & Kashmir (J&K) were 'highly vulnerable' to climate change with little capacity to resist or cope up with the environment. In India, there is a high probability of loss estimated as 10–40% in crop production by 2080–2100 due to global warming (Parry et al., 2004; IPCC, 2014). The IPCC's fourth assessment report suggests that climate change will affect chiefly agricultural yields ranging up to 50% because of rain-based farming. In India, rain-fed agro-ecology accounts for 60% of the net sown area and therefore climate change is expected to cause severe difficulties for Indian farmers. On the other hand, the agriculture sector has a direct impact on the Indian economy.

During 2017, the agriculture sector contributed about 15.4% to Gross Domestic Product (GDP) in India (Indiastat, 2019). With this GDP contribution, the agriculture sector provided $375.61 billion rupees, making India the second largest producer of agricultural products. The total global agricultural output accounts for 7.39% in India (Indiastat, 2019). The agricultural sector's contribution to the Indian economy is much higher than the agricultural contribution of 6.4% to the world's economy. Among the agricultural crops, the contribution of horticulture to Indian GDP is nearly 30% (Chadha, 2015). The horticulture growth rate is high due to the improvement in productivity of horticulture crops which was approximately 28% between 2001–2002 and 2011–2012. India is the second largest producer of fruits (10.9%) and vegetables (8.6%) in the world (FAOSTAT, 2019). It is the third largest producer of (1,396.8 Million nuts) coconuts worldwide (CDB, 2019). India is also the biggest producer, processor, consumer and exporter of cashew.

2. Impact of Climate Change on Fruit Crops

India grows various types of horticultural crops owing to the availability of different climactic zones ranging from extreme temperate to tropical crops. After China, India is the second largest producer of fruits in world with respect to area and production. The impact of horticulture is visible in India as it contributes about 29.5 percent of agriculture GDP from 13.5 percent of area, and hence contributes substantially to the earning obtained from total agricultural exports (Dhillon and Gill, 2015). However, the climate variability has dramatically affected the productivity of crops and their regional distribution in the past decades with cruel impacts on food security. The food production is predominantly susceptible to variability in climate change as the crop yields depend directly on climatic conditions in the tropics and subtropics. Recently, several surveys revealed that the temperature increase has already adversely affected crop growth in different respects. Such adverse effects are mainly evident in horticultural crops such as fruit trees, vegetables, and flowering crops. Many phenological changes are now being frequently observed in horticultural crops as a result of increased temperature. The phenological changes are decreased fruit setting, fruit-size enlargement, insufficient color development, fruit deformity, softening of fruit flesh, increased acidity, fruit skin and leaf burning, lack of head formation in heading crucifers, increased physiological disorders such as tip-burn, blossom end-rot and delay of endodormancy, etc. So, less than 1°C increase in temperature brought about serious phenological disorder in horticultural crops, thus causing substantial yield loss and devaluation of their market prices. In the meantime, it was also observed that the optimum areas for horticultural crop cultivation shifted northwards as the climatic change proceeds. Similar trend was found with the onset and spread of insect pests in crop plants. Every horticulture crop has specific temperature requirements for proper growth, flowering, fruit development and achieving maturity. The typical phenology of each horticultural plant is greatly affected by the change in maximum, minimum and mean temperatures. So, the fruits crops are grown in area where it has the best adaption to the prevailing climates. As a result of climate change and rise in temperature the growing season of crop plants got prolonged (Chmielewski and Rotzer, 2001). Several studies have provided quantitative assessments of the probable impact of climate change on crop production. The broad aspects of phenological changes brought about by climate changes are discussed here in the chapter.

2.1 *Impact on Dormancy and Chilling Requirements*

The rapid climatic changes have altered the adaptability of many temperate fruit crops, causing low productivity problems. There are specific chilling requirements for temperate fruit crops, like apple, pear, peach, almond, cherry, plum, etc. for breaking dormancy and growth (Samish and Lavee, 1982). As the result of climate change, early snow falls from December and January fluctuated and extended up to February and March. The amount of early snow fall contributes nitrogen for plant use, replenish soil moisture and prevent humidity build up and hence determines the number of chilling hour and thereby the time of bud break. However, due to increase in temperature, temperate fruit crops receive insufficient chilling that leads to development of symptoms like delayed foliation, reduced fruit set, increased buttoning and reduced fruit quality. Also, in successful commercial cultivation of many fruit and nut trees requires the fulfillment of a winter chilling requirement, which is specific for every tree cultivar. Lack of chilling during the mild winter results in abnormal patterns of bud-breakage and development in temperate fruit trees (Ameglio et al., 2000). Eventually, warming may affect over-winter chill requirements of temperate tree fruits and require replacement by new cultivars or species. This process finally results in changing crop sizes and maturity stages during harvest that may substantially affect yield and fruit quality. Sometimes uneven early sown leads to hail production which hampers lot of fruit in orchard, if occurred between flowering and fruit development stage. Mainly in temperate region occurrence of hail destroys all the flower buds and injures roughly all developing fruits often creating ugly spots on fruit. The plants use the dormancy mechanism to protect itself from unfavorable climatic condition. Thus, the alteration in climate change will require quiet adaptability of many temperate fruit crop in the near future. The dormancy symptoms of fruits may get prolonged when winter is neither long nor cold enough to break the dormancy adequately; such a condition is referred as 'delayed foliation' (Black, 1952; Ruck, 1975). For a commercially successful cultivation of many temperate fruit trees, the fulfillment of winter chilling is very essential.

2.2 *Impact on Temperature and Crop Shift*

Higher temperature is a crucial factor which speeds up the plant growth and development in major crops. The majority of plant processes associated with maximum yield are high-temperature dependent and require an optimum temperature range for maximum growth and production. The optimum temperature corresponds to the optimum growth through active photosynthesis and other metabolic reactions. In the case of perennial crops, increase in temperature which is more than optimum reduces photosynthesis, thereby affecting productivity, e.g., Banana. At higher temperatures, the biochemical processes and reactions in plants get accelerated up to a threshold where enzyme systems can destroy and die (Chmielewski et al., 2004). The rise in mean temperature from 1.45°C to 2.32°C over the last 20 years adversely affected vernalization of all temperate fruit crops (Ahmed et al., 2011). A study conducted by Fraga and Santos (2021) in Portugal was involved in quantification of the impacts of climate change due to chilling and forcing on the main fresh fruit regions using bias-corrected computed data from several RCM-GCM model chains. They suggested that in future, chilling and forcing may lead to limitations and changes in phenological stages in the most important Portuguese temperate-fruit regions. Even a minor climate sift of 1–2°C could have a vital impact on geographic range of these cultivated crops as the fruit crops are perennial moving production area is just difficult.

In India, the northern hilly zone showed an elevation in temperature approximately by 0.65°C to 2.3°C, while the north-western Himalayas showed about 0.5°C ever since the last century (Bhutiyani and Kale, 2002). A shift in the cultivation area of the temperate crops has been witnessed in these zones. Thus, it indicates that a crop that was productive in one area is now no longer so or it may be the other way round. The consequential effect of climate change is very evident in apple cultivation as the shift in a crop from lower elevations to higher altitudes has occurred in India (Rai et al., 2015). Due to the rise in temperature and decline in rainfall, fruits like apricots and cherries are rapidly disappearing from a few areas of Kashmir Valley, deteriorating both yield and quality of apples. However, the use of proper rootstocks provides a huge number of alternatives to the growers to enhance fruit quality and yield, to attain early fruiting and uniform cropping, to maintain the tree size, to produce opportunity for high-density crop planting, etc. The increase in frost incidence with change in climate is a foremost threat to fruit crops grown in the subtropical regions of north-western India. The occurrence of cold wave from west to eastern India, during 2002–2003, caused a serious damage to fruit plantations. The level of damage reported in mango (40–100%) and litchi (30–80%) was still higher (Samra et al., 2003). In 2007–2008, due to 'western disturbances', the temperature dipped and frost occurred in many parts of India, causing frost and chilling damage to diverse fruit plants, predominantly in mango cultivation (Gill and Singh, 2012). The damage caused was significantly high in frost-prone regions of Punjab where a majority of mango (50–100%) plantations were damaged severely.

2.3 *Impact on Phenology*

Phenology can be defined as the periodic study of events that are happening in life cycles of an organism and how they are influenced

Table 1. Climate change impact in phenology.

Crops	**Climate change factors**	**Impacts**	**Scientist name**
Mango grafted plant	20°C days and 15°C nights	20 weeks to complete the growth cycle	Whiley et al. (1989)
	30°C days and 25°C night	6 weeks only to complete the growth cycle with three times more amount of flush	
Apple and Pear	Increase in Early spring temperature	Early blooming	Grab et al. (2011)
Grapes	Increase in 2–3°C average growing temperature	13–19 days acceleration in bloom, fruiting and harvesting date	Tomasi et al. (2011)
Cherry	Increase of 1.8°C temperature in month of February and march over 25 years	Peak flowering date occurred 5.5 days	Miller et al. (2007)
Mango	Increase in temperature	Early flowing and fruit setting with a smaller number of male flower and a greater number of hermaphrodite flower	Rajatiya et al. (2018a)

by the seasonal and inter-annual climatic variation. In plant species alteration between the duration of juvenile stage and reproductive stage is taking place due to the irregular climate change (Table 1). The change in phenology, i.e., the timing of different physiological activities, is one of the main distinctive effects of climate change (Cleland et al., 2007). A 10-days prior date of the full bloom was observed in apple varieties of 'Boskoop', 'Cox's Orange Pippin' and 'Golden Delicious' on comparing the last 20 years with the earlier 30 years (Blanke and Kunz, 2011). The advancing trends in the blooming dates of many fruit crops indicate the response of climate change.

2.4 *Impact on Flowering, Fruiting and Fruit Dropping*

The most crucial events in all fruit crops are flowering and fruiting which are regulated by the climatic condition. In Himachal Pradesh, the warmer temperature during dormant season induce early flowering in apple varieties of the 'Delicious' group that become vulnerable to spring frost (Bhatia, 2010). Similarly, the climatic shifts affect the fruit-bud differentiation pattern in subtropical fruit crops, which, in turn, affects the time of flowering and fruit production (Ravishanker and Rajan, 2011). The climatic variations disturb the flowering pattern and fruit set in many fruit crops by reducing pollination activity. In temperate fruits, flower induction is deeply influenced by low temperature, however their strong interaction between genotype, photoperiod and temperature that finally decides the flower production. In temperate areas, apple and other fruit crop are susceptible to spring frosts and can also represent the main cause of weather-related damage to crops. During the blooming stage, a single event of temperatures fluctuation can damage flower buds or even kill them (Table 2). While, light frosts result in the deterioration of fruit quality, severe frosts threaten the harvest also. Heavy rains during flowering generally wash out the pollen from the stigma of flowers, resulting in poor or no fruit set. Varu and Viradia (2015) reported that about 80–90% less of mango production occurred due to unseasonal rains followed by heavy dew attack during the flowering season. As a result, there was reduced fruit set, increased fruit drop at pea stage and greater incidence of sooty mold and powdery mildew in mango. During 2008–09, mango yield was drastically low as around 2°C higher temperature was recorded in Gujarat during the flower induction (Parmar et al., 2012). This seems to have a detrimental effect on mango due to poor flowering, eventually affecting the crop yield. Alteration in relative humidity affect major fruit crops hampering the insect activity as a result poor fruit set and excessive fruit drop occurs. High intensity of relative humidity during the flowering period also resulted in a higher infestation of mango hopper and powdery mildew. Transpiration rate and respiration also influence the fruit temperature, which later on influences the normal

Table 2. Climate change impact on blooming.

Crops	Climate change factors	Impacts	Scientist name
Apple	Temperature fluctuation	Alter the pattern of blossoming, bearing and, therefore, fruit yield and the quality of apple deteriorate under Western Himalayan condition of India	Vedwan and Rhoades (2001)
Apple	Greater rise in winter and spring temperatures	Lead to earlier flowering, which coincides with the time of spring frost resulting in a remaining risk of frost damage to apple flowers	Choudhary et al. (2015)
Apple	Change in temperature	Tree spout 2–3 weeks earlier which result in bud opening at the time of March and coincide with frost	Choudhary et al. (2015)
Papaya	High temperature	Led to flower drop in female and hermaphrodite plants and also responsible for changes in sex in hermaphrodite plants.	Reddy et al. (2017)
	Low temperature	If flowering occurs under extremely low temperatures, flower drops in papaya are relatively normal	Reddy et al. (2017)

fruit-ripening processes and lead to development of spongy tissues (Hemanth et al., 2008). In strawberry, the shift in flowering time was observed in comparison to the prevailing climate of the Baltic States as it typically occurs in the middle of June, instead of the middle of May (Bethere et al., 2016). In the case of mango, it was reported that higher temperature caused greater drying of shoots in fruited plants in comparison to non-fruited ones; amongst different varieties, Dashehari mango (77.28%) showed maximum twig drying followed by var. Baneshan (32%) and var. Rajapari (33.93%), whereas lowest twig drying amounting to 19.5% was recorded in var. Langra (Reddy and Singh, 2011).

2.5 *Impact on Crop Duration*

Crop duration and maturity indices are important physiological processes in fruit crops as they affect the fruit quality and final yield. In lowland tropical areas, citrus fruits mature quickly due to high respiration rates in warm temperatures and thus do not have adequate time to accumulate high total soluble solids (TSS) (Zekri, 2011). The acidity of citrus declines rapidly, leading to increase in soluble solids or acid ratio and insipid taste and quick drying of the fruit. It has been reported that the effects of climate change are extremely visible at grape harvest time which starts two weeks earlier than ever before in India (*The Economic Times*, 2016). In Greece, the harvest dates of grape were earlier than in India due to the change in maximum and minimum temperatures (George et al., 2014). In the case of banana, a rise in temperature by 1–2°C beyond 25–30°C increased leaf production, thereby reducing crop duration and improved production (Chaddha and Kumar, 2011).

2.6 *Impact of Effect on Pollination*

Climate change even affected bees and natural pollinators at different levels, interfering their pollinating efficiency (Reddy et al., 2012). Thus, the disturbing climate change scenario has contributed to a significant decrease in the population of pollinating insects. The temperature extremes, i.e., very low or very high are unfavorable to the pollinators, resulting in no fertilization, finally affecting the fruit set. The cross-pollinated fruits, such as walnuts and pistachios, show reduced pollination because of insufficient chilling, leading to reduced crop yields (Gradziel et al., 2007). In temperate fruits, like apple, pear, plum, cherry, etc., optimum temperature for pollination and fertilization is between 20–25°C. The low temperature and foggy conditions had a detrimental effect during pollination in sour cherry in the USA (Zavalloni et al., 2008). A higher temperature during flowering reduces the effective time for pollination by the pollinators, causing poor pollinator activities and desiccation of pollen. Also, the higher temperatures, ranging from 33°C to 36°C during the microsporogenesis or gamete formation in pollen, decrease pollen viability up to 60–85% (Issarakraisila et al., 1993). Even cooler temperatures below 17°C would produce abnormal as well as non-viable pollen grains. The reproductive stage during microsporogenesis appears to be most sensitive to temperatures below 10°C. Hence, cooler conditions also negatively affect pollen germination and pollen-tube growth, which is completely repressed at temperatures below 15°C (Issarakraisila and Considine, 1994). The pre-monsoon showers are also very harmful to fruit plantations as they destroy complete crops of fruits, like grapes, mango and dates plants. The rains washes away the pollen from the stigma of flowers, resulting in poor or no fruit set (Rajan et al., 2020).

2.7 *Impact on Post-Harvest Quality*

Light is the key factor for fruit quality development like colour formation, shape and size development. Especially spectrum of red and blue light influence growth of plant, flowering and yield of plant. The temperature variations directly affect crop photosynthesis and thus have a major impact on the post-harvest quality of fruits by altering important quality parameters, like sugar content, organic acids, antioxidant, peel color and fruit firmness. The expression of anthocyanin biosynthetic genes is induced by low temperatures, so a higher temperature has a negative effect on its biosynthesis. Such a condition arises in red oranges (Lo Piero et al., 2005), apples (Ubi et al., 2006) and grapes (Yamane et al., 2006). In the case of grapes, higher temperatures (above 46°C) cause thick skin of berries. When grown under high temperature, grapes show higher sugar content and lower levels of tartaric acid (Kliewer and Lider, 1970). Similarly, higher night temperatures also reduce anthocyanin accumulation in berry skin because of degradation of anthocyanin pigment as well as the inhibition of RNA transcript which is responsible for its biosynthesis (Mori et al., 2005). In the ripening apples, anthocyanin is apparently induced at temperatures below 10°C (Curry, 1997). However, the synthesis of anthocyanin takes place at higher irradiation at mild temperatures (20–27°C) in mature apples (Curry, 1997; Reay, 1999). High temperature result in poor red color development in the apple peel and reduce its post-harvest quality (Wand et al., 2002, 2005). The increase in temperature from 0.7-1.0°C may shift the area suitable presently for the quality production of Dashehari and Alphonso varieties of mango. Under high temperature, activity of cell wall enzymes reduced causing delay in growth and development during ripening. The fruit development and quality of climate change is presented in Table 3.

Table 3. Development of climate resilient fruit crops for food security.

Name of fruit crop	Climate variables	Salient observations	Quality development	References
Apple (*Malus domestica* Borkh.)	Temperature, humidity, and rainfall	Terpene and volatile contents influenced by the temperature, rainfall and humidity	Flavor development	Vallat et al. (2005)
	Temperature	High temperatures responsible for developing poor red color in apples	Post harvest quality	Wand et al. (2002, 2005)
Var. York Imperial	Drought	Better growth and yield	Growth and yield	NICRA, 2015
Bilberries (*Vaccinium myrtillus* L.)	Overall climate and thermal effect	Anthocyanin concentration in bilberries is influenced by climatic factors	Sensory quality; health-related benefits	Akerstrom et al. (2009)
Fig (*Ficus carica* Linn.)	Overall climate	Secondary compounds are variable due to the influence of climatic conditions	Sensory qualities	Darjazi and Larijani (2012)
Grapes (*Vitis vinifera* L.)	Temperature, solar radiation, rainfall	Presence of phenolic compounds and antioxidant properties are significantly developed	Sensory quality; benefit to human health	Xu et al. (2011)
	Levels of Carbon dioxide	Tartaric acid increase with a rise in carbon dioxide level	Sensory quality	Bindi et al. (2001)
	Temperature	Higher temperatures cause thick skin of berries	Processing and quality	Yamane et al., 2006
	Night temperatures	Higher night temperatures also reduce anthocyanin accumulation in berry skin because of factors such as anthocyanin degradation	Benefit to human health	Mori et al. (2005)

Table 3 contd. ...

...Table 3 contd.

Name of fruit crop	Climate variables	Salient observations	Quality development	References
Hops (*Humulus lupulus* L.)	Overall climate	Concentrations of key compounds depend on climatological conditions with highest levels in poorest weather conditions	Sensory qualities; benefit to human health	Keukeleire et al. (2007)
Pomegranate (*Punica granatum* L.)	Temperature	Seasonal temperature inversely correlated to anthocyanin accumulation	Sensory quality; benefit to human health	Borochov-Neori et al. (2011)
Strawberry (*Fragariax ananassa* Duch.)	Temperature	Higher antioxidant activity and flavonoids on cooler days	Sensory quality; benefit to human health	Wang and Zheng (2001)
Red oranges (*Citrus sinensis* (L.) Osbeck.) Var.'Blood orange')	Temperature	Expression of anthocyanin biosynthetic genes induced by low temperature	Sensory quality; benefit to human health	Lo Piero et al. (2005)
Citrus sinensis (L.) Osbeck.) Var. Mosambi	Drought	Sustain in high temperature and better growth	Better quality	NICRA (2015)
Apricot (*Prunus armeniaca* L.)	Drought stress	High yield	Better quality	NICRA (2015)
Banana (*Musa acuminata* L.) Var. Karpuravalli (ABB genome)	Drought	Sustain in drought condition and better growth	Sensory quality; benefit to human health	NICRA (2015)
Banana (*Musa acuminata* L.) Var. Shrimanti, Grand Naine	Temperature	Sustain in high temperature and better growth	Sensory quality; benefit to human health	NICRA (2015)

Table 3 contd. ...

...Table 3 contd.

Name of fruit crop	Climate variables	Salient observations	Quality development	References
Mango (*Mangifera indica* L.) Var. Arka,, Neelachal, Sinduri/Jalore seedless	Drought	Sustain in high temperature; better growth; ripening with flavor; high storage quality	Good quality; benefit to human health	NICRA (2015)
Guava (*Psidium guajava* L.) Var. Allahabad Safeda, Lucknow-49	Salinity stress	Better growth and storage quality; maintain the ascorbic acid content	Good quality; benefit to human health	NICRA (2015)

2.8 *Impact on Irrigation Water*

Agriculture always demands water for irrigation; it is considered more sensitive to climate change. The elevated temperature increases the demand for more irrigation, which is another restraint affecting the productivity of fruit crops. A change in the field-level climate may alter the need, amount and timing of irrigation, especially the fruit crops cultivated under rain-fed conditions as these are drastically affected. The fruit crop requirement of annual irrigation will increase, not because of higher evaporation, but because the trees develop at a faster rate during the annual period due to higher atmospheric CO_2 levels. Gradually the change in Indian's monsoon had resulted into severe droughts and intense flooding in various parts of India. So, several fruits crops were affected and unable to thrive in severe climatic conditions like drought and flood. About 80% of the reduction in apple yield was predictable due to water shortage for irrigation and 20% due to high evaporation rate in the apple-growing areas of Himachal Pradesh, India (Singh et al., 2016). It is predicted that around 2025, all irrigated areas of India would need more irrigation water and also the global net irrigation requirements would increase, irrespective of climate change by 3.5–5% (Pathak et al., 2014).

2.9 *Impact on Physiological Disorders*

Furthermore, climate change has increased the occurrence of physiological disorders in plants. High moisture and temperature stress results in increased sunburn and cracking of apples, apricot, and cherries. Three types of sunburn were reported to be caused by heat and or light stress (Felicetti and Schrader, 2008). The first is 'sunburn necrosis' and is induced by heat burn that occurs when the temperature of the fruit surface reaches 52°C in about 10 min. This causes the death of cells in the peel followed by dark brown or black necrotic spots. In the second case, 'sunburn browning', which is the most common type, leads to yellow, brown, or dark tan spots on the sun-exposed side of the fruit. The third and final type of sunburn is induced when fruits are suddenly exposed to full sunlight, as they are not acclimated to sun exposure. The apples with sunburn necrosis show higher relative electrical conductivity than the fruits with no sunburn or with sunburn browning (Schrader et al., 2001). Increased relative electrical conductivity in fruits with sunburn necrosis showed that the membrane integrity was damaged, allowing the leakage of electrolytes. The increased temperature at maturity level will lead to fruit cracking and burning, as in the case of litchi (Mark and Marin, 2016). In custard apple, the maximum black spots (35.63%) was reported on the skin of the fruit due to the high-speed wind recorded in the same year (Varu et al., 2010).

2.10 Impact on Pest and Disease Incidence

The rise in temperature as a consequence of climate change has indirectly affected the pest and disease incidences in crop species. All stages of the insect life cycle are affected by climate, i.e., life span, molting, fecundity, dispersal, mortality and genetic adaption. The higher temperature in spring season results in a faster reproduction rate of insects, thereby boosting the pest population (Patterson et al., 1999). Insect are categories as harmful (fruit fly, stone weevil, fruit borers, aphids, thrips, lemon butter fly, scales and chaffer beetles) and beneficial insects (Ladybird beetles/Ladybugs, Ground beetles, Green Lacewings, Spiders, Syrphid flies/Hoverflies, Praying mantids, Minute pirate bug, Aphid midge). Due to climate change some beneficial insect may be eradicated from nature (predators) which increases the harmful pest population in crop field. In other case due to climate change some harmful insect may mutate quickly and cause epidemic or pandemic condition in some location. The crop protection chemicals are also affected as their efficacy reduces due to changes in temperature and precipitation. In addition, there is a direct impact of climate change on apple productivity; it has also aggravated the susceptibility to various diseases and pest attacks, resulting in further losses in yield (Gautam et al., 2013). In apples, with every 1°C rice in temperature reduction in mortality of sting bug by reported by 15% in winters (Kiritani, 2006). It was also reported by the Harrington et al. (2007), that 5 extra generation of woolly aphids are now seen in apples after the overall rise in 2°C temperature. While, in Citrus increased shoots generation during spring and summer had resulted into increase population of leaf hopper that are main vector of citrus variegated chlorosis (Milanez et al., 2002). Increased summer precipitation, particularly heavy storms, increases the incidences of *Rhynchosporium* leaf blotch and Septoria leaf spot diseases (Royle et al., 1986). With increased temperature and elevated CO_2 it was suggested that pests, such as aphids and weevil larvae, will respond positively (Newman, 2004; Staley and Johnson, 2008). The change in climatic parameters also increased the threat of new incursions. In, Cavendish Banana group one new strain of fusarium wilt race-4 in Asian countries have been reported due to climate change beside the two old strains, i.e., race 1 and 2 for which the resistant was developed (Padmanaban et al., 2009). Now, the mutant stain race-4 is a great concern and the entry of this new mutant can ruin the total banana plantation in India. Sigatoka disease in Maharashtra is now a devastating disease, where it was never considered a problem previously due to the climate change. In the case of mango, hot and dry climate reduces the risk of fungal diseases like anthracnose and powdery mildew because sunlight, low humidity and temperature extremes (below 18°C or more than 35°C) rapidly inactivate the spores (Alfonso and Brent, 2014). Correlation analysis of weather factors and inflorescence pests showed that relative humidity had a negative correlation with hopper population and a highly negative correlation with flower bug, thrips/leaf and thrips/inflorescence of mango. The minimum temperature and evening relative humidity had a significant negative correlation with flower bug and thrips/inflorescence population, while displaying a highly negative correlation with thrips/leaf population (Bhut and Jethva, 2015). Hence, climate change has a varied and detrimental effect on fruit crops. The rise in mean temperature, long spells of drought during the summer season, delay in the start of winter season and reduced rainfall or snowfall have condensed the total area supposed to be slightly suitable for apple and other temperate fruit cultivation. The critical chill units required for apple production have been showing a decline. A trend analysis study indicates that snowfall is decreasing at the rate of 82.7 mm/annum in Himachal Pradesh, India. Consequently, the apple cultivation region is moving further up in elevation due to the warmer climate (Gautam et al., 2014). In India, a total decrease of around 2–3% in apple yield had been reported in districts of Shimla, Kullu, Lahul and Spiti during mid-2000s and a maximum decline of about 4% was witnessed in marginal farms (Bhagat et al., 2009). IPCC projected that the average air-temperature rise will be a maximum of 4°C during the 21st century. That's why it has become very urgent to develop effectual and promising technologies to mitigate the severe risks induced by climate change in fruit crop production. Nowadays, climate change adaptation and mitigation strategies are being incorporated more deeply in governmental structures and policies with the rising outline of climate action in national political agendas. Several countries are appointing inter-ministerial committees to oversee climate action and comprehensive national systems to monitor, evaluate and report on progress in mitigating climate change.

3. Adaptation of Fruit Crop to Climate Change

Global climate change is an outcome of a complex phenomenon that seems to be beyond one's control. It is considered to be a challenge and prioritizes action against it. In addition, its science is specifically specializing, but our approach to this effect has more or less been generalizing in nature. Adaptation aids in the preparation for the prospective effects of climate change by reducing their impacts on ecosystems and people's well-being. Adaptation to climate change is nothing but to embrace the actions that will include adjusting practices, processes and making strategies for minimizing the risk of climate change on fruits crops in an integrated approach. Diversification in crop production is a key strategy applied to minimize risk and build resilience in the farming systems (Sthapit, 2012). Monocultures may produce bumper harvests during favorable weather as well as market conditions, but they also expose the producers to the risk of complete crop failure and loss. There is an enormous diversity in agricultural practices with respect to cultural, institutional and economic factors and their interactions because of the variable climatic and environmental conditions. This provides us various opportunities to develop a correspondingly large array of possible adaptations for the existing agricultural systems and often supporting climate risk management Rajatiya et al. (2018b). Adaptation strategies could encompass innumerable factors like the creating awareness about the new climate resilient technology, adopting cultivars develop to cope up with the extreme climatic conditions, upgrading agronomic practices as per the requirement of particular ecology, adopting management practices that can reducing the stress on the land and water requirements, establishing better water conservation programs, shifting to new newer forms of agriculture, etc. An essential component of this approach is to implement the adaptation frameworks that are very relevant, robust and can easily be operated by all stakeholders. Alternatively, agricultural production systems with high biodiversity bring high stability in yield and also limit pest and disease outbreaks while increasing resilience to disturbances of climate change (Frison et al., 2011). Fruit trees add resilience to the farming systems as they can better withstand climate adversity than annual crop plants. Depending on the fruit crop species, they can also provide numerous use values as seen in the case of timber, fodder, firewood, nitrogen fixation in soil and protection from windbreaks. The reproductive stages of fruit trees are most susceptible to climate change due to implications on the quantity and quality of fruits produced (Ramos et al., 2011). They bring change or adaptation in varieties of longer-lived fruit trees and this is an emerging challenge because of the rapidly occurring climate changes. However, fruit trees have long productive lives, generally ranging from over two to four decades, and hence any change in variety can happen over this long period of time (Lobell et al., 2006).

Numerous strategies of adaptation and mitigation measures have been reported by various researchers to minimize the climate change effect (Li et al., 2002; Zhong, 2003; Cao and Sun, 2007; Wei et al., 2007; Duan et al., 2008; Xue et al., 2008; Sun et al., 2009; Chen, 2010). Although it's very hard to foresee the exact impacts of climate change on fruit crops, we have to get accustomed with the fluctuating circumstances and prepare in the future. The traditional agronomical methods extensively adopted as adaptation strategies to address the adverse impacts of climate change on productivity and quality of crops are:

1) Raising awareness within the farming communities to adapt to the changing climate and for providing better information on challenges and solutions.
2) Development of cost-effective and climate-resilient technologies suitable for farmers.
3) Changing varieties or cultivars and altering the planting and harvest dates to achieve an effective, low-cost option. However, it may increase the risk that the farmer's product will put in a different market window with lower prices.
4) Adopting strategies like altering or shifting planting dates in order to combat the increasing temperature and water stress periods during the main crop-growing season.
5) Adopting dormancy avoidance strategy which is a method that can prevent the plants from entering dormancy condition and aids in bud burst without requiring proper chilling temperature. Luedeling et al. (2011) demonstrated that the defoliation of the trees just after the harvest disrupts the annual plant crop cycle and allow to resume without chilling requirements This can be practiced in production of temperate fruits and possible in our country India for bud break in case of apple, Japanese plum, apricot, and pear, etc. (Haokip et al., 2019).
6) Adopting heat treatment strategy to influence the phonological phase of fruit trees in temperate crops. Reports of short-term high temperature treatment were reported in pears and apples inducing bud break (Haokip et al., 2019).
7) Adopting crop-based adaptations by using climate-ready crops that could induce flowering at higher temperatures or rootstock having low vigor but a strong root system.
8) Promoting development of lateral roots and small roots as an adaptive strategy to improve water absorption by increasing more absorbent surface (Abobatta, 2021). It has been reported previously that when

drought period prolongs in Arabian coffee plants (*Coffea arabica* L.), the leaf area is reduced and minimum new shoots generates for survival under stress conditions (DaMattta et al., 2018).

9) Introducing adaptation strategies through cropping pattern, i.e., cropping systems, intercropping, crop diversification and relocation of crops in alternative areas.
10) Adaptations based on using variable crop species or tolerant/resistant cultivars and rootstock against climate change, like drought, high temperature, etc.
11) Modifying the crop management practices, such as zero tillage or minimum tillage practices to improve soil drainage, using sustainable liquid fertilizer, changing inland use management practices, etc.
12) Use of tolerant rootstocks and prompting rootstock breeding for abiotic stress management. For grape production grapes rootstocks from *V. berlandieri* × *V. rupestris* and *V. berlandieri* × *V. riparia* are reported to be drought tolerant (Mitra, 2018).
13) Adopting screening methodology for breeding approaches, i.e., a set of germplasm are evaluated in different environments, so a stable genotype can be identified and it can be exploited at the particular ecological condition (Sarkar et al., 2021). Evaluation of the wild species and folk varieties or landraces should also be probed thoroughly as could act a source of resistant genes.
14) Developing and introducing transgenics in the fruit crops. Currently, various genomic and proteomic research in fruit trees on stress response have been reported which provides a more comprehensive understanding of environmental stress resistance. The studies conducted on overexpression of antioxidant enzyme (APX or POX or SOD) are providing a viable approach to improving resistance to environmental stresses (Mitra, 2018).
15) Weed management and maintaining crop residues in the field for efficient use of resources.
16) Improving existing irrigation systems or implementing new ones, like drip irrigation systems.
17) Providing irrigation during critical stages of the crop growth and thus conserving soil's moisture reserves.
18) Using deep rooted genotypes are for enhancing drought tolerance of the crop (Uga et al., 2013).
19) Introducing drought resistant varieties in water scarce areas. In regions where water availability is scare for grapevine productivity can use drought resistant rootstocks '110R', '140Ru' and '1103P' will be very beneficial (Chen, 2000). Also, we could adopt drought

tolerant coconut cultivars like 'West Coast Tall', 'Laccadive Ordinary', 'Andaman Ordinary', etc. in low rainfall areas (Mitra, 2018).

20) Adopting new farming techniques and resource-conserving technologies, for example, bagging of fruits, fertigation, etc. In mango, bagging of mango fruits at marble stage with brown paper or securing bags give maximum fruit retention. Bagging with newspaper bags gave the highest fruit weight and also reduced the occurrence of spongy tissue (Haldankar et al., 2015). Again, bagging of pomegranate fruits with prgmen bags reduces fruit cracking and sunburn-like physiological disorders (Mohamed, 2016).
21) Mulching of cultivation beds with reflective silver-color film is a commonly used method to improve the skin/peel coloring of apples as it increases sunlight reflection from the bottom. Mulching helps to conserve the soil moisture, soil micro-climate, improve microbial activities and soil health. The use of plastic mulch has increased yield in papaya (64.24%), mango (45.23%), banana (33.95%), Indian jujube (ber-27.06%), guava (25.93%), pineapple (14.63%) and litchi (12.61%) when compared to control of the plant (Patil et al., 2013).
22) Plastic mulching in combination with drip irrigation is a common practice to attain high-quality and high-yield production in citrus orchards.
23) In greenhouse cultivation of vegetables, a range of devices like efficient ventilator, shading, fogger cooling, heat pumps, photo-selective film, etc. are developed for practical use to minimize rise in interior temperature.
24) Anti-traspirants, like chitosan, kaolin, etc. in agriculture are used to reflect the heat radiation from plant parts so that water losses through transpiration and temperature of fruit and leaf surface are reduced (Parashar and Ansari, 2012). The anti-transpirant kaolin is also an effective treatment for reducing sunburn in pomegranate fruit (Ehteshami et al., 2011).
25) Applications of organic solutes like Jasmonic acid, Salicylic acid, and Proline can increase the tree tolerance for different abiotic stress (Khan et al., 2015; Ali and Baek, 2020). Some studies showed foliar application of alcoholic sugars like Sorbitol, and Mannitol, etc. can also be effective in increasing plant adaptability (Nosarzewski et al., 2012).
26) Among the frost reduction approaches, the chemical called the Bordeaux mixture is extensively used for reduction of frost damages on grapes grown in a moderately cold climate as compared to other frost-reduction approaches (Yadollahi, 2011).

27) Modifying the fertilizer application schemes to enhance nutrient availability and use of other soil amendments to improve soil fertility for enhancing nutrient uptake are other methods.
28) Wind breaks or shelterbelts should be prepared for modifying the microclimate of the orchard as they provide shelter to pollinating insects and protect orchards against natural disasters, like wind erosion. It was reported that the minimum mortality percentage of fruit plants affected by frost in an orchard surrounded by windbreaks was 2.97–30.81%, whereas in the absence of this barrier, maximum mortality reached up to 91.43% (Yadollahi, 2011).
29) Weather forecasting by use of GIS and crop insurance schemes should be introduced to the farmers.
30) In some cases, excessive soil moisture, because of heavy rains, becomes a major problem and such a condition can be overcome by growing crops on raised beds.
31) Water conservation measures need to be encouraged among the farmers to increase resilience to climate change and these activities should be adopted in a widespread manner.
32) Significant efforts at water conservation are necessary in regions where a major portion of the total water resource is used for agricultural purposes. Water-saving measures, such as rainwater harvesting, crop rotations make efficient use of available water and adjust in sowing dates as per the temperature and rainfall patterns, use of suitable crop varieties for new weather conditions (e.g., crop varieties with shorter cycles, more resilient to water stress), the espousal of water conservation practices that favor in-filtration and the reuse of waters should be practiced on arable land that reduces water run-off and acts as a windbreaker.
33) Beyond the farm level, measures such as modernizing the irrigation infrastructure, can be applied.
34) Assessment of the 'vulnerability' of all the major regions and/or fruit commodities should be done to identify the current 'at-risk' production sites and/or industries.
35) Improving the supply chain of the fruits to the markets (Fraisse, 2022).
36) Identification should be done of the processes and practices that in the long term will be a threat to horticultural regions and cropping systems as a consequence of climate change adaptation. So, developing adaptation strategies that are appropriate practically and economically is necessary, in consultation with framers and researchers.

37) Reviewing and/or developing, where necessary, Good Agriculture Practices (GAP) for fruit cultivation, include adaptation and mitigation components.
38) Adoption of 'conservation agriculture' by keeping the land covered with vegetation all round the year prevents soil erosion and rainwater runoff on soil developed on limestone.
39) Restoration of biodiversity and forest plantation in all barren lands.
40) River banks to be enclosed with agroforestry, with shrubs and trees, having high CO_2 demand.
41) High-density orchards and garden plants practices must be encouraged to keep the soil covered with shadow horticultural crops.
42) Propose legal bans against any construction or non-farming activities in productive land.
43) Promote the construction of highways and railway tracks along the river banks.
44) Topsoil restoration is a significant issue to be addressed in relation to the adjoining environment situations, including slope gradient, natural vegetation, effective soil depth, rainfall intensity, clay types, nature of land use and soil biodiversity.
45) Assessment of the economic benefits of silvi-horticulture as well as the benefits that might accrue through adaptation and mitigation.
46) For adaption we could also adopt precision farming methods as they improve ground cover within the orchard thereby maintaining an ideal microclimate.
47) Documentation of the effects of climate change for major overseas production regions, particularly in those countries that are chief competitors to India's production.

Along with the adoption of modified crop management practices, the challenges posed by climate change could be tackled by developing new climate-resilient tolerant varieties. Numerous institutions have produced hybrid varieties which are tolerant of heat, salinity and drought-stress conditions. Such hybrids must be planted efficiently to combat the consequence of climate change. Sufficient efforts should be focused on the development of new varieties well suited to diverse agro-ecological regions under changing climatic conditions. In contrast to annual crops, where the application of adaptation strategies can be visualized comparatively quickly by using a broad range of cultivars and altering the planting dates or season; planting and reorganizing orchards require contemplation of more long-term planning for climate change. Hence, before resorting to any adaptation strategy, detailed research on the impact of climate change on perennial crops is very essential.

4. Mitigation Strategies of Fruit Crop to Climate Change

Mitigation strategies are considered as human interventions to reduce the effect or impact of human activities on the climate system (Fig. 2). The mitigation tools are more frequently used in fruits and vegetables as the climate-related issues are increasing, whereas the adaptive strategies could be considered for more long-term stress management. Hence, we concluded that in future to cope up or combat with the changing climate we have to explore the unmapped rootstocks and scions, develop advance genome sequencing technologies, identify advantageous genomic sequences, identify speedy mapping traits, familiarize with biotechnological tools for gene editing, depend on quicker phenotyping, and work on the functional validation of the putative genes as major candidates regulating development under stress conditions in fruit crops. It predominately includes strategies that reduce GHG sources or emissions and hence enhance GHG sinks. Field crops are a good source of GHGs, cause due to soil disturbance, emission of methane and nitrous oxide by burning straw, and emissions from fossil fuels used in field management practices including direct (fuel) or indirect (chemicals) ways, whereas the perennial trees, fruit orchards, and grasslands act as efficient sinks of atmospheric carbon (Sharma et al., 2021). Hence its very essential to create sustainable agricultural systems that could minimize GHGs emissions and sequester carbon within the atmosphere. 'Climate mitigation' primarily comprises of steps to reduce the emissions or enhance the sinks of GHGs which are responsible for climate change with the help of conscious practices, which can permanently eradicate or reduce the long-term risk and hazard of climate change to human life. So, mitigation includes the process in which the emission of GHGs is reduced or is sequestered. The improved agronomic practices for enhanced nutrient use efficiency, water use efficiency, reduction of GHGs and eco-friendly disease and pest

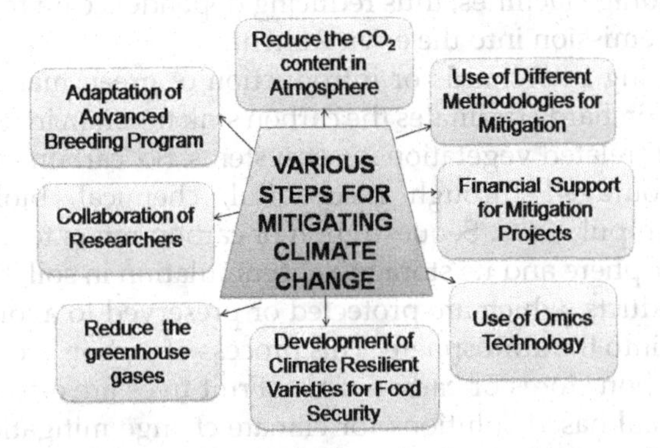

Fig. 2. Different mitigation strategies to cope with climate change and to sustain food security.

management strategies are also parts of mitigation. Application of the right amount of fertilizer, use of more biological control measures, extension of water-saving technology like dripping irrigation or sparkling irrigation, water-fertilizer coupling effect technologies could be helpful in decreasing GHGs emissions. Several mechanical, as well as chemical efforts have been made to minimize CO_2, CH_4 and N_2O emissions. Planting trees and enhancing agroforestry for high CO_2 demand is one of the efforts that will reduce GHG's effects. A research by Forster et al. (2021) calculated the potential GHG mitigation in UK by national planting strategy of 30,000 ha yr^{-1} from 2020 to 2050 using a dynamic life cycle assessment and recorded that the commercial forest could mitigate 1.64 Pg CO_2e by 2120. They found that forest growth rate was most important for determining cumulative mitigation, irrespective of whether trees were harvested by 2120. Greenhouse gas (GHGs) emission from burning fossil, liquid, solid and gaseous fuels including other sources is being chemically sequestered in some other forms. Minimizing the load on petroleum and coal, fossil fuel sources are being used through different biofuel extraction methods from *Jatropa* and *Pongamia* species and its utilization. The use of biofuels—fuels with lower carbon content, i.e., natural gas, CNG, cow-dung gas—will improve the efficiency of electricity generation, transmission and distribution throughout the world. Many architectural manipulations were employed by developing greenery over buildings through lawns and similar constructions. The reduction in CO_2 and CH_4 emissions were taken into consideration throughout the whole agriculture processes. So, there is a need to quantify the GHG emissions from all improved agronomic practices before cataloguing or tagging any technology as 'green' or 'low-carbon' technology. In processing technology, biogas production is extremely useful as it reuses or recycles various farm wastes and can be used together with possible renewable energies, like hydropower and wind energy. The biogas produced can also be used in transportation activities or storage facilities, thus reducing dependence on fossil fuel and reducing CO_2 emission into the environment.

Intercropping in orchards or introduction of green manure crops or cover crops in orchards facilitates the carbon sink function in different fruit crops and their related vegetation in ecosystems. So, carbon sequestration is being encouraged through mechanical, chemical, biological and pedogenic manipulations. Sequestration of carbon refers to CO_2 removal from the atmosphere and its storage or accumulation in soil, biomass and harvested products, which are protected or preserved to avoid CO_2 to be released back into the atmosphere. This process of carbon accumulation is referred as carbon stores or carbon sinks. Fruit trees are also an essential part of perennial-based solutions for climate change mitigation. In 2012, it was reported that perennial crops can sequester around 320–1,100 kg of soil carbon per hectare when compared to 0–450 kg of carbon in annual crops and hence are more likely to provide better yields than annual crops at higher temperatures (Glover et al., 2007; Sthapit and Scherr, 2013). The horticultural crops are perennial, such as plantations, fruit trees and tree spices which are potential candidates for carbon sequestration. Though currently most of these do not fall under carbon trade, there is a lot of scope for these tree species to be used in carbon sequestration and climate regulation system. Similar studies conducted on coconut plantations suggested that annual carbon sequestration in coconut above the ground biomass varies between 15–35 Mg CO_2 ha^{-1} yr^{-1} depending on cultivar, agro-climatic zone, soil type and management (Kumar, 2013). Cocoa-arecanut intercropping also is a good system for carbon sequestration with a potential to sequester 5–7 Mg CO_2 ha^{-1} yr^{-1} (Kumar, 2009; Balasimha and Kumar, 2010). The mitigation of climate change damage for various tropical fruits is done through zoning of crop suitability for proper allocation of land resources and amiable environment for compatible fruit cultivars. It encourages breeding for a high and wide range of stress-resistant varieties through conventional breeding techniques and biotechnology. Such improved cultivars are highly recommended or incorporated with protected horticulture production for better and safer economic returns.

Mitigation strategies also include the policy and institutional capability required for mitigation of climate change adversities (Fig. 2). The adaptation policy proposal consists of more activities, i.e., further research focusing on plant breeding, photosynthetic capacity, biological nitrogen fixation rate, level of stress resistance, protected horticultural crops, and precision horticulture for augmentation of crop adaptability to climate change. Development of climate monitoring, forecasting and pre-warning capability in all provinces will contribute to adaptability to climate change. Therefore, improvement in public awareness of climate change problems and the need for betterment of adaptation and mitigation strategies is highly recommended. As a result, there is a critical need for a decision support system (DSS) to facilitate improved management of agriculture at the farm level and for sustainable and climate-resilient crop production. A farmer needs to answer several questions related to crop production and marketing before deciding on the technology best suited to the farm condition. The crop models are of immense use in the climate change perspective as well as from the crop management point of view. The crop models developed also need to be calibrated and validated for the study purpose before conducting a large-scale analysis. Linking the models to GIS and remote-sensing data also offers enormous scope for regional estimations of GHG emissions and for significantly reducing the yield loss. The major sectors for application of crop simulation models include crop management, agro-ecological zoning, estimating potential production, yield gap analysis and developing breeding strategies. Strategic and anticipatory decision-making in the system also helps in finding crop potential zones for land use planning and hi-tech horticulture. Crop insurance and weather-based horti-advisory are also dependent on crop simulation models. Crop modeling is considered an important tool in research and development of perennial plantation crops because it takes a lot of time for conducting research experiments. Similar to any other approach, simulation modeling also has limitations and uncertainties attached to it.

5. Research Development for Climate Resilience in Fruit Crop

Harmful effects of climate change include increased temperature followed by changes in time, intensity and pattern of rainfall, which in turn lead to higher frequency and extent of natural calamities, like floods, cyclones, droughts and augmented soil salinity. Thus, the hazardous damage caused by climate change inevitably brings about a shift in the optimum planting time of various crops, particularly in rain-fed agricultural areas. The critical impacts of climate change (associated with high temperature, frost, GHGs emissions, etc.) can be easily visualized on the physiological processes of fruit crops, geographic shift of production areas, changes in cropping practices, changes in the frequency of disease and pest spread, crop production and yield, product quality, etc. This enhances the urgent need to initiate the development of potential cultural practices and other countermeasures to cope with climate change.

To alleviate the adverse effects of climate change on the productivity and quality of fruits produced, several strategies and technological countermeasures have been planned and developed. Still, more contemporary strategies can be produced for reducing or managing the chilling requirements of temperate fruit crops. We should work towards the development of environmentally-friendly chemicals to break the dormancy period of such temperate fruits. The adaptation technologies against increased temperature and other climatic abnormalities also include breeding and biotechnological intervention for increased production. The breeding interventions include measures like systematic breeding or phenotyping of fruit crop genetic wealth against temperature increase, moisture stress and genetic enhancement for tolerance towards biotic and abiotic stress. Marker-assisted selection, molecular characterization and development of transgenic or climate-resilient crops having resistance to various traits in relation to biotic and abiotic stress are some common measures adopted in biotechnological interventions for improving climate resilience. Gene pyramiding against stress and *in-vitro* conservation of rare and useful species can be used as the future thrust areas of research for climate change. So, there is a critical need right now to conduct focused and precise research to generate adequate information on impacts of climate change and thus derive suitable adaptation and mitigation strategies for a particular region. The literature review on the basis of PRISMA-P (Preferred Reporting Items for Systematic review and Meta-Analysis Protocols) showed that small-scale farmers in the past 30 years have adopted climate-resilient crops and varieties to counter abiotic stresses, such as drought, heat, flooding and salinity in lower- and middle-income countries (Acevedo et al., 2020). On the basis of the collected literatures, the use of climate-resilient varieties has been always recommended as an excellent way for farmers to cope with or adapt to climate change. Various researchers have reported a series of pathways and interventions that can contribute to ever higher adoption rates of climate-resilient crops.

6. Conclusion and Prospects

Agricultural production is highly affected due to climate change. Extreme weather conditions, such as heat waves, droughts, cyclones and floods lead to reduction of agricultural food production and poverty, especially in rural communities. Climate change significantly leads to yield reduction to the tune of about 30% in most crops apart from lower productivity and failure of crop. The increase in global population and change in food habits in developing countries will highly impact the natural resources and lead to food insecurity. To cope with climate change, farmers need to modify farm management practices including change in planting time, supplementing irrigation, intercropping, adopting conservation agriculture, accessing short- and long-term crop and planting more climate-resilient crop varieties. On the basis of the literature, inclusion of a series of pathways and interventions can contribute to higher adoption rates of climate-resilient crops. Climate resiliency at farm level is of utmost importance in order to achieve food security and improve livelihood support to the rural communities, especially the communities that depend on local agricultural produce to ensure household income and achieve daily adequate caloric intake and balanced nutrition. Farmers are adopting various strategies under climate-smart agriculture to build highly resilient and sustainable agricultural systems. Environment friendly and ecologically sound chemicals must be used for breaking the rest period or dormancy. In tropical climate, dormancy can be induced artificially by defoliating after harvesting (Griesbach, 2007). Also, the application of sprays of hydrogen cyanamide has been effective in breaking dormancy, thereby promoting blooming (Erez et al., 2008; Ashebir et al., 2010; Chabchoub et al., 2010). Weather-based monitoring strategies must be adopted for rapid and effective diagnosis of insects, pests and diseases using the GIS system. Recent achievements on the development of newer technologies and crop genetic improvement that include variety adaptation, crop diversification, biodiversity identification and underutilized crops manipulation have identified cultivars that can adapt to high temperature, drought and floods.

References

Abobatta, W. F. (2021). Fruit orchards under climate change conditions: adaptation strategies and management. *Journal of Applied Biotechnology and Bioengineering*, 8(3): 99–102.

Acevedo, M., Pixley, K., Zinyengere, N., Meng, S., Tufan, H., Cichy, K., Bizikova, L., Isaacs, K., Ghezzi-Kopel, K. and Porciello, J. (2020). A scoping review of adoption of climate-resilient crops by small-scale producers in low- and middle-income countries. *Nature Plants*, 6: 1231–1241.

Ahmed, N., Lal, S., Das, B. and Mir, J.I. (2011). Impact of climate change on temperate fruit crops. pp. 141–150. *In*: Dhillon, W.S. and Aulakh, P.S. (eds.). *Impact of Climate Change on Fruit Crops*. Narendra Publishing House, New Delhi.

Akerstrom, A., Forsum, A., Rumpunen, K., Jaderlund, A. and Bang, U. (2009). Effects of sampling time and nitrogen fertilization on anthocyanidin levels in *Vaccinium myrtillus* fruits. *Journal of Agricultural and Food Chemistry*, 57: 3340–3345.

Alfonso, D.R. and Brent, M.S. (2014). Agricultural adaptation to climate change in the Sahel: expected impacts on pests and diseases afflicting selected crops. pp. 53–54. *In*: *African and Latin Americal Resilience to Climate Change Project*. USAID, Washington.

Ali, M. and Baek, K.H. (2020). Jasmonic acid signaling pathway in response to abiotic stresses in plants. *International Journal of Molecular Sciences*, 21(2): 621.

Ameglio, T., Alves, G., Bonhomme, M., Cochard, H., Ewres, F. et al. (2000). Winter functioning of walnut: Involvement in branching processes. pp. 230–238. *In: L'Arbre, Biologieet Development*. Montreal (CAN), Isabelle Quentin.

Anonymous, *Economic Times*. (2016). Climate Change Advances Wine Grape Harvest by Two Weeks. New Delhi.

Ashebir, D., Deckers, T., Nyssen, J., Bihon, W., Tsegay, A., Tekie, H., Poesen, J., Haile, M., Wondumagegneheu, F., Raes, D., Behailu, M. and Deckers, J. (2010). Growing apple (*Malus domestica*) under tropical mountain climate conditions in northern Ethiopia. *Experimental Agriculture*, 46: 53–65.

Ayala, A., Barahona, D.F., Huss, M., Pellicciotti, F., McPhee, J. and Daniel Farinotti. (2020). Glacier runoff variations since 1955 in the Maipo River basin, in the semiarid Andes of central Chile. *The Cryosphere*, 14: 2005–2027.

Balasimha, D. and Kumar, N.S. (2010). Net primary productivity, carbon sequestration and carbon stocks in areca-cocoa mixed cropping system. pp. 215–226. *In*: *Proceedings of the 16th International Cocoa Research Conference*, Bali.

Baumert, K.A., Herzog, T. and Pershing, J. (2009). *Navigating the Numbers, Greenhouse Gas Data and International Climate Policy*. World Resources Institute, Washington, DC, USA.

Bethere, L., Tija, S., Juris, S. and Bethers, U. (2016). Impact of climate change on the timing of strawberry phenological processes in the Baltic States. *Estonian Journal of Earth Sciences*, 65(1): 48–58.

Bhagat, R.M., Rana, R.S. and Kalia, V. (2009). *Global Climate Change and Indian Agriculture*. ICAR, pp. 48–53, Aggarwal, P.K. (ed.). New Delhi, India.

Bhatia, H.S. (2010). Evaluation of new apple cultivars under changing climate in Kullu Valley of Himachal Pradesh. pp. 1–6. *In*: *Proceeding of National Seminar on Impact of Climate Change on Fruit Crops* (ICCFC, 2010), 6–8 October, PAU, Ludhiana.

Bhut, J.B. and Jethva, D.M. (2015). Impact of weather factor on incidence of inflorescence pests of mango. pp. 291–296. *In*: *National Seminar on Water Management and Climate Smart Agriculture*, 13–14 February, JAU, Junagadh.

Bhutiyani, M.R. and Kale, V.S. (2002). *Climate Change in the Last Century are the Himalaya Warming*. Sapper, 13: 37–46.

Bindi, M., Fibbi, L. and Miglieta, F. (2001). Free air CO_2 enrichment (FACE) of grapevine (*Vitis vinifera* L.): II. Growth and quality of grape and wine in response to elevated CO_2 concentrations. *European Journal of Agronomy*, 14: 145–155.

Black, M.W. (1952). The problem of prolonged rest in deciduous trees. pp. 1122–1131. *In*: *Proceedings of 13th International Horticultural Congress*. London.

Blanke, M.M. and Kunz, A. (2011). Effects of climate change on pome fruit phenology and precipitation. *Acta Horticulturae*, 922: 381–386.

Borochov-Neori, H., Judeinstein, S., Harari, M., Bar-Yaaakov, I., Patil, B.S. et al. (2011). Climate effects on anthocyanin accumulation and composition in the pomegranate (*Punica granatum* L.) fruit arils. *Journal of Agricultural and Food Chemistry*, 59: 5325–5334.

CAIT, Climate Analysis Indicators Tool. (2011). CAIT version 8.0. Available from: http://cait.wri.org.

Cao, M.H. and Sun, Y.Z. (2007). Guangdong fruits and climate damages. *Modern Agricultural Sciences and Technology*, 4: 39–41.

CCPI, Climate Change Performance Index. (2019). *The New Climate Institute and the Climate Action Network*. website published by Germanwatch.

CDB, Coconut Development Board. (2019). *All India Final Estimates of Area, Production and Productivity of Coconut*.

Chabchoub, M.A., Aounallah, M.K. and Sahli, A. (2010). Effect of hydrogen cyanamide on bud break, flowering and fruit growth of two pear cultivars (*Pyrus communis*) under Tunisian conditions. *Acta Horticulturae*, 884: 427–432.

Chadda, K.L. and Kumar, S.N. (2011). Climate change impacts on production of horticultural crops. pp. 3–9. *In*: Dhillon, W.S. and Aulakh, P.S. (eds.). *Impact of Climate Change on Fruit Crops*.

Chadha, K.L. (2015). Global climate change and Indian horticulture, Chapter 1. *In*: Choudhary et al. (eds.). *Climate Dynamics in Horticultural Science: Impact, Adaptation, and Mitigation*, vol. 2, Apple Academic Press, Inc.

Chandio, A.A., Jiang, Y., Rehman, A. and Rauf, A. (2020). Short and long-run impacts of climate change on agriculture: An empirical evidence from China. *International Journal of Climate Change Strategies and Management*, 12(2): 201–221.

Chen, Q. (2012). Adaptation and mitigation of impact of climate change on tropical fruit industry in China. *Acta Horticulturae*, 101–104.

Chen, J.F. (2000). The status of research on grape rootstock varieties and its prospect. *Guoshu Xuebao*, 17: 38–146.

Chmielewski, F.M. and Rotzer, T. (2001). Response of tree penology to climate change across Europe. *Agricultural Forest Meteorology*, 108: 101–112.

Chmielewski, F.M., Muller, A. and Bruns, E. (2004). Climate changes and trends in phenology of fruit trees and field crops in Germany 1961–2000. *Agricultural Forest Meteorology*, 121: 69–78.

Choudhary, M.L., Patel, V.B., Siddiqui, M.W. and Mahsi, S.S. (2015). Climate Dynamics. In Horticultural Science Volume-I: Principles and Applications. Apple Academic Press, Toronto.

Cleland, E.E., Chuine, I., Menzel, A., Mooney, H.A. and Schwartz, M.D. (2007). Shifting plant phenology in response to global change. *Ecological Evolution*, 22: 357–365.

Curry, E.A. (1997). Temperatures for optimal anthocyanin accumulation in apple skin. *Journal of Horticultural Science and Biotechnology*, 72: 723–729.

DaMatta, F.M., Avila, RT, Cardoso, A.A, et al. (2018). Coffee tree growth and environmental acclimation. Achieving Sustainable Cultivation of Coffee. Breeding and Quality Traits

Darjazi, B.B. and Larijani, K. (2012). The effects of climatic conditions and geographical locations on the volatile flavor compounds of fig (*Ficus carica* L.) fruit from Iran. *African Journal of Biotechnology*, 11: 9196–9204.

Dhillon, W.S. and Gill, P.P.S. (2015). Climate change and fruit production, Chapter 2. *In*: Choudhary et al. (eds.). *Climate Dynamics in Horticultural Science: Impact, Adaptation, and Mitigation*. vol. 2, Apple Academic Press, Inc.

Duan, H.L., Qian, H.S., Yu, F. and Song, Q.H. (2008). Temperature suitability of longan and its changes in south China area. *Acta Ecologica Sinica*, 28: 5303–5313.

Ehteshami, S., Sarikhani, H. and Ershadi, A. (2011). Effect of kaolin and gibberellic acid application on some qualitative characteristics and reducing the sunburn in pomegranate fruits (*Punica granatum* L.) cv. 'Rabab Neiriz'. *Plant Products Technology, Agricultural Research*, 11(1): 15–23.

EPA, Environmental Protection Agency. (2022a). Causes of Climate Change. Climate Change Science. U.S.

EPA, Environmental Protection Agency. (2022b). Impacts of Climate Change. Climate Change Science. U.S.

Erez, A., Yablowitz, Z., Aronovitz, A. and Hadar, A. (2008). Dormancy breaking chemicals' efficiency with reduced phytotoxicity. *Acta Horticulturae*, 772: 105–112.

FAOSTAT, Food and Agriculture Organization Statistics. (2019). Data accessed: 28 August 2019.

Fahey, D.W., Doherty, S.J., Hibbard, K.A., Romanou, A. and Taylor, P.C. (2017). Physical drivers of climate change. *In*: *Climate Science Special Report: Fourth National Climate Assessment*, Volume I. U.S. Global Change Research Program, Washington, DC, p. 80.

Felicetti, D.A. and Schrader, L.E. (2008). Photo-oxidative sunburn of apples characterization of a third type of apple sun-burn. *International Journal of Fruit Science*, 8(3): 160–172.

Forster, E.J., Healey, J.R., Dymond, C. and Styles, D. (2021). Commercial afforestation can deliver effective climate change mitigation under multiple decarbonisation pathways. *Nature Communications*, 12: 3831.

Fraga, H. and Santos, J.A. (2021). Assessment of climate change impacts on chilling and forcing for the main fresh fruit regions in Portugal. *Frontier of Plant Science*, 12: 689121.

Frison, E.A., Cherfas, J. and Hodgkin, T. (2011). Agricultural biodiversity is essential for a sustainable improvement in food and nutrition security. *Sustainability*, 3: 238–253.

Fraisse, C.W. (2022). Climate adaptation and mitigation in fruit and vegetable supply chains. Non-Technical Summary report of University of Florida under National Institute of Food and Agriculture.

Gautam, H.R., Bhardwaj, M.L. and Kumar, R. (2013). Climate change and its impact on plant diseases. *Current Science*, 105: 1685–1691.

Gautam, H.R., Sharma, I.M. and Kumar, R. (2014). Climate change is affecting apple cultivation in Himachal Pradesh. *Current Science*, 106: 498–499.

George, K., Mavromatis, T., Stefanos, K., Nikolaos, M.F. and George, V.J. (2014). Viticulture-climate relationships in Greece: The impacts of recent climate trends on harvest date variation. *International Journal of Climatology*, 34(5): 1445–1459.

Gill, P.P.S. and Singh, N.P. (2012). Decline of mango diversity in sub-montane and Kandi zone of Punjab—An overview. *Indian Journal of Ecology*, 39(2): 313–315.

Glover, J.D., Cox, C.M. and Reganold, J.P. (2007). Future farming: A return to roots? pp. 82–89. *Scientific American*, August, 2007.

Grab, S. and Craparo, A. (2011). Advance of apple and pear tree full bloom dates in response to climate change in the southwestern Cape, South Africa: Agric. *For. Meteorol.*, 151: 406–413.

Gradziel, T.M., Lampinen, B., Connell, J.H. and Viveros, M. (2007). Winters' almond: An early-blooming, productive, and high-quality pollinizer for nonpareil. *Hort. Science*, 42: 1725–1727.

Griesbach, J. (2007). *Growing Temperate Fruit Trees in Kenya*. World Agroforestry Center (ICRAF), Nairobi, Kenya.

Haldankar, P.M., Parulekar, Y.R., Kireeti, A., Kad, M.S., Shinde, S.M. and Lawande, K.E. (2015). Studies on influence of bagging of fruits at marble stage on quality of mango cv. Alphonso. *Journal of Plant Studies*, 4(2): 12–20.

Haokip, S.W., Shankar, K. and Lalrinngheta, J. (2020). Climate change and its impact on fruit crops. *Journal of Pharmacognosy and Phytochemistry*, 9(1): 435–438.

Harrington, R., Clark, S.J., Welham, S.J., Verrier, P.J., Denholm, C.H. et al. (2007). Environmental change and the phenology of European aphids. *Global Change Biology*, 13(8): 1550–1564.

Hayhoe, K., Wuebbles, D.J., Easterling, D.R., Fahey, D.W., Doherty, S., Kossin, J., Sweet, W., Vose, R. and Wehner, M. (2018). Our changing climate. *In*: Reidmiller, D.R., Avery, C.W., Easterling, D.R., Kunkel, K.E., Lewis, K.L.M., Maycock, T.K. and Stewart, B.C. (eds.). *Impacts, Risks, and Adaptation in the United States: Fourth National Climate Assessment*, volume II. U.S. Global Change Research Program, Washington, DC. pp. 76.

Hemanth, K.N., Ravishankar, K.V., Narayanaswamy, P. and Shivashankara, K.S. (2008). Influence of temperature on spongy tissue formation in Alphonso Mango. *International Journal of Fruit Science*, 8(3): 226–234.

Indiastat, India's most comprehensive e-resource of socio-economic data. (2019). Indiastat.com.

IPCC. (2007). Climate Change: Mitigation. *Contribution of Working Group III to the Fourth Assessment Report of the Intergovernmental Panel on Climate Change* [Metz, B., Davidson, O.R., Bosch, P.R., Dave, R. and Meyer, L.A. (eds.)], Cambridge University Press, Cambridge, United Kingdom and New York, NY, USA.

IPCC. (2009). *Climate Change: The Scientific Basis*. Cambridge University Press.

IPCC. (2014). *Climate Change: Mitigation of Climate Change, Fifth Assessment Synthesis Report of Intergovernmental Panel on Climate Change*.

IPCC. (2019). Intergovernmental panel on climate change, *Special Report on Climate Change, Desertification, Land Degradation, Sustainable Land Management, Food Security, and Greenhouse Gas Fluxes in Terrestrial Ecosystems*. SPM Approved Draft, Summary for Policy Makers.

IPPC, IPPC Secretariat. (2021). *Scientific Review of the Impact of Climate Change on Plant Pests—A Global Challenge to Prevent and Mitigate Plant Pest Risks in Agriculture, Forestry and Ecosystems*. Rome, FAO on Behalf of the IPPC Secretariat.

Issarakraisila, M., Considine, J.A. and Turner, D.W. (1993). Effects of temperature on pollen viability in mango cv. Kensington. *Acta Horticulturae*, 341: 112–124.

Issarakraisila, M. and Considine, J.A. (1994). Effects of temperature on pollen viability in mango cv. 'Kensington'. *Annals of Botany*, 73: 231–240.

Keukeleire, J., Janssens, I., Heyerick, A., Ghekiere, G., Cambie, J. et al. (2007). Relevance of organic farming and effect of climatological conditions on the formation of r-acids, a-acids, desmethylxanthohumol, and xanthohumol in hop (*Humulus lupulus* L.). *Journal of Agricultural and Food Chemistry*, 55: 61–66.

Khan, M.I.R., Fatma, M., Per, T.S. et al. 2015. Salicylic acid–induced abiotic stress tolerance and underlying mechanisms in plants. *Frontiers in Plant Science*, 6: 462.

Khan, T., Qiu, J., Banjar, A., Alharbey, R., Alzahrani, A.O. and Mehmood, R. (2021). Effect of climate change on fruit by co-integration and machine learning. *International Journal of Climate Change Strategies and Management*, 13(2): 208–226.

Kliewer, M.W. and Lider, L.A. (1970). Effects of day temperature and light intensity on growth and composition of *Vitisvinifera* L. fruits. *Journal of the American Society for Horticultural Science*, 95: 766–769.

Kiritani, K. (2006). Predicting impacts of global warming on population dynamics and distribution of arthropods in Japan. *Population Ecology*, 48(1): 5–12.

Kumar, N.S. (2009). Carbon sequestration in coconut plantations. *In*: Aggarwal, P.K. (ed.). *Global Climate Change and Indian Agriculture—Case Studies from ICAR Network Project*, ICAR Pub., New Delhi.

Kumar, S.N. (2013). Modelling climate change impacts, adaptation strategies and mitigation potential in horticultural crops, Chapter 3. *In*: Singh, H.P. et al. (eds.). *Climate-resilient Horticulture: Adaptation and Mitigation Strategies*. Springer Publications, India.

Li, Y.L., Su, Z. and Tu, F.X. (2002). The effects of climate factors on yields of lichee and longan in Guangxi. *Journal of Guangxi Academy of Sciences*, 18: 135–140.

Lobell, D.B., Field, C.B., Cahill, K.N. and Bonfils, C. (2006). Impacts of future climate change on California perennial crop yields: Model projections with climate and crop uncertainties. *Agricultural and Forest Meteorology*, 141: 208–218.

Lo Piero, A.R., Puglisi, I., Rapisarda, P. and Petrone, G. (2005). Anthocyanins accumulation and related gene expression in red orange fruit induced by low temperature storage. *Journal of Agricultural Food Chemistry*, 53: 9083–9088.

Luedeling, E., Girvetz, E.H., Semenov, M.A. and Brown, P.H. (2011). Climate change affects winter chill for temperate fruit and nut trees. *PLoS One*, 2011: 6(5).

Mark, D.C.J. and Marin, R.A. (2016). Carbon sequestration potential of fruit tree plantation in southern Philippines. *Journal of Biodiversity and Environmental Sciences*, 8(5): 164–174.

Milanez, J.M., Parra, J.R.P., Custodio, I.A., Magri, D.C., Cera, C. and Lopes, J.R.S. (2003). Feeding and survival of citrus sharpshooters (Hemiptera: Cicadellidae) on plants. *Florida Entomologist*, 86: 154–157.

Miller Rushing, A.J., Toshio, K., Primack, R.B., Ishii, Y., Sang, D.L. and Hiroyoshi, H. (2007). Impact of global warming on a group of related species and their hybrids: cherry tree (Rosaceae) flowering at Mt. Takao, Japan. *American Journal of Botany*, 94(9): 1470–1478.

Mitra, S.K. (2018). Climate change: impact, and mitigation strategies for tropical and subtropical fruits. *Acta Horticulture*, 1216: 1–12.

Mohamed, A.W. (2016). Effect of bagging type on reducing pomegranate fruit disorders and quality improvement. *Egyptian Journal of Horticulture*, 41(2): 263–278.

Mori, K., Sugaya, S. and Gemma, H. (2005). Decreased anthocyanin biosynthesis in grape berries grown under elevated night temperature condition. *Scientia Horticulturae*, 105: 319–330.

NAS, National Academy of Sciences. (2020). Climate change: Evidence and causes: Update 2020. The National Academies Press, Washington, DC. pp. 7.

NASA. (2022). The Effects of Climate Change. Earth Science Communications Team at NASA's Jet Propulsion Laboratory.

Newman, J.A. (2004). Climate change and cereal aphids: the relative effects of increasing CO_2 and temperature on aphid population dynamics. *Global Change Biology*, 10: 5–15.

NICRA. (2015). *Climate Resilient Crop Varieties for Sustainable Food Production under Aberrant Weather Conditions*. ICAR, Central Research Institute for Dry Land Agriculture, Hyderabad, Bulletin No. 4, pp. 1–56.

Nosarzewski, M., Downie, A.B., Wu, B. et al. (2012). The role of sorbitol dehydrogenase in *Arabidopsis thaliana*. *Functional Plant Biology*, 39(6): 462–470.

Padmanaban, B., Thangavelu, R., Gopi, M. and Mustaffa, M.M. (2009). First report on the occurrence of a virulent strain of fusarium wilt pathogen (race-1) infecting cavendish (AAA) group of bananas in India. *Journal of Biological Control*, 23(3): 277–283.

Parashar, A. and Ansari, A. (2012). A therapy to protect pomegranate (*Punica granatum* L.) from sunburn. *Pharmacie Globale*, 3(5): 1–3.

Parmar, V.R., Shrivastava, P.K. and Patel, B.N. (2012). Study on weather parameters affecting the mango flowering in south Gujarat. *Journal of Agrometeorology*, 14: 351–353.

Parry, M.L., Rosenzweig, C., Iglesias Livermore, A.M. and Fischer, G. (2004). Effects of climate change on global food production under SRES emissions and socio-economic scenarios. *Global Environmental Change*, 14: 53–67.

Pathak, S., Pramanik, P., Khanna, M. and Kumar, A. (2014). Climate change and water availability in Indian agriculture: Impacts and adaptation. *Indian Journal of Agricultural Sciences*, 84(6): 671–679.

Patil, S.S., Kelkar, T.S. and Bhalerao, S.A. (2013). Mulching: A soil and water conservation practice. *Research Journal of Agriculture and Forestry Sciences*, 1(3): 26–29.

Patterson, D.T., Westbrook, J.K., Joyce, R.J.V., Lingren. P.D., Rogasik, J. et al. (1999). Weeds, insects and disease: Climate change: impact on agriculture. *Climate Change*, 43: 711–727.

Puri, M.G., Murai, A.M., Gholape, S.M., Shigwan, A.S. and Mesare, S.N. (2017). Climate smart agriculture: an approach to sustainably increasing agricultural productivity. pp. 39–42. *In: Proceedings of National Conference on Climate Change Adaption*. 24–25 February, Hyderabad, India.

Rai, R., Joshi, S., Roy, S., Singh, O., Samir, M. and Chandra, A. (2015). Implications of changing climate on productivity of temperate fruit crops with special reference to apple. *Journal of Horticulture*, 2(2): 1–6.

Rajan, R., Ahmad, M.F., Pandey, K., Aman, A. and Kumar, V. (2020). Climate change and resilience in fruit crops. pp. 337–3354. *In: Climate Change and its Effects on Agriculture*, BIOTEC BOOKS Publisher.

Rajatiya, J.H., Varu, D.K., Farheen, H. and Solanki, M.B. (2018a). Correlation of climatic parameters with flowering characters of mango. *International Journal of Pure Applied Bioscience*, 6(3): 597–601.

Rajatiya, J., Varu, D.K., Gohil, P., Solanki, M., Halepotara, F., Gohil, M., Mishra, P. and Solanki, R. (2018b). Climate change: impact, mitigation and adaptation in fruit crops. *International Journal of Pure and Applied Bioscience*, 6(1): 1161–1169.

Ramos, C., Intrigliolo, D.S. and Thompson, R.B. (2011). Global change challenges for horticultural systems. *In:* Araus, J.L. and Slafer, G.A. (eds.). *Crop Stress Management and Global Climate Change*. CAB International.

Ravishanker, H. and Rajan, S. (2011). Possible impact of climate change on mango and guava productivity. pp. 151–156. *In*: Dhillon, W.S. and Aulakh, P.S. (eds.). *Impact of Climate Change on Fruit Crops*.

Reay, P.F. (1999). The role of low temperature in the development of the red blush on apple fruit (Granny Smith). *Scientia Horticulturae*, 79: 113–119.

Reddy, A.G.K., Kumar, J.S., Maruthi, V., Venkatasubbaiah, K. and Rao, C.S. (2017). Fruit production under climate changing scenario in India: a review. *Environment and Ecology*, 35(2B): 1010–1017

Reddy, R.P.V., Verghese, A. and Rajan, V.V. (2012). Potential impact of climate change on honeybees (*Apis* spp.) and their pollination services. *Pest Management in Horticultural Ecosystems*, 18: 121–127.

Reddy, Y.N. and Singh, O. (2011). Role of heat shock proteins in adaptivity of plants to higher temperatures and alleviation of heat shock in mango. pp. 57–64. *In*: Dhillon, W.S. and Aulakh, P.S. (eds.). *Impact of Climate Change on Fruit Crops*.

Royle, D.J., Shaw, M.W. and Cook, R.J. (1986). Pattern of development of *Septorianodorume* and *S. tritici* in some winter wheat crops in Western Europe, 1983–84. *Plant Pathology*, 35: 466–476.

Ruck, H.C. (1975). Deciduous fruit tree cultivars for tropical and sub-tropical regions. *Hort. Rev 3*, Commonwealth Burr. Horticulture and Plantation Crops, East Malling, UK.

Samish, R.M. and Lavees, S. (1982). The chilling requirement of fruit trees. pp. 372–388. *In: Proc. of XVI International Horticultural Congress*. Brussels.

Samra, J.S., Singh, G. and Rama Krishna, Y.S. (2003). *Cold Wave of 2002–2003: Impact on Agriculture*. Natural Resource Management Division, ICAR, Krishi Bhavan, New Delhi.

Sarkar, T., Roy, A., Choudhary, S.M. and Sarkar S.K. (2021). Impact of climate change and adaptation strategies for fruit crops. Chapter 4. *In*: Islam, M.N. and van Amstel, A. (eds.). *India: Climate Change Impacts, Mitigation and Adaptation in Developing Countries*. Springer Climate.

Schrader, L.E., Zhang, J. and Duplaga, W.K. (2001). Two types of sunburn in apple caused by high fruit surface (peel) temperature. *Plant Health Progress*, 10: 1094.

Sharma, S., Rana, V.S., Prasad, H., Lakra, J. and Sharma, U. (2021). Appraisal of carbon capture, storage, and utilization through fruit crops. *Frontier in Environmental Science*, 9: 1–10.

Singh, H.P. (2010). Impact of climate change on horticultural crops. pp. 1–8. *In*: *Challenges of Climate Change – Indian Horticulture*. Westville Publishing House, New Delhi.

Singh, N., Sharma, D.P. and Hukam, C. (2016). Impact of climate change on apple production in India: A review. *Current World Environment*, 11(1): 251–259.

Srivastava, S.K. (2021). New challenges on natural resources and their impact on climate change in the Indian. Chapter 1. pp. 1–15. *In*: Islam, M.N. and van Amstel, A. (eds.). *Climate Change Impacts, Mitigation and Adaptation in Developing Countries*. Springer Climate, U.S. Environmental Protection Agency.

Staley, J.T. and Johnson, S.N. (2008). Climate change impacts on root herbivores. *In*: Johnson, S.N. and Murray, P.J. (eds.). *Root Feeders: An Ecosystem Perspective*. Wallingford, UK: CABI.

Sthapit, B.R., Ramanatha Rao, V. and Sthapit, S.R. (2012). *Tropical Fruit Tree Species and Climate Change*. Bioversity International, New Delhi, India.

Sthapit, S.R. and Scherr, S.J. (2013). Tropical fruit tree species and climate change. pp. 15–26. *In*: *Bioversity International*. New Delhi, India.

Sun, J., Chen, S.J. and Xin, J.W. (2009). Analysis on the effect of climate change on mango production in Changjiang City. *Journal of Anhui Agricultural Sciences*, 37: 4962–4963.

Tomasi, D., Jones, G.V., Giust, M., Lovat, L. and Gaiotti, F. (2011). Grapevine phenology and climate change: relationships and trends in the veneto region of Italy for 1964–2009. *Am. J. Enol. Vitic.*, 62: 329–339.

Ubi, B.W., Honda, C., Bessho, H., Kondo, S., Wada, M., Kobayashi. S. and Moriguchi, T. (2006). Expression analysis of anthocyanin biosynthetic genes in apple skin effect of UV–B and temperature. *Plant Science*, 170: 571–578.

Uga, Y., Sugimoto, K., Ogawa, S., Rane, J., Ishitani, M., Hara, N., Kitomi, Y., Inukai, Y., Ono, K., Kanno, N., Inoue, H., Takehisa, H., Motoyama, R., Nagamura, Y., Wu, J., Matsumoto, T., Takai, T., Okuno, K. and Yano, M. (2013). Control of root system architecture by DEEPER ROOTING 1 increases rice yield under drought conditions. *Nat. Genet.*, 45: 1097–1102.

UNFCCC, *United Nations Framework Convention on Climate Change Report*. (2019). Climate action and support trends, based on national reports submitted to the UNFCCC secretariat under the current reporting framework.

Vallat, A., Gu, H. and Dorn, S. (2005). How rainfall, relative humidity and temperature influence volatile emissions from apple trees *in situ*. *Phytochemistry*, 66: 1540–1550.

Varu, D.K., Viradia, R.R., Chovatia, R.S. and Barad, A.V. (2010). Response of different genotypes of custard apple to weather parameters. pp. 24–36. *In: AGRESCO Report-2010*, JAU, Junagadh.

Varu, D.K. and Viradia, R.R. (2015). Damage of mango flowering and fruits in Gujarat during the year 2015. *Survey Report of Department of Horticulture*, JAU, Junagadh.

Vedwan, N. and Rhoades, R.E. (2001). Climate change in the Western Himalayas of India: a study of local perception and response. *Climate Resilient*, 19: 109–117.

Wand, S.J.E., Steyn, W.J., Mdluli, M.J., Marais, S.J.S. and Jacobs, G. (2002). Over tree evaporative cooling for fruit quality enhancement. *South Africa Fruit Journal*, 2: 18–21.

Wand, S.J.E., Steyn, W.J., Mdluli, M.J., Marais, S.J.S. and Jacobs, G. (2005). Use of evaporative cooling to improve 'Rosemarie' and 'Forelle' pear fruit blush color and quality. *Acta Horticulturae*, 671: 103–111.

Wang, S.Y. and Zheng, W. (2001). Effect of plant growth temperature on antioxidant capacity in strawberry. *Journal of Agricultural and Food Chemistry*, 49: 4977–4982.

Wei, J.H. and Mo, R. (2007). Agricultural meteorological disasters and prevention counter measures of affecting on the high-quality fruit project in Baise City. *Guangxi Agricultural Sciences and Technology*, 38: 212–214.

Whiley, A.W., Rasmussen, T.S., Saranah, J.B. and Wolstenholme, B.N. (1989). Effect of temperature on growth, dry matter production and starch accumulation in ten mango (*Mangifera indica* L.) cultivars. *Journal of Horticultural Science and Biotechnology*, 64: 753–765.

Wuebbles, D.J., Fahey, D.W., Hibbard, K.A., DeAngelo, B., Doherty, S., Hayhoe, K., Horton, R., Kossin, J.P., Taylor, P.C., Waple, A.M. and Weaver, C.P. (2017). Executive summary: Climate science special report. pp. 12–34. *In*: Wuebbles, D.J., Fahey, D.W., Hibbard, K.A., Dokken, D.J., Stewart, B.C. and Maycock, T.K. (eds.). *Fourth National Climate Assessment*, volume I. U.S. Global Change Research Program, Washington.

Xu, C., Zhang, Y., Zhu, L., Huang, Y. and Lu, J. (2011). Influence of growing season on phenolic compounds and antioxidant properties of grape berries from vines grown in subtropical climate. *Journal of Agricultural and Food Chemistry*, 59: 1078–1086.

Xue, J.J., Lu, H.C., Wang, H.S. et al. (2008). Investigation of chilling damage to Litchi and Longan in Nanning City and measures for remedy flowering. *China Fruits*, 5: 66–71.

Yadollahi, A. (2011). Evaluation of reduction approaches on frost damages of grapes grown in moderate cold climate. *African Journal of Agricultural Research*, 6(29): 6289–6295.

Yamane, T., Jeong, S.T., Goto-Yamamoto, N., Koshita, Y. and Kobayashi, S. (2006). Effects of temperature on anthocyanin biosynthesis in grape berry skins. *American J. Enology Viticulture*, 57: 54–59.

Zavalloni, C., Andresen, J.A., Black, J.R., Winkler, J.A., Guentchev, G. et al. (2008). A preliminary analysis of the impacts of past and projected future climate on sour cherry production in the Great Lakes Region of the USA. *Acta Horticulturae*, 803: 123–130.

Zekri, M. (2011). Factors affecting Citrus Production and quality. *Citrus Industry*. Online at crec.ifas.ufl.edu.

Zhong, S.Q. (2003). Causes of low flowering success of longan and litchi in Guangxi. *Chinese Journal of Agrometeorology*, 24: 55–57.

Part II
Warm Temperate Fruits

3

Citrus Production in Climate Change Era

Waleed Fouad Abobatta

1. Introduction

1.1 Cultivation and Distribution

Citrus tree is one of the important fruits worldwide, particularly under pandemic conditions of Covid-19 due to the health benefits of the fruit. There is a significant relationship between its yield and climate conditions in different growth stages. Therefore, climate change conditions affect drastically growth and productivity of citrus orchards, particularly in arid and semi-arid regions, which are considered the main commercial production areas of citrus (Wu et al., 2018). Citrus spp. are considered the main genus of citrus trees beside Fortunella and Poncirus genus (Liu et al., 2012; Uzun et al., 2009).

The main areas for citrus production in the Northern Hemisphere is the USA and Mediterranean basin, while in the Southern Hemisphere, Brazil is considered the biggest citrus producer followed by Argentina, South Africa, South Asia region and Australia (United States Department of Agriculture (USDA, 2019)). Currently, there are huge areas planted with citrus spp., particularly under warm and semi-arid regions (Pommer et al., 2009).

Citrus Department, Horticulture Research Institute, Agriculture Research Center, 9 Cairo University St., 1212, Orman, Giza, Egypt.
Email: wabobatta@yahoo.com

According to the Food and Agriculture Organization of the United Nations (FAO), the major cultivation and production areas are concentrated in regions with mild winters, as in the Northern Hemisphere, while China, Brazil, the USA, India, Mexico, Egypt and Spain are considered the world's leading citrus fruit-producing countries. There has been rapid growth in the annual production of the citrus crop in the last decade—from 116.13 million metric tons in 2008 approximately to 124.24 million metric tons in 2016, particularly in oranges, lemons, limes, grapefruit and tangerines which are considered the most important commercial species of citrus fruits. Sweet orange (*Citrus sinensis* (L.) Osb.) represents the major citrus crop worldwide, contributing more than half of the citrus production about (66.97 million metric tons in 2016) in the form of oranges, followed by 32.97 million metric tons of tangerines (*Citrus* spp.), 15.98 million metric tons of lemons (*Citrus limon Burm.* F.) and limes (*Citrus latifolia* Tan.), and 8.32 million metric tons of grapefruit (*Citrus paradise*). Therefore, under the realities of climate change, citrus growers need to manage the situation to maintain their production, thus needing to understand the effects of climate change on citrus productivity (FAO, 2017).

As the universal climate is deteriorating, there is an increase in different phenomena like decrease in precipitation, rise in temperature, increase of drought and soil salinity. Consequently, there is an increase in adverse impacts on the agricultural production with climate variances causing temporal and spatial changes in citrus cultivation (Verner et al., 2016).

2. Health and Medicinal Benefits

There is a gradually increased demand for citrus fruits and their products since the last century with the global consumption of citrus fruits increasing yearly due to their numerous health benefits. Citrus fruits are considered an important food item and are an essential part of our daily diet as it plays a key role in providing various vitamins, particularly vitamin C, numerous phytochemicals, energy, fibers and nutrients. Citrus fruit is consumed fresh or as a processed product; also its peel is the main byproduct during processing in various industries, such as for pectin production and so on (Wang et al., 2020).

There is a strong relationship between regular consumption of citrus fruit and products and the prevention of chronic diseases. For maintenance of human health, daily consumption of citrus fruit provides adequate amounts of vitamin C to adults as per the recommended dietary allowance. Citrus fruit has low protein and very little fat content; the carbohydrate content includes sucrose, glucose and fructose, Citrus fruits also contain several phytochemicals which are considered anti-cancer agents, besides providing a refreshing fragrance, thirst-quenching ability, distinct aroma and delicious taste (Abobatta, 2019a; Liu et al., 2012).

In the current work, we have tried to handle the rise in abiotic stresses, such as temperature, drought and salinity, which challenge citriculture, mainly under the Mediterranean climate. Consequently, we discuss the impacts of climate change on the growth and productivity of citrus in major production regions worldwide and explore the different strategies to adapt citrus orchards to the impact of potential changes in climate conditions.

3. Botany and Classification

Citrus spp. is an important evergreen economic fruit tree grown in different regions of the world and there is a significant relationship between its yield and climate conditions during its different growth stages. Citrus is one of the warm-climate fruit trees; it originated in the humid subtropical regions of Southeast Asia and spread to other regions of the world (Wu et al., 2018).

Citrus, belonging to the family Rutaceae, has numerous evergreen species of genus Citrus containing different commercial varieties, besides genus Fortunella, in addition to Poncirus which, the lonely deciduous genus in citrus, is considered the most tolerant of low temperatures dipping as far as –20°C (Luro et al., 2017; Liu et al., 2012).

Currently, more than 7 million hectares are cultivated with citrus in 140 countries, mostly in warm temperate regions, while, major growing regions are located between latitude 40° North-South where temperature is generally more than 3°C (Court et al., 2017) while the minimum temperature is above –4°C (Narouei-Khandan et al., 2016). Commercial production is concentrated in about 30 countries around the world, the main citrus producers being the USA, Brazil, Argentina, Spain, Egypt, Turkey, Italy, Morocco, South Africa, South Asian region and Australia (USDA, 2019).

In the Mediterranean climate, citrus sprouts one to three times yearly, while in subtropical conditions, the trees produce new flushes throughout the year. Regular sprouting sustains a complex tree structure with regular vegetative and reproductive growth as follows:

1. At spring flush, flowers and subsequently fruits bloom.
2. The other flush at summer-fall generates branches on which spring flushes will develop (Malik et al., 2015).
3. Sometimes due to the warm climate and sufficient nutrition, trees sprout in fall as the third flush (Abobatta, 2019b).

Flowers develop from buds on spring flushes, originating from the previous season's shoots, while the spring sprouts bear the fruits in different citrus varieties except for lemon cultivars which bear fruits three times yearly or more (Albrigo and Chica, 2011).

Indeed, there are numerous types in the genus of citrus that differ from each other in their leaves, flowers, twigs and fruits; so there are various botanical classifications of citrus species. Tanaka (1977) classifies 162 species of citrus. Table 1 shows the botanical classification of citrus genus according to Hodgson (1967) and as cited in Novelli et al. (2006).

The citrus genus is classified into four main groups—acid members, oranges, mandarins and pummelos.

Table 1. Classification of citrus genus according to Hodgson (1967).

No.	Group type	Scientific name	Common name
1	Acid members	*C. medica* (Citron)	Citron
		C. limon	Lemon
		C. jambhiri	Rough lemon
		C. limettioides	Sweet lime
		C. limetta (Lemon × lime)	Sweet lime
2	Oranges	*C. aurautium*	Sour orange
		C. sinensis	Sweet orange
3	Mandarins	*C. reticulata*	Mandarin orange
		C. unshiu	Satsuma mandarin
		C. nobilis (*C. reticulata* × *C. sinensis*)	Mandarin
4	Pummelos	*C. paradisi*	Grape fruit
		C. maxima/ C. grandis	Shaddock or pummelo.

3.1 *Citriculture and Climate Change*

Citrus is adapted to grow in various soil types and climatic conditions of subtropical, warm and semi-arid regions, while the Mediterranean climate is considered preferable. In recent years, several reputed organizations have warned about the threats of climate change to the agriculture industry worldwide and of course, citriculture too is likely to be affected (Lobell et al., 2011). In the climate change era, there is a need to understand the effects of climate change on citrus growth and productivity and the impact of the potential climate changes in the next decades in major citrus-producing regions.

There is a significant relationship between climatic conditions and citrus performance as well as the geographical distribution of production regions due to global warming and continuous fluctuations in climate.

Citrus cultivation is more susceptible to climate change and faces persistently challenging numerous abiotic situations, like temperature fluctuation (Zhu et al., 2011), water deficit (Zhang et al., 2018; Wu et al., 2017), soil salinity (Zhang et al., 2017; Wu and Zou, 2013) and flooding (Mahdavian et al., 2020; Legua et al., 2018), which increase deterioration in productivity and fruit quality. For that, exposure of citrus tree to a combination of abiotic stresses at the same time increases deleterious effects than any individual factor in citriculture in terms of yield and fruit quality (Santos et al., 2017). The adverse effects of climate change affect various metabolic activities (Fig. 1), like cell division, stem elongation, root system architecture, gas exchange, photosynthesis and increasing oxidative damage in addition to increasing pest infestation, disease frequency and inflated production cost. Ultimately, climate change seriously obstructs citriculture worldwide (Zandalinas et al., 2017).

Nevertheless, there is variation in the capability of citrus varieties to maintain reasonable growth and productivity under abiotic stresses, like drought and salinity, due to different factors, including planting systems, growth stage and tree age which affect water and nutrient uptake, stomatal opening, photosynthesis and transpiration (Rodrigues-Gamir et al., 2011).

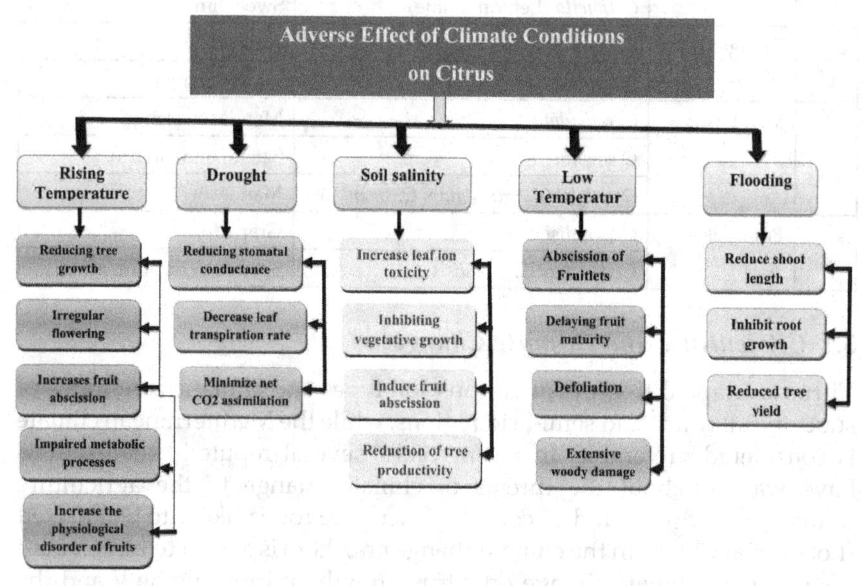

Fig. 1. Schematic diagram of the adverse effects of climate change on citrus growth and productivity.

3.2 Temperature and Citrus Trees

Citrus as a warm-climate fruit tree which grows between 12.8–35°C without freezing nights; also the soil temperature affects the root growth which requires 10°C at least to recover metabolism activity. While the optimal temperature for citrus growth is between 25–35°C, maximum photosynthesis performance occurs at 30°C (El-Aidy et al., 2018). On the other side, cool weather is a restrictive factor in citrus cultivation. The low temperature for a short period reduces the growth rate and may even stop the growth when freezing for a long period is deleterious for the whole tree. Therefore, citrus-producing is commercially viable in areas positioned between 40° North and South latitude where the minimum temperature is generally around –4°C (Pereira et al., 2017).

Temperature fluctuation affects the vegetative growth and productivity of citrus trees. Low and high temperatures cause numerous damages to new sprouts of citrus trees, delays flowering patterns, increases dropping of flowers, aborts fruit set and produces poor fruit quality. On the other side, some varieties, like Satsuma mandarin, can tolerate temperatures as low as 3°C for a short time, while the genus of Poncirus is the only deciduous citrus rootstock to tolerate freezing. Therefore *Poncirus trifoliata* is preferred as rootstocks for citrus cultivation in cold areas. Also, the hybrid of *Poncirus trifoliata* enhances the hardiness of scion, making it cold-tolerant. Therefore, its use as a rootstock is preferred in cold areas (Zabihia et al., 2016).

3.2.1 Influence of High Temperature

As mentioned above, citrus growth increases with rise in temperatures from 12°C up to 30°C, while the growth slows down subsequently to stop completely at 40°C or above (Kumar et al., 2011). Due to global warming, rise in temperature of more than 2.5°C is expected in the next decades with decrease in rainfall by about 10–15% by 2050. This however, affects negatively the growth and productivity of various fruit trees (Dosio et al., 2018; Pathak et al., 2012). For example, harsh climate conditions that occurred during the 2019–20 season, particularly the rise in temperature restricted flowering and fruit set, leading to a serious decrease in citrus production, especially in Spain and Morocco (USDA, 2020).

Previous scientific studies have reported adverse effects of exceeding temperatures in the last few decades on citrus growth and productivity. In addition, there are expected changes in physiological, hormonal and molecular behavior of citrus tree as a response to unfavorable high temperatures that will affect growth and yield in the next decades (Abobatta, 2019b; Balfagon et al., 2018; Vives-Peris et al., 2017; Zhang and Sonnewald, 2017; De Ollas et al., 2018; Devireddy et al., 2018). There are significant changes in the behavior of trees as a result of fluctuations in

Fig. 2. Effect of high temperature on tree growth (Field image by Dr. Abobatta, 2021).

temperature during the growth season (Fig. 2), affecting the growth and productivity of citrus trees, depending on the intensity and duration of temperature (Shanker et al., 2018).

3.2.2 Adverse Effects of High Temperature

Rising temperatures negatively affect the various growth stages and productivity of citrus trees, particularly during flowering and fruit set. It increases abortion of fruit set. Also, a high temperature alters citrus fruit growth and directly affects the quality and total yield.

The numerous adverse effects of high temperature are as follows:

- Reduction in tree growth.
- Irregular flowering by way of timing and longevity.
- Increase in fruit abscission.
- Thermal degradation of photosystem under high temperature.
- Impaired metabolic processes.
- Increase in the physiological disorder of fruits (e.g., sunburn, cracking and splitting).

For instance, during the 2015–16 season, there was a reduction in orange production globally due to increase in temperature; so, there is was a yield reduction in the major production areas, like the USA, Brazil, Spain, Egypt and South Africa (USDA, 2019; Fares et al., 2017).

3.2.3 Warm Nights

Warm nights during winter represent a serious threat for citrus production by inhibiting flowering, delaying flowering and reducing crop yield (Kumar et al., 2011).

3.2.4 Effects of Low Temperature

Low temperature is one of the main limiting aspects of citrus cultivation globally, as it restricts the growth of citrus and reduces tree productivity. So, temperature below 13°C restricts tree vigor and delays fruit maturity, while frost destroys the yield and could have a deleterious impact on the entire tree (Huchche et al., 2010). There are different injuries due to low temperature on citrus trees including damage to new flushes, increase dropping of fruitlets, delay in fruit maturity; additional hours under freezing temperature could destroy the vegetative growth (Wang et al., 2016). Therefore, cold temperature is considered the chief determining factor in citrus cultivation.

There are various negative effects of low temperature on citrus cultivation:

* Abscission of fruitlets and flowers at –1.7°C.
* Delay in fruit maturity and damage to fruits at –2°C.
* Defoliation at –4.4 to –5.6°C.
* Browning of main branches and stem at –6.7°C.
* Frost causes extensive damage to wood of young trees or may even lead to death of the tree.

3.2.5 Heat Waves

Owing to climate change, there are increasing heat waves both by way of duration and intensity. Heat waves are expected to increase considerably in the coming decades, affecting dramatically the fruit set, fruit quality and leading to significant losses in the yield worldwide (Lesk et al., 2016). In 2019, heat waves hit the Mediterranean region accompanied with a high temperature of more than 30°C and high solar radiation accompanied with drought conditions, which negatively affected citrus production and loss of about 15–20% of citrus production, particularly in Spain, Turkey, and Morocco (FAO, 2019).

There are adverse effects of heat waves on the productivity of different citrus varieties, primarily the seedless cultivars, such as Navel orange, Shamouti orange, a number of mandarin and lemon cultivars. Moreover, the influence of heat waves is severe during the flowering and fruit set. Also the month of June sees a severe drop in the yield (Salvi et al., 2013).

3.3 *Drought*

Due to global warming and decrease in rainfall, drought has led to severe abiotic stress in agriculture during the last few decades. According to the global warming scenario, there are fears of facing more drought periods combined with frequent hot days and the temperature rising more than

30°C. There is a prospect of several regions worldwide becoming susceptible to drought in the coming era, though the severity and longevity would vary from region to other due to other factors. The arid and semi-arid regions will be more susceptible for negative effects of severe drought (Balfagon et al., 2021; McDowell et al., 2018; Michaelides et al., 2018).

Drought has a major impact on citrus production due to increasing evaporation and shortage of water with limited rainfall. The response of citrus tree to drought stress is complex and is related to other biotic and abiotic stress conditions (Abobatta, 2020; Farooqi et al., 2020). Drought stress has adverse effects on citrus production and may restrict the cultivation of citrus in some regions worldwide (Vincent et al., 2020), while the combination of drought with rising temperature may increase the negative impact on citrus, both in tree vigor and productivity (Fig. 3).

In addition, drought affects the fruit quality and quantity. So, knowledge is necessary about how the potential climate change would affect the availability of water resources in citrus production globally and how to sustain citrus production (Rodriguez et al., 2019).

Various factors affect the growth and productivity of citrus under drought conditions and water shortages that including:

- The physiological growth stage.
- The duration of water shortage.
- The degree of water stress.
- Evapotranspiration.
- Transpiration intensity.
- Age and vigor of trees.

Indeed, there are different elements that affect citrus tree's water requirements and these depend on the variety, the season, temperature,

Fig. 3. Impact of drought on lemon growth in the Egyptian desert (Field image by Dr. Abobatta, 2020).

solar radiation, geographical region, rootstock-scion combination and soil characteristics.

3.3.1 *Influence of Drought on Citrus Cultivation*

Generally shortage of water supply impairs the reproductive processes of plants, leading to a reduction in yield and impacting citrus growth-varying ability, which depends on the time of occurrence (it is more deleterious during summer or autumn flushing, whereas there is little effect in spring flushing (Chang et al., 2020; Vincent et al., 2020; Abobatta, 2019c)). The variation in drought pattern during fruit growth is more pronounced in semi-arid regions, where lower precipitation accompanied with temperatures above the optimal requirement cause harmful effects on citrus productivity (Lamaoui et al., 2018). Moreover, drought increases fruit abscission and induces irregular fruit growth, which decrease the yield and fruit quality (Agusti and Primo-Millo, 2020).

Drought causes various physiological disorders of citrus tree:

- Reduces stomatal conductance.
- Decreases leaf transpiration rate.
- Minimizes net CO_2 assimilation.

3.4 *Flooding*

Actually, there are no citrus species which are tolerant and able to adapt to permanent soil-flooding conditions. Citrus is subjected to seasonal flooding in some citrus-producing regions, like the coasts of the Mediterranean, Florida, Brazil and some areas of China, affecting dramatically the growth (reduced shoot length and inhibited root growth) and tree yield (Fares et al., 2017). Currently, there are some tolerant rootstocks against temporary waterlogging, like Carrizo citrange (C. *Sinensis* × *P. trifoliata*) or citrumelo CPB-4475 (*C. paradise* × *P. trifoliata*), which have the capacity to grow under temporary flooding conditions (Ferrarezi et al., 2020; Arbona et al., 2009). On the other hand, sensitive rootstocks, such as of Cleopatra mandarin deplete under flooding or excessive irrigation particularly in clay soil (Fig. 4). Therefore, under flood-prone conditions, tolerant varieties are the most important for growers (Hossain et al., 2009).

Nevertheless, due to climate change conditions, there is a likelihood of intensive rains in certain regions. There are two factors that affect citrus growth under flood-prone conditions and determine the citrus ability to recovery after stress:

* Flooding period
* Tolerant variety

Fig. 4. Effect of excessive irrigation on tree growth (Modified from Abobatta, 2021).

4. Soil Salinity

Citrus, classified as sensitive plants to salinity conditions in either water or soil, experience different morphological and physiological effects of salinity stress and may face yield loss (Vincent et al., 2020; Mesquita et al., 2015). Soil salinity is considered as one of the major abiotic restraints in citrus cultivation, particularly under irrigation conditions in arid and semi-arid regions with high concentrations of salts. Also, salinity and drought in the calcareous soil of the Mediterranean region can lead to major problems (Abd-Elgawad, 2020).

Under repeated droughts accompanied by high temperatures, there is increase in average annual evapotranspiration. Consequently, there is a rising demand for water and excessive irrigation, particularly in arid and semi-arid regions, so that the water-table gets raised and salts get located close to the root zone. Thus, increase in soil salinity inhibits citrus growth and productivity (Aouad et al., 2015) due to

- Water deficit.
- Disturbance in plant metabolism.
- Nutritional imbalance.
- Specific ion toxicity.

It has been well documented that increase in salinity above 1.4 dS/m decreases citrus production and the yield loses about 13 per cent for every increase of 1 dS/m, while growth and productivity of citrus are impaired at soil salinity of about 2 dS/m without any concomitant expression of leaf symptoms (Islam et al., 2019; Brito et al., 2017; Raga et al., 2014; Syvertsen and Garcia-Sanchez, 2014; Kumar et al., 2012; Abadi et al., 2010). There are severe effects of salinity on citrus growth and productivity, as shown

Fig. 5. Field image of adverse effects of salinity on orange tree (field image by Dr. Abobatta, 2020).

in Fig. 5, particularly in regions with limited water. Use of underground water for irrigation increases soil salinity (Cimen and Yesiloglu, 2016). The impact of salinity on citrus growth is related to other factors, including soil characteristics, drought, high temperature, water availability, besides the agricultural practice employed. In this context, citrus production impairs and maybe destroys the whole tree under high-salinity conditions (Colmenero-Flores, 2020).

4.1 *Some Adverse Effects of Salinity on Citrus Tree*

* Increase in leaf ion toxicity.
* Inhibiting of vegetative growth.
* Induction of fruit abscission.
* Reduction of tree productivity.
* Production of poor fruit quality.

4.2 *Salinity Tolerance of Citrus*

Due to climate change conditions, there is an increment in soil salinity in major citrus production regions. Therefore, more attention is to be paid to generating salinity-tolerant hybrids genotypes, be they rootstocks or grafted varieties, to sustain citrus production, particularly under arid and semi-arid conditions seen in the Mediterranean climate (Etehadpour et al., 2020).

The tolerance to salinity in citrus varies with the rootstocks and grafted varieties. There are citrus species which can tolerate or resist salinity, like mandarin (*Citrus reticulata*) and pummelos (*C. maxima*), and the related genotypes, like Troyer Citrange, Citrumelo rootstocks and trifoliate citrus hybrids (Hussain et al., 2012). Previous studies verified the tolerance based on some mechanisms, like early induction, salt avoidance mechanism, ion exclusion from shoot organs, tight control of transpiration, rapid inhibition of photosynthesis, primary metabolism, biosynthesis of protective molecules, checking the charges in gas exchange and chlorophyll fluorescence (Forner-Giner et al., 2020; Brito et al., 2016; Aouad et al., 2015). The same trend was noticed by Sa et al. (2017) in lemon and mandarin hybrids; Brito et al. (2017) on trifoliate citrus hybrids; Fadli et al. (2014) on some Citrumelo cultivars; Silva et al. (2014) and Brito et al. (2016) on trifoliate hybrid citrus and on Sunki hybrid rootstock; Silva et al. (2019) on Tahiti acid lime grafting on Sunki mandarin hybrids.

5. Influence of Climate Change on Citrus Growth and Productivity

Citrus cultivation requires specific climate conditions including air temperature, humidity, solar radiation and light intensity, whereas yield and fruit quality are very sensitive to temperature, drought and salinity. Climate change would contribute to fluctuations in the timing of flowering, fruit set and fruit maturity. Also the increasing impact of biotic stress would affect negatively the citrus production. Effects of climate change on citrus varies widely due to geographical, spatial and climate type, while a change in citrus-producing regions globally would depend on availability of water for irrigation, particularly for growing sweet orange (Dalal et al., 2017; Dolkar et al., 2017).

There are the following different trends:

1. Arid and semi-arid regions, particularly the coastal areas face salinity challenges.
2. In the Mediterranean climate, there is frequency of droughts which tend to be lengthy too.
3. In humid subtropical regions, freezing is less frequent.

5.1 *Effect of Climate Change on Flowering and Fruit Set*

Citrus flowering occurs once yearly in the commercial fruit-producing regions (with the exception of lemon and lime) and that too during spring after the cool winter is over. This is seen in temperate climates like that in the Mediterranean when flowers are produced within three to four weeks. However, flowering intensity and duration are related to climate conditions, particularly the temperature during autumn. In tropical and subtropical climates, most of the citrus plant species bloom into flowers throughout the year (Distefano et al., 2018; Avila et al., 2012).

According to previous research, environmental factors play an essential role in the initiation and induction of flowering buds on citrus trees. The majority of flowers form on new shoots that originate from vegetative growth of the previous season. Global warming has an adverse effect on flowering. In this context, temperature affects flowering intensity and longevity, therefore, the warm weather during bud initiation stage impairs the forming of the flower buds. Therefore, warm nights during autumn and winter affect negatively and delay flowering. Thus predicting climate conditions and proper management of flushing are effective to control flowering and fruit set (Jadhav et al., 2020; Nawaz et al., 2019; Sharma et al., 2017; Malik et al., 2015).

On another hand, the citrus tree needs sufficient hours of cool temperature, ranging between 15–20°C to form flower parts within the bud and sprout inflorescence. Increase of temperature more than 30°C weakens the development of the flower buds and has an adverse effect on the type of inflorescence and leaves (Balfagon et al., 2018).

Regarding fruit set, fluctuation in temperatures (higher or lower) affects negatively the fruit set, reduces fertilization, apportions fruit set and increases fruitless dropping, particularly in seedless varieties. So, higher temperatures would inevitably reduce the crop of different citrus varieties (Bennici et al., 2019; Irenaeus and Mitra, 2014; Sugiura et al., 2012; Kumar et al., 2011).

5.2 *Impact on Productivity*

The development of citrus fruit is considerably related to climatic variables. Consequently, there is an economic effect on citrus productivity due to fluctuations in climate conditions, like temperature, soil moisture, solar radiation and wind speed during the growth season, leading to significant losses in the fruit yield (Lobell et al., 2011). Climate change conditions increase the projected drop rate of citrus fruit, affect fruit quality and reduce the tree yield. Thus rising temperature and drought are considered the main reasons for this projection, which may reach 17 per cent (FAO, 2017).

The effect of climate conditions on citrus production has been described in previous researches which state that the response depends on the frequency of stress rate and the phenological stages. Therefore, the expected increase in climate variables, such as drought and high temperature particularly during fruit growth phases I and II, will have a dramatic impact on citrus productivity (Agusti and Primo-Millo, 2020; Ouyang et al., 2019; Abobatta, 2019b; Zandalinas et al., 2018), decrease the growth of fruit (Nawaz et al., 2020a), affect the external and internal fruit characteristics (Lado et al., 2018; Juan and Jiezhong, 2017), delay fruit maturity (Manera et al., 2012) and increase the physiological disorders (Nawaz et al., 2019).

5.3 *Effect on Fruit Growth*

There are various factors that affect fruit growth and these include time, location and the climate conditions which determine the longevity of different phenophases of fruit development. In this context, there is a remarkable influence of climate conditions on all the growth phases of citrus fruit (Dalal et al., 2018; Chang and Lin, 2020). For instance, in tropical regions, the citrus tree begins to bloom earlier and the fruit grows quickly during the cell-enlargement phase but the fruit reaches the ripening phase later with low fruit quality, like suppressed peel color (Singh et al., 2015).

Rising temperature and drought stress over the last few decades has altered citrus fruit growth and restricted the total crop, producing poor fruit quality as shown in Fig. 6 and increasing physiological disorders, like fruit cracking, fruit splitting and creasing (Rodriguez et al., 2019; El-Aidy et al., 2018; Juan and Jiezhong, 2017).

Fig. 6. Effect of high temperature on fruit growth (Field image by Dr. Abobatta, 2021).

5.4 *Effect on Maturity and Harvesting*

It has been described that the fluctuation of climate alters the fruit maturity which may be earlier in some cultivars, like New Hall orange and later in another variety, like Murccot mandarin (Micheloud et al., 2020). Cold night temperature increases anthocyanin and carotenoid in pericarps which accelerate the peel coloring (Brotons et al., 2013) on lemon fruits (*Citrus lemon* L. Burm. f.) (Manera et al., 2013) and on grapefruit (*Citrus paradisi* Mac f.).

The rising day temperatures and warm nights during the fruit maturation stage could diminish the fruit quality, cause poor coloration of the citrus fruit, delay peel coloring of lemon (*Citrus limon* L. Burm. f) (Erena et al., 2019) and the harvesting period (Manera et al., 2012). Also, Chang and Lin (2020) reported that rising temperatures inhibit the development of Kumquat fruit and affect fruit quality. On the other hand, there are adverse effects of frequent rainfall during the fruit maturity stage as it affects negatively the fruit quality and increases infection by fungal pathogens, like fruit rot, particularly in late mature varieties, like Valencia orange and Kinnow mandarin (Khalid et al., 2018).

Indeed, the negative effects of climate change on fruit quality include among others:

* The fluctuations of climate change alter fruit maturity.
* Reduce fruit quality.
* Increase sunburn, particularly in easy-peeling varieties, like mandarins and Fairchild's orange.
* Spread physiological disorders, including fruit cracking, splitting, creasing, etc.
* Reduce fruit rigidity, decrease total sugar acidity ratio and increase pulp softening.

6. Climate Change and Biotic Stresses

Indeed, there are various challenges facing the citrus industry regarding climate change and which cause economic losses, biotic stresses, etc. (Ullah et al., 2015). In this context, citrus is suspected of being attacked by different pests and pathogens, and the magnitude of this phenomenon is noticeable in various citrus production regions worldwide, as shown in Fig. 7.

Rising temperature has increased the biotic stress and accelerated spread of pests and diseases that include, among others, fruit fly

Fig. 7. Impact of climate change by spreading pests and diseases (field image by Dr. Abobatta, 2021).

(Diptera: Tephritidae), citrus thrips (*Scirtothrips citri* (Moulton)), citrus rust mite (*Phyllocoptruta oleivora*), red scale (*Aonidiella aurantii*), Nematodes (*Tylenchulus semipenetrans*), citrus Tristeza virus, citrus gummosis (*Phytophthora citrophthora*), citrus psorosis virus, citrus canker (*Xanthomonas campestris* pv. Citri), alternaria stem-end rot, alternaria brown spot, root rot disease, dry root rot disease (*Fusarium solani*), and greening (Huanglongbing) disease (Nawaz et al., 2020b; Xu et al., 2013).

7. Adaptation Strategies to Alleviate the Influence of Climate Change

Climate change will affect various agricultural activities and can be acute for citrus orchards because of rising temperatures and extended periods of drought, particularly in arid and semi-arid regions (Korres et al., 2016). Plants use different strategies to adapt to threats of climate change and these include indirect effects through increasing evapotranspiration and closing stomatal, which in turn, affects various metabolic processes:

* Compatible solute accumulation and osmotic protection (Sharma et al., 2019).
* Accumulated intracellular proline (Brito et al., 2019).
* Synthesis and accumulation of glycine betaine within the cell (Gupta and Huang, 2014).
* Activation of antioxidant enzymes and synthesis of antioxidant compounds.
* Cell and tissue water conservation (Vives-Peris et al., 2017).

Currently, more attention is being paid to new adaptation strategies which address these challenges to sustain citrus cultivation in different production regions. Though different adaptation strategies are used by citrus growers in coping with climatic changes to preserve productivity and reduce adverse effects of climate change on citrus (Garcia-Tejero et al., 2013; Martínez-Ferri et al., 2013). In addition, due to environmental stress, there is a change in citrus tree behavior as it tries to adapt to environmental conditions in order to reduce unwanted negative impacts on growth, yield and fruit quality.

7.1 *Some Adaptation Strategies*

7.1.1 Canopy Management

Under climate threat, tree size must be controlled to adapt to the diverse environmental conditions, whereas, the size of citrus trees is subjected to the scion/rootstock combination and environment conditions in order to increase the efficiency of tree productivity, improve fruit quality and facilitate harvesting (Donadio et al., 2018). Therefore, canopy management, high density planting and the use of dwarf rootstock lead to more productivity and enhancing fruit quality under climate change conditions (Ziogas et al., 2021; Nawaz et al., 2020b).

7.1.2 Top Netting

As a result of climate change, there is increase in dramatic climatic phenomena represented through extreme heat waves, rising temperatures, excessive sunlight, strong winds during various fruit-growth stages which are result in reduced productivity of citrus trees combined with poor fruit quality. Therefore, top netting practice presents a practical solution for harsh climate conditions. It could play an important role in protecting citrus trees against dramatic environmental events through modified microclimate. Shading, absorbing some spectral regions and controlling light intensity are imperative for preserving fruit sugar content and rind color (Dovjek et al., 2020, 2021; Cronje et al., 2020; Wang et al., 2019).

7.1.3 Breeding Salinity-tolerant Hybrids Genotypes

Due to climate changes, more attention is given to generating salinity-tolerant hybrid genotypes, be they rootstocks or grafted varieties, with positive responses to drought and salinity conditions to sustain citrus production, particularly under arid and semi-arid conditions as in the Mediterranean climate (De Souza et al., 2017). So, looking for genotypes adapted to stress conditions and acclimatizing with other abiotic stresses is considered an essential strategy to protect the citrus industry worldwide (Ziogas et al., 2021).

7.1.4 Citrus Management to Mitigate Climate Change

It is well documented that citrus production is vulnerable and sensitive to risks of climate change. With the acceleration of climate change since the second half of the last century, there is a gradual increase in abiotic stress confronting citrus cultivation, particularly in the Mediterranean climate (Steiner et al., 2018; Zandalinas et al., 2018; Hailai et al., 2010). On the one hand, production of citrus requires further techniques to suit the current climate conditions, like the use of rootstocks adapted to diverse soil and climate conditions, use of protected cultivation as a part of the production system of citrus, besides modifying the schedule of various agricultural practices through proper timing (Morais et al., 2020). On the other hand, citrus tree uses various mechanisms in response to threats of stress by adapting and surviving under conditions which include stomatal regulation, accumulation of proteins and organic acids (Abobatta, 2019c; Forni et al., 2017).

8. Conclusion

Citrus tree grows in different ecological zones while its commercial production is concentrated in about 30 countries around the world, particularly in the Mediterranean region. There is a significant relationship between climatic conditions and citrus production. Due to fluctuations in the climatic conditions, citrus is more susceptible to numerous abiotic stresses, like rising temperature, drought, soil salinity, flooding, low temperature and heat waves. In addition, there is widespread attack by pests and pathogens as a result of climate fluctuations which affect citrus cultivation, increasing deterioration in productivity and reducing fruit quality. Therefore, there is a change expected in citrus-producing regions globally due to climate conditions and the availability of irrigation water, particularly for sweet orange.

Citrus tree uses various metabolic processes, like synthesis and accumulates compatible solutes and resorts to osmotic protection in order to adapt to the impact of fluctuations in climatic conditions. Currently, there is more attention given to adaptation strategies which address these challenges so as to sustain the citrus industry and reduce the adverse effects of climate change on citrus in different regions. These adaptation strategies include canopy management, using top netting practice and breeding salinity-tolerant genotypes.

References

Abadi, F.S.G., Mostafavi, M., Eboutalebi, A., Samavat, S. and Ebadi, A. (2010). Biomass accumulation and proline content of six citrus rootstocks as influenced by long-term salinity. *Research Journal of Environmental Sciences*, 4(2): 158–165.

Abd-Elgawad, M.M. (2020). Managing nematodes in Egyptian citrus orchards. *Bulletin of the National Research Centre*, 44(1): 1–15.

Abobatta, W.F. (2019a). Nutritional benefits of citrus fruits. *Am. J. Biomed. Sci. & Res.*, 3(4). Doi 10.34297/AJBSR.2019.03.000681.

Abobatta, W.F. (2019b). Influence of climate change on citrus growth and productivity (effect of temperature). *Adv. Agri. Tech. Plant Sciences*, 2(4): 180036.

Abobatta, W.F. (2019c). Drought adaptive mechanisms of plants—A review. *Adv. Agr. Environ. Sci.*, 2(1): 42–45. Doi: 10.30881/aaeoa.00021.

Abobatta, W.F. (2020). Plant responses and tolerance to combined salt and drought stress. pp. 17–52. *In*: Hasanuzzaman, M. and Tanveer, M. (eds.). *Salt and Drought Stress Tolerance in Plants*. Springer, Cham.

Agusti, M. and Primo-Millo, E. (2020). Flowering and fruit set. pp. 219–244. *In*: *The Genus Citrus*. Woodhead Publishing.

Albrigo, L.G. and Chica, E.J. (2011). Citrus shoot age requirements to fulfill flowering potential. *Proc. Fla. State Hort. Soc.*, 124: 56–59.

Aouad, A.E., Fadli, A., Aderdour, T., Talha, A., Benkirane, R. and Benyahia, H. (2015). Investigating salt tolerance in citrus rootstocks under greenhouse conditions using growth and biochemical indicators. *Biolife*, 3(4): 827–837. Doi:10.17812/blj.2015.3413.

Arbona, V., Lopez-Climent, M.F., Perez-Clemente, R.M. and Gomez-Cadenas, A. (2009). Maintenance of a high photosynthetic performance is linked to flooding tolerance in citrus. *Environ. Exp. Bot.*, 66: 135–142. https://doi.org/10.1016/j.envexpbot.2008.12.011.

Avila, C., Guardiola, J.L. and Nebauer, S.G. (2012). Response of the photosynthetic apparatus to a flowering-inductive period by water stress in citrus. *Trees*, 26(3): 833–840.

Balfagon, D., Zandalinas, S.I. and Gomez-Cadenas, A. (2018). High temperatures change the perspective: integrating hormonal responses in citrus plants under co-occurring abiotic stress conditions. *Physiologia Plantarum*. Doi:10.1111/ppl.12815.

Bennici, S., Distefano, G., Las Casas, G., Di Guardo, M., Lana, G., Pacini, E., Malfa, S. and Gentile, A. (2019). Temperature stress interferes with male reproductive system development in clementine (*Citrus clementina* Hort. ex. Tan.). *Annals of Applied Biology*, 175(1): 29–41.

Brito, C., Dinis, L.T., Moutinho-Pereora, J. and Correia, C.M. (2019). Drought stress effects and olive tree acclimation under a changing climate. *Plants*, 8(7): 232. https://doi.org/10.3390/plants8070232.

Brito, M.E.B., Sa, F.V.S., Filho, W.S.S., Silva, L.A. and Fernandes, P.D. (2016). Gas exchange and fluorescence of citrus rootstocks varieties under saline stress. *Revista Brasileira de Fruticultura*, 38(2): e-951. http://dx.doi.org/10.1590/0100-29452016951.

Brito, M.E.B., Sa, F.V.S., Silva, L.A., Filho, W.S.S., Gheyi, H.R., Moreira, R.C.L., Fernandes, P.D. and Figueuredo, L.C. (2017). Saline stress on to growth and physiology of trifoliate citrus hybrids during rootstock formation. *Biosci. J.*, Uberlandia, 33(6): 1523–1534.

Brotons, J.M., Manera, J., Conesa, A. and Porras, I. (2013). A fuzzy approach to the loss of green colour in lemon (*Citrus lemon* L. Burm. f.) rind during ripening. *Computers and Electronics in Agriculture*, 98: 222–232. https://doi.org/10.1016/j.compag.2013.08.011.

Chang, Y.C. and Lin, T.C. (2020). Temperature effects on fruit development and quality performance of Nagami Kumquat (*Fortunella margarita* [Lour.] Swingle). *The Horticulture Journal*, UTD-120. https://doi.org/10.2503/hortj.UTD-120.

Chang, Y.C., Chang, Y.S., Chen, I.Z. and Lin, L.H. (2020). *Phenological Characteristics of Potted Kumquat under Protected Culture*, 38(2): 130–145.

Cimen, B. and Yesiloglu, T. (2016). Rootstock breeding for abiotic stress tolerance in citrus. pp. 527–563. *In*: Shanker, A.K. and Shanker, C. (eds.). *Abiotic and Biotic Stress in Plants—Recent Advances and Future Perspectives*. InTech, Rijeka.

Colmenero-Flores, J.M., Arbona, V., Morillon, R. and Gomez-Cadenas, A. (2020). Salinity and water deficit. pp. 291–309. *In*: *The Genus Citrus*. Woodhead Publishing.

Court, C.D., Hodges, A.W., Rahmani, M. and Spreen, T.H. (2017). *Economic Contributions of the Florida Citrus Industry in 2015–16*, final sponsored project report to the Florida Department of Citrus, University of Florida-IFAS, Food & Resource Economics, Gainesville, FL, May 2017. Available at http://flcitrusmutual.com/citrus-101/citrushistory.aspx.

Cronje, P.J.R., Botes, J., Prins, D.M., Brown, R., North, J., Stander, O.P.J., Hoffman, E.W., Zacarias, L. and Barry, G.H. (2020). The influence of 20% white shade nets on fruit quality of Nadorcott mandarin. *Acta Hortic.*, 1268: 279. https://doi.org/10.17660/ActaHortic.2020.1268.36.

Dalal, R.P.S. and Raj Singh, A.K. (2017). Prevailing weather condition impact on different phenophases of Kinnow Mandarin (*Citrus nobilis Lour* Citrus deliciosa Tenore*). *Int. J. Pure App. Biosci.*, 5(2): 497–505.

De Ollas, C., Arbona, V., Gomez-Cadenas, A. and Dodd, I.C. (2018). Attenuated accumulation of jasmonates modifies stomatal responses to water deficit. *J. Exp Bot.*, 69: 2103–2116.

De Souza, D.J., Silva, A.E.M., Filho, C.M.A., Morillon, R., Bonatto, D., Micheli, F. and Gesteira, S.A. (2017). Different adaptation strategies of two citrus scion/rootstock combinations in response to drought stress. *PLoS ONE*, 12(5): e0177993. https://doi.org/10.1371/journal.pone.0177993.

Devireddy, A.R., Zandalinas, S.I., Gomez-Cadenas, A., Blumwald, E. and Mittler, R. (2018). Coordinating the overall stomatal response of plants: rapid leaf-to-leaf communication during light stress. *Sci. Signal*, 11: eaam9514.

Distefano, G., Gentile, A., Hedhly, A. and La Malfa, S. (2018). Temperatures during flower bud development affect pollen germination, self-incompatibility reaction and early fruit development of clementine (*Citrus clementina* Hort. ex Tan.). *Plant Biology*, 20(2): 191–198.

Dolkar, D., Bakashi, P., Wali, V.K., Khushu, M.K., Singh, M., Bhushan, B. and Shah, R.A. (2017). Effect of climate variability on growth, yield and quality of Kinnow mandarin. *Journal of Agrometeorology*, 19(5): 67–47.

Donadio, L.C., Lederman, I.E., Roberto, S.R. and Stucchi, E.S. (2018). Dwarfing-canopy and rootstock cultivars for fruit trees. *Rev. Bras. Frutic. Jaboticabal.*, 41(3): (e-997). Doi: http://dx.doi.org/10.1590/0100-29452019997.

Dosio, A., Mentaschi, L., Fischer, E.M. and Wyser, K. (2018). Extreme heat waves under 1.5°C and 2°C global warming. *Environ. Res. Lett.*, 13(2018): 054006.

Dovjek, I., Nemera, D.B., Wachsmann, Y., Shlizerman, L., Ratner, K., Kamara, I., Morozov, M., Charuvi, D., Shahak, Y., Cohen, S. and Sadka, A. (2020). Top netting as a practical tool to mitigate the effect of climate change and induce productivity in citrus: Summary of experiments using photo-selective nets. *Acta Hortic.*, 1268: 265–270. https://doi.org/10.17660/ActaHortic.2020.1268.34.

Dovjek, I., Nemera, D.B., Cohen, S., Shahak, Y., Shlizerman, L., Kamara, I., Florentin, A., Ratner, K., McWilliam, S.C., Puddephat, I.J., FitzSimons, T.R., Charuvi, D. and Sadka, A. (2021). Top photoselective netting in combination with reduced fertigation results in multi-annual yield increase in Valencia Oranges (*Citrus sinensis*). *Agronomy*, 11(10): 2034. https://doi.org/10.3390/agronomy11102034.

El-Aidy, A.A., Alam-Eldein, S.M. and Esa, M.W. (2018). Effect of organic and bio-fertilization on vegetative growth, yield, and fruit quality of Valencia orange trees. *J. Product. & Dev.*, 23(1): 111–134.

Erena, M., Brotons, J.M., Conesa, A., Manera, F.J., Castaner, R. and Porras, I. (2019). Influence of climate change on natural degreening of lemons (*Citrus limon* L. Burm. f). *J. Agr. Sci. Tech.*, 21: 169–179.

Etehadpour, M., Fatahi, R., Zamani, Z., Golein, B., Naghavi, M.R. and Gmitter, F. (2020). Evaluation of the salinity tolerance of Iranian citrus rootstocks using morph-physiological and molecular methods. *Scientia Horticulturae*, 261: 109012.

Fadli, A., Chetto, O., Talha, A., Benkirane, R., Morillon, R. and Benyahia, H. (2014). Characterization in greenhouse conditions of two salt tolerant Citrumelo (*Citrus paradisi Macf.* x *Poncirus trifoliata* (L.) Raf.) cultivars. *Journal of Life Sciences*, 8: 955–966. Doi: 10.17265/1934-7391/2014.12.005.

FAO, Food and Agriculture Organization of the United Nation. (2017). *Citrus Fruit – Fresh and Processed Statistical Bulletin 2016*. http://www.fao.org/3/a-i8092e.pdf.

FAO. (2017). *Citrus Fruit Fresh and Processed Statistical Bulletin 2016*. FAO: Rome, Italy, 2017. http://www.fao.org/3/a-i8092e.pdf.

FAO. (2019). *The State of Food and Agriculture (2019). Moving Forward on Food Loss and Waste Reduction*. Rome. Licence: CC BY-NC-SA 3.0 IGO.). http://www.fao.org/3/ca6030en/ca6030en.pdf.

Fares, A., Bayabil, H.K., Zekri, M., Mattos-Jr, D. and Awal, R. (2017). Potential climate change impacts on citrus water requirement across major producing areas in the world. *Journal of Water and Climate Change*, 8(4): 576–592.

Farooqi, Z.U.R., Ayub, M.A., Rehman, M.Z., Sohail, M.I., Usman, M., Khalid, H. and Naz, K. (2020). Regulation of drought stress in plants. pp. 77–104. *In*: *Plant Life under Changing Environment*. Academic Press.

Ferrarezi, R.S., Rodriguez, K. and Sharp, D. (2020). How historical trends in Florida all-citrus production correlate with devastating hurricane and freeze events. *Weather*, 75(3): 77–83.

Forner-Giner, M.A., Continella, A. and Grosser, J.W. (2020). Citrus rootstock breeding and selection. pp. 49–74. *In*: Gentile, A., La Malfa, S. and Deng, Z. (eds.). *The Citrus Genome: Compendium of Plant Genomes*. Springer, Cham.

Forni, C., Duca, D. and Glick, B.R. (2017). Mechanisms of plant response to salt and drought stress and their alteration by rhizobacteria. *Plant and Soil*, 410(1-2): 335–356.

García-Tejero, I.F., Arriaga, J., Duran-Zuazo, V.H. and Muriel-Fernández, J.L. (2013). Predicting crop-water production functions for long-term effects of deficit irrigation on citrus productivity (SW Spain). *Archives of Agronomy and Soil Science*, 59(12): 159–1606.

Gupta, B. and Huang, B. (2014). Mechanism of salinity tolerance in plants: Physiological, biochemical and molecular characterization. *Int. J. Genom.*, 701596: 18.

Hailai, D., Huaisui, Q., Mingxia, L. and Yaodong, D. (2010). Changes of citrus climate risk in subtropics of China. *J. Geogr. Sci.*, 20(6): 818–832. Doi: 10.1007/s11442-010-0813-6.

Hodgson, R.W. (1967). Horticultural varieties of citrus. *History, World Distribution, Botany and Varieties*, 431–591.

Hossain, Z., Lopez-Climent, M.F., Arbona, V., Perez-Clemente, R.M. and Gomez-Cadenas, A. (2009). Modulation of the antioxidant system in citrus under waterlogging and subsequent drainage. *J. Plant Physiol.*, 166: 1391–1404. https://doi.org/10.1016/j.jplph.2009.02.012.

Huchche, A.D., Panigrahi, P. and Shivankar, V.J. (2010). Impact of climate change on citrus in India. *In*: Singh, H.P., Singh, J.P. and Lal, S.S. (eds.). *Challenges of Climate Change – Indian Horticulture*. Delhi, India: Westville Publishing House, 2010, 224 pp.

Hussain, S., Luro, F., Costantino, G., Ollitrault, P. and Morillon, R. (2012). Physiological analysis of salt stress behavior of citrus species and genera: Low chloride accumulation as an indicator of salt tolerance. *South African Journal of Botany*, 81: 103–112.

Irenaeus, T. and Mitra, S.K. (2014). Understanding the pollen and ovule characters and fruit set of fruit crops in relation to temperature and genotype—A review. *Journal of Applied Botany and Food Quality*, 87: 157–167. Doi:10.5073/JABFQ.2014.087.023.

Islam, F., Abou Leila, B. and Gaballah, M. (2019). Effect of antioxidants on citrus leaf anatomical structure grown under saline irrigation water. *Plant Archives*, 19(1): 840–845.

Jadhav, A.K., Sharma, R.M. and Dubey, A.K. (2020). Physiology of flowering in Citrus species. *Indian Journal of Horticulture*, 77(1): 65–71.

Juan, L. and Jiezhong, C. (2017). Citrus fruit-cracking: causes and occurrence. *Horticultural Plant Journal*, 3(6): 255–260. https://doi.org/10.1016/j.hpj.2017.08.002.

Khalid, M.S., Malik, A.U., Khan, A.S., Saleem, B.A., Amin, M., Malik, O.H., Khalid, S. and Rehman, A. (2018). Geographical location and agroecological conditions influence Kinnow mandarin (*Citrus nobilis* × *Citrus deliciosa*) fruit quality. *Intl. J. Agric. Biol.*, 20: 647–654.

Korres, N.E., Norsworthy, J.K., Tehranchian, P., Gitsopoulos, T.K., Loka, D.A., Oosterhuis, D.M., Gealy, D.R., Moss, S.R., Burgos, N.R., Miller, M.R. and Palhano, M. (2016). Cultivars to face climate change effects on crops and weeds: A review. *Agron. Sustain. Dev.*, 36: 12. https://doi.org/10.1007/s13593-016-0350-5.

Kumar, K., Rashid, R., Bhat, J.A. and Bhat, Z.A. (2011). Effects of high temperature on fruit crops. *Elixir Appl. Botany*, 39: 4745–4747.

Kumar, P., Sharma, S. and Dalal, R.S. (2012). Citrus decline in relation to soil-plant nutritional status—a review. *Agricultural Reviews*, 33(1): 62–69.

Lado, J., Gambetta, G. and Zacarias, L. (2018). Key determinants of citrus fruit quality: Metabolites and main changes during maturation. *Scientia Horticulturae*, 233: 238–248.

Lal, S., Singh, D.B., Sharma, O.C., Mir, J.I., Sharma, A., Raja, W.H., Kumawat, K.L. and Rather, S.A. (2018). Impact of climate change on productivity and quality of temperate fruits and its management strategies. *IJARSE*, 7(4): 1833–1844.

Lamaoui, M., Jemo, M., Datla, R. and Bekkaoui, F. (2018). Heat and drought stresses in crops and approaches for their mitigation. *Frontiers in Chemistry*, 6: 26.

Legua, P., Martinez-Cunenca, M.R., Bellver, R. and Forner-Giner, M.A. (2018). Rootstock's and scion's impact on lemon quality in southeast Spain. *International Agrophysics*, 32(3): 325–333. doi: 10.1515/intag-2017-0018.

Lesk, C., Rowhani, P. and Ramankutty, N. (2016). Influence of extreme weather disasters on global crop production. *Nature*, 529: 84–98. Doi:10.1038/nature16467.

Liu, Y-Q., Heying, E. and Tanumihardjo, S.A. (2012). History, global distribution, and nutritional importance of citrus fruits. *Comprehensive Reviews in Food Science and Food Safety*, 11: 530–545.

Lobell, D.B., Schlenker, W. and Costa-Roberts, J. (2011). Climate trends and global crop production since 1980. *Science*, 333: 616–620.

Luro, F., Curk, F., Froelicher, Y. and Ollitrault, P. (2017). Recent insights on citrus diversity and phylogeny. pp. 16–28. *In*: Zech-Matterne, Véronique and Fiorentino, Girolamo (eds.). *AGRUMED: Archaeology and History of Citrus Fruit in the Mediterranean: Acclimatization, Diversifications, Uses*. new ed. [online]. Naples: Publications du Centre Jean Bérard. Doi: 10.4000/books.pcjb.2107 (generated 11 May 2020). Available on the Internet. https://books.openedition.org/pcjb/2169.

Mahdavian, M., Sarikhani, H., Hadadinejad, M. and Dehestani, A. (2020). Putrescine effect on physiological, morphological, and biochemical traits of carrizo citrange and volkameriana rootstocks under flooding stress. *International Journal of Fruit Science*, 20(2): 164–177.

Malik, N.S.A., Perez, J.L. and Kunta, M. (2015). Inducing flushing in citrus cultivars and changes in polyphenols associated with bud break and flowering. *J. Horticulture*, 2: 148. Doi:10.4172/2376-0354.1000148.

Manera, F.J., Brotons, J.M., Conesa, A. and Porras, I. (2012). Influence of temperature on the beginning of degreening in lemon peel. *Sci. Hortic.*, 145: 34–38.

Manera, F.J., Brotons, J.M., Conesa, A. and Porras, I. (2013). Relation between temperature and the beginning of peel color change in grapefruit (*Citrus paradisi* Macf.). *Scientia Horticulturae*, 160: 292–299.

Martinez-Ferri, E., Muriel-Fernandez, J.L. and Díaz, J.R. (2013). Soil water balance modelling using SWAP: an application for irrigation water management and climate change adaptation in citrus. *Outlook on Agriculture*, 42(2): 93–102.

McDowell, N.G., Michaletz, S.T., Bennett, K.E., Solander, K.C., Xu, C., Maxwell, R.M. and Middleton, R.S. (2018). Predicting chronic climate-driven disturbances and their mitigation. *Trends in Ecology and Evolution*, 33(1): 15–27.

Mesquita, E.F., Sa, F.V.S., Bertino, A.M.P., Cavalcante, L.F., Paiva, E.P. and Ferreira, N.M. (2015). Effect of soil conditioners on the chemical attributes of a saline-sodic soil and on the initial growth of the castor bean plant. *Semina: Ciencias Agrarias.*, 36: 2527–2538. https://doi.org/10.5433/1679-0359.2015v36n4p2527.

Michaelides, S., Karacostas, T., Sánchez, J.L., Retalis, A., Pytharoulis, I., Homar, V., Romero, R., Zanis, P., Giannakopoulos, C., Buhl, J., Ansmann, A., Merino, A., Melcon, P., Lagouvardos, K., Kotroni, V., Bruggeman, A., Lopez-Moreno, J.I., Berthet, C., Katragkou, E., Tymvios, F., Hadjimitsis, D.G. and Mamouri, R.-E. (2018). Reviews and perspectives of high impact atmospheric processes in the Mediterranean. *Atmospheric Research*, 208: 4–44.

Micheloud, N.G., Castro, D.C., Buyatti, M.A., Gabriel, P.M. and Gariglio, N.F. (2020). Factors affecting phenology of different Citrus varieties under the temperate climate conditions of Santa Fe, Argentina. *Rev. Bras. Frutic. Jaboticabal*, 40(1): (e-315).

Morais, A.L.D., Zucoloto, M., Malikouski, R.G., Babosa, D.H.S.G., Passos, O.S. and Altoe, M.S. (2020). Vegetative development and production of 'Tahiti' acid lime clone selections grafted on different rootstocks. *Rev. Bras. Frutic. Jaboticabal.*, 42(3): (e-585). Doi: http://dx.doi.org/10.1590/0100-29452020585.

Narouei-Khandan, H.A., Halbert, S.E., Worner, S.P. and van Bruggen, A.H. (2016). Global climate suitability of citrus huanglongbing and its vector, the Asian *citrus psyllid*, using two correlative species distribution modeling approaches, with emphasis on the USA. *European Journal of Plant Pathology*, 144(3): 655–670.

Nawaz, R., Abbasi, N.A., Hafiz, I.A., Khalid, A., Ahmad, T. and Aftab, M. (2019). Impact of climate change on kinnow fruit industry of Pakistan. *Agrotechnology*, 8(186): 2.

Nawaz, R., Abbasi, N.A., Hafiza, I.A. and Khalid, A. (2020a). Impact of climate variables on growth and development of Kinnow fruit (*Citrus nobilis* Lour x *Citrus deliciosa* Tenora) grown at different ecological zones under climate change scenario. *Scientia Horticulturae*, 260: 108868. https://doi.org/10.1016/j.scienta.2019.108868.

Nawaz, R., Abbasi, N.A., Hafiz, I.A. and Khalid, A. (2020b). Increasing level of abiotic and biotic stress on Kinnow fruit quality at different ecological zones in climate change scenario. *Environmental and Experimental Botany*, 171: 1–13. https://doi.org/10.1016/j.envexpbot.2019.103936.

Novelli, V.M., Cristofani, M., Souza, A.A. and Machado, M.A. (2006). Development and characterization of polymorphic microsatellite markers for the sweet orange (*Citrus sinensis* L. Osbeck). *Genetics and Molecular Biology*, 29(1): 90–96.

Ouyang, Z.G., Mi, L.F., Duan, H.H., Hu, W., Chen, J.M., Peng, T. and Zhong, B.L. (2019). Differential expressions of citrus CAMTAs during fruit development and responses to abiotic stresses. *Biologia Plantarum*, 63(1): 354–364.

Pathak, H., Aggarwal, P.K. and Singh, S.D. (eds.). (2012). *Climate Change Impact, Adaptation and Mitigation in Agriculture: Methodology for Assessment and Applications*. Indian Agricultural Research Institute, New Delhi, 302 pp.

Pereira, F.F.S., Sanchez-Roman, R.M. and Gonzalez, A.M.G.O. (2017). Simulation model of the growth of sweet orange (*Citrus sinensis* L. Osbeck) cv. Natal in response to climate change. *Clim. Chang.*, 143: 101–113. https://doi.org/10.1007/s10584-017-1986-0.

Pommer, C.V. and Barbosa, W. (2009). The impact of breeding on fruit production in warm climates of Brazil. *Revista Brasileira de Fruticultura*, 31: 612–634.

Raga, V., Bernet, G.P., Carbonell, E.A. and Asins, M.J. (2014). Inheritance of rootstock effects and their association with salt tolerance candidate genes in a progeny derived from Volkamer lemon. *Journal of the American Society for Horticultural Science*, 139(5): 518–528.

Rodríguez-Gamir, J., Ancillo, G., González-Mas, M.C., Primo-Millo, E., Iglesias, D.J. and Forner-Giner, M.A. (2011). Root signaling and modulation of stomatal closure in flooded citrus seedlings. *Plant Physiol. Biochem.*, 49: 636–645. https://doi.org/10.1016/j.plaphy.2011.03.003.

Rodriguez, J., Anoruo, A., Jifon, J. and Simpson, C. (2019). Physiological effects of exogenously applied reflectants and anti-transpirants on leaf temperature and fruit sunburn in citrus. *Plants*, 8(12): 549.

Sa, F.V.D.S., Brito, M.E., Figueiredo, L.C.D., Melo, A.S.D., Silva, L.D.A. and Moreira, R.C. (2017). Biochemical components and dry matter of lemon and mandarin hybrids under salt stress. *Revista Brasileira de Engenharia Agrícola e Ambiental*, 21(4): 249–253.

Salvi, B.R., Damodhar, V.P., Mahaldar, S.R. and Munj, A.Y. (2013). Effect of high temperatures on fruit drop and spongy tissue in Alphonso mango. *Ann. Plant Physiol.*, 27(l): 5–10.

Santos, J.A., Costa, R. and Fraga, H. (2017). Climate change impacts on thermal growing conditions of main fruit species in Portugal. *Climate Change*, 140: 273–286. https://doi.org/10.1007/s10584-016-1835-6.

Shanker, T.G., Navneet, P., Swapnil, P. and Vishal, N. (2018). Impact of temperature stress on growth and development of fruit crops and its mitigation strategies. *International Journal of Innovative Horticulture*, 7(1): 1–9.

Sharma, A., Shahzad, B., Kumar, V., Kohli, S.K., Sidhu, G.P. S., Bali, A.S., Handa, N., Kapoor, D., Bhardwaj, R. and Zheng, B. (2019). Phytohormones regulate accumulation of osmolytes under abiotic stress. *Biomolecules*, 9(7): 285.

Sharma, N., Sharma, S. and Niwas, R. (2017). Thermal time and phenology of citrus in semi-arid conditions. *Journal of Pharmacognosy and Phytochemistry*, 6(5): 27–30.

Silva, L.A., Brito, M.E.B., Sa, F.V.S., Moreira, R.C.L., Filho, W. S.S. and Fernandes, P.D. (2014). Physiological mechanisms in citrus hybrids under saline stress in hydroponic system. *Revista Brasileira de Engenharia Agricola e Ambiental*, 18: S1–S7.

Silva, L.A., Brito, M.E.B., Fernandes, P.D., Sa, S.F.V., Moreira, R.C.L., Sales, G.N.B., Almeida, J.F. and Filho, W.S.S. (2019). Growth and fluorescence of Tahiti acid lime/rootstock on Sunki mandarin hybrids under salinity. *Biosci. J. Uberlandia*, 35(4): 1131–1142. http://dx.doi.org/10.14393/BJ-v35n4a2019-42192.

Steiner, J.L., Briske, D.D., Brown, D.P. and Rottler, C.M. (2018). Vulnerability of Southern Plains agriculture to climate change. *Climatic Change*, 146(1): 201–218.

Sugiura, T., Sumida, H., Yokoyama, S. and Ono, H. (2012). Overview of recent effects of global warming on agricultural production in Japan. *JARQ*, 46(1): 7–13.

Syvertsen, J.P. and Garcia-Sanchez, F. (2014). Multiple abiotic stresses occurring with salinity stress in citrus. *Environmental and Experimental Botany*, 103: 128–137.

Tanaka, T. (1977). Fundamental discussion of citrus classification. *Studia Citrologia*, 14: 1–5.

Ullah, R., Shivakoti, G.P. and Ali, G. (2015). Factors effecting farmers' risk attitude and risk perceptions: the case of Khyber Pakhtunkhwa, Pakistan. *Int. J. Disast. Risk Reduct.*, 13: 151–157.

USDA, United States Department of Agriculture. (2019). *Citrus: World Markets and Trades, 2019.* http://www.agrirowad.com/UploadFiles/AdminFiles/citrus0dd5bf.pdf.

USDA. (2020). *Citrus Fruit Data and Analysis, 2020.* United States Department of Agriculture. https://www.fas.usda.gov/data/search?f%5B0%5D=field_commodities%3A11.

Uzun, A.Y.D.I.N., Yesiloglu, T., Aka-Kacar, Y.I.L.D.I.Z., Tuzcu, O. and Gulsen, O. (2009). Genetic diversity and relationships within citrus and related genera based on sequence related amplified polymorphism markers (SRAPs). *Scientia Horticulturae*, 121(3): 306–312.

Verner, D., Treguer, D., Redwood, J., Christensen, J., Mcdonnell, R., Elbert, C. and Konishi, Y. (2016). *World Bank Document.* http://documents.worldbank.org/curated/en/318211538415630621/pdf/130406-WP-P159856-Tunisia-WEB2.pdf.

Vincent, C., Morillon, R., Arbona, V. and Gómez-Cadenas, A. (2020). Citrus in changing environments. pp. 271–289. *In: The Genus Citrus*. Woodhead Publishing.

Vives-Peris, V., Gomez-Cadenas, A. and Perez-Clemente, R.M. (2017). Citrus plants exude proline and phytohormones under abiotic stress conditions. *Plant Cell Reports*, 36(12): 1971–1984.

Wang, F., Huang, Y., Wu, W., Zhu, C., Zhang, R., Chen, J. and Zeng, J. (2020). Metabolomics analysis of the peels of different colored citrus fruits (*Citrus reticulata* cv.'Shatangju') during the maturation period based on UHPLC-QQQ-MS. *Molecules*, 25(2): 396.

Wang, Y., Fu, X., He, W., Chen, Q. and Wang, X. (2019). Effect of plastic film mulching on fruit quality of *Citrus grandis* cv. 'Hongroumiyou' and 'Sanhongmiyou'. *In: IOP Conference Series: Earth and Environmental Science*. vol. 358, No. 2, p. 022030, IOP Publishing.

Wu, G.A., Terol, J., Ibanez, V., Lopez-Garcia, A., Perez-Roman, E., Borreda, C., Domingo, C., Tadeo, F.R., Carbonell-Caballero, J., Alonso, R., Curk, F., Du, D., Ollitrault, P., Roose, M.L., Dopazo, J., Gmitter Jr, F.G., Rokhsar, D.S. and Talo, M. (2018). Genomics of the origin and evolution of citrus. *Nature*, 554: 311–328.

Wu, H.H., Zou, Y.N., Rahman, M.M., Ni, Q.D. and Wu, Q.S. (2017). Mycorrhizas alter sucrose and proline metabolism in trifoliate orange exposed to drought stress. *Scientific Reports*, 7: 42389.

Wu, Q.S. and Zou, Y.N. (2013). Mycorrhizal symbiosis alters root H+ effluxes and root system architecture of trifoliate orange seedlings under salt stress. *J. Anim. Plant Sci.*, 23: 143–148.

Xu, M., Liang, M., Chen, J., Xia, Y., Zheng, Z., Zhu, Q. and Deng, X. (2013). Preliminary research on soil conditioner mediated citrus Huanglongbing mitigation in the field in Guangdong, China. *European Journal of Plant Pathology*, 137(2): 283–293.

Zabihia, H., Vogeler, I., Amina, Z.M. and Gourabi, B.R. (2016). Mapping the sensitivity of citrus crops to freeze stress using a geographical information system in. Ramsar, Iran. *Weather and Climate Extremes*, 14: 17–23.

Zandalinas, S.I., Balfagon, D., Arbona, V. and Gómez-Cadenas, A. (2017). Modulation of antioxidant defense system is associated with combined drought and heat stress tolerance in citrus. *Front. Plant Sci.*, 8. https://doi.org/10.3389/fpls.2017.00953.

Zandalinas, S.I., Balfagon, D., Arbona, V. and Gomez-Cadenas, A. (2018). Regulation of citrus responses to the combined action of drought and high temperatures depends on the severity of water deprivation. *Physiologia Plantarum*, 162(4): 427–438.

Zhang, F., Zou, Y.N. and Wu, Q.S. (2018). Quantitative estimation of water uptake by *mycorrhizal extraradical* hyphae in citrus under drought stress. *Scientia Horticulturae*, 229: 132–136.

Zhang, H. and Sonnewald, U. (2017). Differences and commonalities of plant responses to single and combined stresses. *Plant J.*, 90: 839–855.

Zhang, Y.-C., Wang, P., Wu, Q.-H., Zou, Y.-N., Bao, Q. and Wu, Q.-S. (2017). Arbuscular mycorrhizas improve plant growth and soil structure in trifoliate orange under salt stress. *Archives of Agronomy and Soil Science*, 63(4): 491–500. Doi: 10.1080/03650340.2016.1222609.

Zhu, G.F., Li, X., Su, Y.H., Lu, L. and Huang, C.L. (2011). Seasonal fluctuations and temperature dependence in photosynthetic parameters and stomatal conductance at the leaf scale of *Populus euphratica* Oliv. *Tree Physiology*, 31(2): 178–195.

Ziogas, V., Tanou, G., Morianou, G. and Kourgialas, N. (2021). Drought and salinity in citriculture: optimal practices to alleviate salinity and water stress. *Agronomy*, 11(7): 1283.

4

Climate Change Impacts and Adaptation Strategies of Common Fig (*Ficus carica* L.) Cultivation

M Moniruzzaman,[1,2,*] *Zahira Yaakob,*[2] *Nurina Anuar,*[2] *Jameel M Al-Khayri*[3] and *Islam El-Sharkawy*[1]

1. Introduction

The genus *Ficus* is one of 40 genera of the Moraceae family and accommodates over 800 species (Woodland, 1997). Fig plants (*Ficus carica* L.) are shrubs or small trees (Fig. 1); the leaves are described as hand-shaped (Li, 1578), the foliage is single, alternate and large, deeply lobed with three or seven lobes, rough and hairy on the adaxial surface, and soft and hairy on the abaxial side; the bark is smooth and gray (Lansky et al., 2008). The plant bears auxiliary aerial root systems extending to the ground from the branches or trunks. Ripe fruits are soft with thin and tender skins. Fruits are consumed either fresh or dried. Fig fruits and leaves

[1] Center for Viticulture & Small Fruit Research (CVSFR), College of Agriculture and Food Sciences, Florida A&M University, Tallahassee, FL 32308, USA.

[2] Research Centre for Sustainable Process Technology (CESPRO), Faculty of Engineering and Built Environment, Universiti Kebangsaan Malaysia, Bangi 43600, Malaysia.

[3] Department of Agricultural Biotechnology, College of Agriculture and Food Sciences, King Faisal University, P.O. Box 420, Al-Hassa 31982, Saudi Arabia.

Emails: zahira@ukm.edu.my; nurina@ukm.edu.my; jkhayri@kfu.edu.sa; islam.elsharkawy@famu.edu

* Corresponding author: monirbge@gmail.com

bear enormous food and medicinal values (Barolo et al., 2014; Palmeira et al., 2019; Raafat and Wurglics, 2019; Abdel-Aty et al., 2019). Based on nutritional values, figs have much more overall content of energy, dietary fiber and minerals (potassium, calcium and iron) as compared to other fruits, i.e., apples, bananas, dates, grapes, oranges, strawberries (Flaishman et al., 2007). Ficus species are a rich source of different chemicals, such as alkaloid, coumarin, flavonoid, phenolic, pyrimidine, sterol, triterpenoid and anthocyanin. These biologically active compounds from the leaves, roots, bark, latex and fruits demonstrate antibacterial, antiviral, acetyl cholinesterase inhibition (Oliveira et al., 2010), antifungal (Raskovic et al., 2016), anti-helminthic (de Amorin et al., 1999; Mavlonov et al., 2008), anti-carcinogenic (Lansky et al., 2008; Rubnov et al., 2001; Wang et al., 2008), anti-inflammatory and antioxidant activities (Lazreg-Aref et al., 2012; Loizzo et al., 2014). Therefore it has been concluded that figs could be a potential source of novel drugs (Zhang et al., 2012b; Zhang et al., 2012a).

Thus, the fig is an economically important fruit crop (Kitajima et al., 2018; Stover et al., 2007). The fig economy is mainly distributed around the Mediterranean basin and the annual production of fig in Mediterranean basin is 1,064,784 metric tons (mt) from 311,080 hector (ha) of plantations. Turkey, Egypt, Iran, Greece, Algeria and Morocco account for ~ 70% of world annual production. Commercial fig production is reported in 14 US states. The USA produces about 4% of global fig production (FAOSTAT, 2012) for consumption. On the other hand, wild fig is a key role player (critical food source) in many forest ecosystems. Around 1,200 species of world vertebrates exclusively feed on wild figs. Moreover, many fig species are being used as ornamental or road-side plant in cities.

Northern Europe and Mediterranean regions are the primary zone for fig fruit production (Fig. 2). Climate change causes largely changing spatial distribution of rainfall and increasing global temperature in these regions. In northern Europe, precipitation patterns shift with a decrease in summer, when bulk amount of water is needed as it is the major growing period (Jacob et al., 2014). In the Mediterranean region, there is an expected overall reduction of annual precipitation to 39.1 ± 55.1 mm; consequently, an increase of air temperature of 1.57 ± 0.27°C between the years 2000 and 2050 (Saadi et al., 2015). In consequence, the annual reference evapotranspiration increase is 6.7% (92.3 ± 42.1 mm). In western Europe, prolonged summers with more frequent heat waves and unprecedented temperature extremes is expected (IPCC, 2013). Eventually this will cause greater water scarcity in summer. The Mediterranean region has already suffered 0.4°C temperature rise than global average temperature (Cramer et al., 2018). Due to the increased temperature, summer precipitation is predicted to decrease by up to 30%, and increase up to 18%, demanding more irrigation in some regions until the end of the century. As a result, agricultural practice will be more affected by drought stress. Drought can

Fig. 1. Different parts of common fig. (a) Fig plants in experimental orchard, (b) fruiting plants in orchard, (c) individual leaf of fig tree, (d–e) tree ripe fig fruits, (f) section of ripe fig fruit (Moniruzzaman et al., 2020).

afflict plants at any growth stage and can affect the productivity, depending on the degree, intensity and duration. Drought has been identified as the most detrimental environmental stress-affecting agriculture worldwide and responsible for the greatest loss of yield (Ruggiero et al., 2017; Hu and Xiong, 2014). In fact, previous studies estimated that drought stress results in an average yield loss of at least 50% in rice (Venuprasad et al., 2007), chickpea (Varshney et al., 2014), and of 18–32% in potato (Hijmans, 2003). Similar result is expected in the case of fig crop, though there is no similar study in the case of fig.

Global food security is one of the biggest mind-numbing concerns nowadays. Food insecurity is also associated with lower fruit and

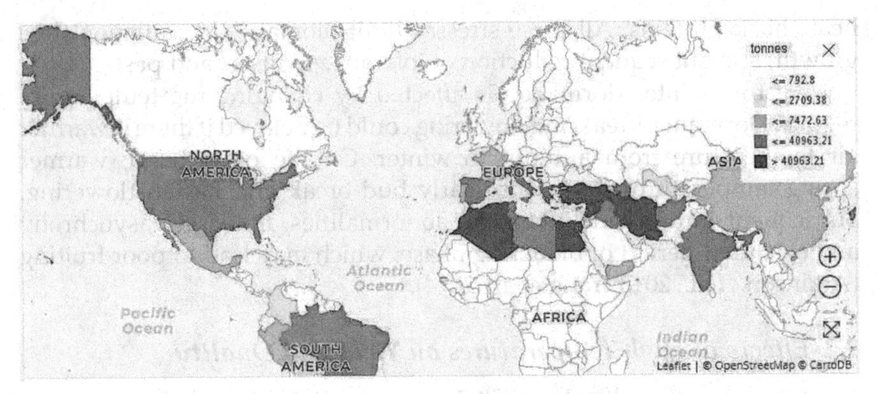

Fig. 2. Fig cultivation areas in the world.

vegetable intake (Leung et al., 2014; Litton and Beavers, 2021). Inadequate consumption of fruits and vegetables is associated with the risk of chronic diseases, including type-2 diabetes and cardiovascular disease, as well as with increased mortality (Wang et al., 2016; Wang et al., 2014; Oyebode et al., 2014; Seligman et al., 2010). However, production and availability of fruits have been negatively affected by global climate change, causing vulnerable food security and the situation would be even worse if appropriate measures are not taken in advance. To make awareness and determine the actions for adequate fig production, in this chapter, we have discussed the climate change effects and adaptation strategies for fig cultivation in adverse climate.

2. Impacts of Climate Change

The very common phenomena of climate change are increased atmospheric temperature, decrease in precipitation, season change and frequent occurrence of natural catastrophe.

2.1 *Effects of High Temperatures on Tree Physiology*

Higher temperature causes excess tree water consumption because of increased leaf-to-air vapor pressure deficit. As a consequence, the intrinsic tree hydraulic conductivity in xylem and leaves could exceed, especially when stomata poorly regulate the transpiration. This may lead to rupture of xylem vessel in association with low-stem hydric potential. It causes loss of tree turgor, death of some organ (leaves, twigs, fruit) and even the whole tree. Fig tree operates C3 carbon cycle and the most favorable temperature for photosynthesis range at 25–35°C. The elevated (> 40°C) temperature leads to heat stress, because no efflux regulation of latent heat results from transpiration. It hampers photosynthesis and leads to synthesis of heat shock proteins. All these stresses limit biomass accumulation (tree growth) and subsequent reduction of tolerance to insect and pest.

Fig tree winter dormancy is affected by changing the temperature regime. Dormancy break and flowering could be delayed if there is warmer air temperature from autumn to winter. On the other hand, warmer spring temperatures may cause early bud-break and hasten flowering. Subsequent impacts include floral abnormalities, flowering asynchrony and extended period of blooming phase, which may lead to poor fruiting (Atkinson et al., 2013; Yang et al., 2014).

2.2 *Effects of High Temperatures on Yield and Quality*

Plant reproductive development is sensitive to elevated temperatures. Increase in temperatures, associated with climate change, may reduce the yield and quality of fruits (Hasanuzzaman et al., 2013; Challinor et al., 2014); even moderate increases in temperature alter the phenological dates, bloom-to-fruit maturity period and plant water requirements (Bita and Gerats, 2013; Hatfield and Prueger, 2015). A shorter bloom-to-fruit maturity period negatively impacts the potential fruit size. Fruit texture and shape may also deteriorate because of shorter phage of cell multiplication and increased individual cell growth. Fruit eating quality, acidity to sugar ratio, is also affected by warm temperatures at harvest maturity, as the malic respiration rate is stimulated by warm temperature. Moreover, fruit skin color is also affected by altered temperature during fruit maturity, decreasing the commercial value of fruits. Higher temperatures increase the accumulation of growing degree days and reduce chilling hours during winter-time dormancy, both affecting the flowering time of fruit trees (El Yaacoubi et al., 2014; Santos et al., 2017) and subsequent reduction of fruit yield and fruit quality.

2.3 *Effects of High Temperatures on Fig WASP and Fig Pollination/Population*

Mutual obligation and ancient relationship between fig trees and fig wasps seem to be running a fatal interference due to human-induced global warming. Several species of fig wasps is less tolerant to the fast-escalating temperatures across their range in the equatorial tropics. The warm temperatures are shortening WASP's life-spans by modulating cellular machinery at the molecular level. The consequence is shorter life spans that end with broken and incomplete life cycles. As a result, fig wasp populations have started to decline. Thus, eventually, there won't be enough wasp populations to offer fig-pollination services, which will lead to decline in fig fruit production (Smyrna variety) and fig tree populations (because of the unavailability of viable seeds).

Fig tree is considered one of the most important groups of plants because of its ecological and even cultural values. Wild fig fruits are a critical food source of many forest ecosystems. Around 1,200 species of world vertebrates exclusively feed on wild figs. So, declining fig tree population will eventually interrupt the pre-existing food chain, ultimately collapsing the ecosystem. On the other hand, fig tree plantation could cause rain-forest regeneration because many fig-eating animals also dispense the seeds of several others key plant species of a forest. Moreover, several fig species are being planted on road-sides and in many cities as CO_2 adsorbent. Hence, declining population of fig tree would render futile the restoration of forests and combatting of the greenhouse gas emissions, reversing the impacts of climate change and improving livelihoods.

2.4 *Effect of Elevated Temperatures on Pests and Diseases*

Insect and pests life cycles and reproduction rates are also influenced by climatological framework change, besides crop yields. Insect mortality rates in winter drops due to increase in global temperature. This will facilitate quick and severe infestations (Hullé et al., 2010; Van Damme et al., 2015; Bisbis et al., 2018). Plant-insect interactions are affected by extra atmospheric CO_2 probably by changing the host plant's feeding quality (Dáder et al., 2016). Wind blow and direction influence strongly the pest and disease appearance and spread (Collier et al., 2008).

Besides securing food supply, an important aspect of human nutrition is food safety. As a consequence of global warming, alternaria mold might occur more frequently in cool regions, e.g., Poland and less in warm regions, e.g., in Spain, ultimately altering mycotoxin content in fruits (Van de Perre et al., 2015).

2.5 *Effect of Increasing Atmospheric CO_2*

Carbon dioxide (CO_2) accumulation in the atmosphere will increase between 442 ppm (RCP2.6) and 540 ppm (RCP8.5) by mid-century (IPCC, 2013). Increase amounts of fertilizer application may be required because of the predicted extra CO_2 concentration in atmosphere that leads to larger nitrogen uptake by the crop (Dong et al., 2018c; Dong et al., 2018a). Climate change may also cause loss of nitrogen through leaching or gaseous losses. This may shift fertilizer demand for crop production (Bindi and Howden, 2004).

The CO_2 fertilization input cost may reduce due to greater atmospheric CO_2. It may be beneficial for crop production, especially during summer, depending on the existing climate situation. However there might be the question of product quality (Dong et al., 2018b).

2.6 *Effects of Water Deficit on Seed/Fruit Growth, Yield and Quality*

Scarcity of water for agriculture use is one of the adverse effects of global climate change. Efficient and sustainable use of water resources is a timeless demand now. Employing the modern irrigation systems (e.g., drip and sprinkler systems), especially in the newly reclaimed sandy soils, could manage drought effects in crops and ensure better water use. Fig trees' response to drought stress and supplemental irrigation was determined (Tapia et al., 2003). Prolonged drought stress causes severe injuries to the plant (Fahad et al., 2017). Drought-stress incidences result in massive leaf abscission and reduced fruit yield and its quality (Doaa et al., 2015; Türk and Aksoy, 2011; Gholami et al., 2012; Karimi et al., 2012). Furthermore, all supplemental irrigation treatments gave a good positive correlation between irrigation rates and vegetative growth or productivity (Jafari et al., 2015).

3. Adaptation of Fig Cultivation in Adverse Climate

Fig trees prefer a total yearly rainfall of 500–550 mm, especially 40–45% humidity and average temperatures of 18–20°C yearly (Polat and Caliskan, 2017). However, fig crop has been confronting or will be severely confronted with adverse climate in the coming future because of climate change effects. Adaptation of fig to adverse climate requires integration of multiple strategies, i.e., genetic resource utilization (fig genetic diversity) and genetic improvement, cultivation system management (sequence of crops and management techniques used on a particular agricultural field) and farmer behavior and farming system improvement (Fig. 3).

3.1 *Fig Genetic Improvement for Changing Environments*

Genetic improvement is the science of applying genetic and plant-breeding principles as well as physiology and biotechnology to improve crops for human use. Traditional improvement programs are the prime focus for increasing crop yield. However, this methodology is not optimal because the yield is a character with low heritability and high genotype (G) × environment (E) interaction (Richards et al., 2002). The scenario is even more complex if agronomic management (M) is considered as a third influential factor, which would make it even more difficult to understand this complex interaction in the field (G × E × M) (Tuberosa et al., 2014).

In order to develop suitable cultivars that can cope with the climate change effect, focus should be on stress-tolerance-related traits and genotype-phenotype relationships need to be understood. However, most of the stress-tolerance-related traits are complex because many

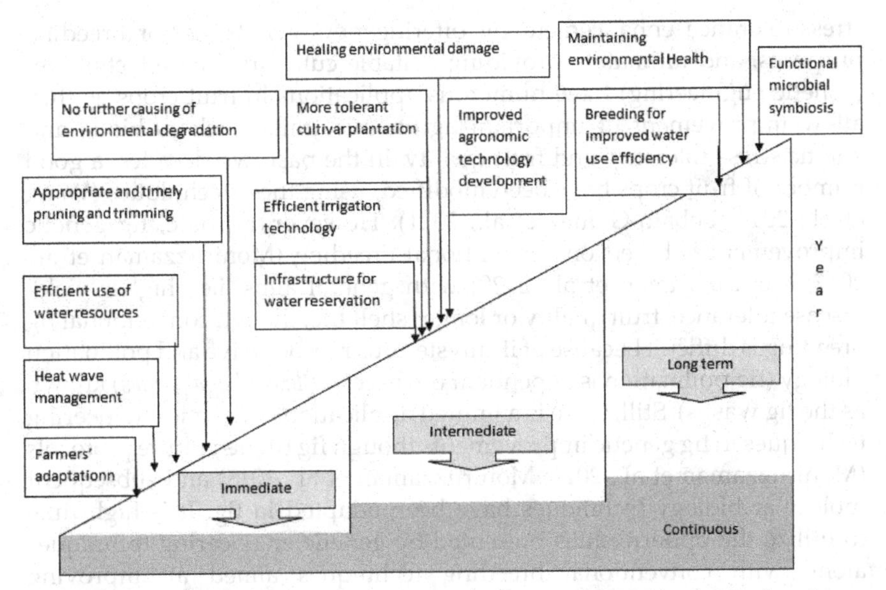

Fig. 3. Schematic representations of adaptation strategies of fig cultivation under adverse climate condition.

genes are involved (Diab et al., 2004; Tondelli et al., 2006; Araus et al., 2008) and genotype-phenotype relationships are not fully understood yet (Fischer, 2007). This demonstrates the urgency of developing technologies and methodologies that could facilitate understanding of the vulnerability or adaptation of a genotype in a given environment, contributing at the same time to improving the predictive capacity of marker-assisted breeding (Camargo and Lobos, 2016; Yang et al., 2014; Reynolds and Langridge, 2016; Dhondt et al., 2013). Hence, in order to develop cultivars that are well adapted to environmental fluctuations or constraints, breeders will be forced to evaluate the maximum number of possible traits that allow characterization of a genotype's performance under different conditions. The only way to satisfy all these demands is through the use of high-dimensional phenotypic data (high-throughput field phenotyping) or 'phenomics' (Houle et al., 2010). This new approach would allow a reduction in cost of breeding programs (Brennan et al., 2007) and, by allowing the early selection of genetic material of interest, increase the chances of releasing more cultivars (Reynolds et al., 2009) in a short period (Lobos and Hancock, 2015; Camargo and Lobos, 2016). In addition, biotechnological tools can be used to help crop adaptation to climate and socioeconomic changes. Recent advancements in recombinant DNA technology allows transfer of the desired genes from any organism, plant or microorganism into fruit crops, even precisely editing of existing genes, thereby extending the opportunities for fruit yield and stress-tolerance enhancement by offering new genotypes for breeding purposes, and ultimately providing suitable cultivars. These techniques (genetic engineering) have numerous applications in fruit crops as they allow improvement of important agronomic traits, such as biotic and abiotic stress tolerance and fruit quality. In the past two decades, a good number of fruit crops have been modified, using these techniques (Wang et al., 2019; Lobato-Gómez et al., 2021). However, till date, fig genetic improvement is based on conventional breeding (Moniruzzaman et al., 2017; Moniruzzaman et al., 2020) with general aims, i.e., higher yield, disease tolerance, fruit quality or longer shelf life, though conventional fig breeding is difficult because of its mysterious reproductive and pollination biology (fig pollination is dependent on insects (*Blastophaga psenes*) known as the fig wasps). Still, there is a limited application of genetic engineering techniques in fig genetic improvements though fig tissue-culture protocols (Moniruzzaman et al., 2021; Moniruzzaman et al., 2015) and subsequent molecular biology techniques have been adapted in fig. It is high time to utilize the opportunities presented by genetic engineering techniques along with conventional breeding techniques aimed at improving water-use efficiency along with other agronomic traits for getting improved fig cultivars that could withstand the climate change challenges.

3.2 *Efficient Agronomic Practices Could Improve Tolerance against Adverse Climate*

Efficient agronomic practices could prevent soil and water-resource degradation (Zalidis et al., 2002), maximize the use of available water, control pest and disease; thus improving fruit productivity and quality. In rain-fed conditions, Mediterranean fig fruit production could largely improve when rainfall occurs in spring and early summer than the total amount of overall precipitation. Water use during drought periods could be effectively managed by considering soil properties (texture, porosity, organic matter, impeding layers in the soil profile, among others), mulching and crop residue management that promote water infiltration and accumulation in the soil (Bescansa et al., 2006; Lampurlanés et al., 2016). In the case of irrigation, it will be necessary to improve water-transport infrastructure to reduce water losses, improve irrigation system distribution and on-farm application, and adopt irrigation-efficient technology (Sowers et al., 2011). Integrated pest management and good agricultural practice could prevent fig bark beetle (*H. scabricollis*) and root nematodes infection (Cutajar and Mifsud, 2017). Fig tree trimming and pruning are necessary to stimulate new growth or to control tree size. Pruning should be carried out when the tree is dormant, that is, between leaf fall and bud burst (usually between November and early

March). Timely pruning could control transpiration, thus improving water-utilization capacity.

3.3 *Functional Symbiosis for Adverse Climate Challenge*

Plants are not always autonomous individuals. The internal tissue of plants provide specialized environment for diverse communities of symbiotic microbes, which largely influence plant growth and productivity (Brachmann and Parniske, 2006; Trappe, 2005). Interestingly, it is known that plant-fungi symbiosis could act as an early warning system or network for environmental constraints by transmitting signals between plants, even between plants of different species (Babikova et al., 2013). It is also reported that mutualistic fungi contribute to or are responsible for plant adaptation to environmental stress (Molina-Montenegro et al., 2015). On top of that, functional symbioses in turn pose fewer risks to human health and have little or no residual effects, having already been accepted in organic farming (Kumar and Singh, 2014).

Fig trees possess endophytes and several studies are reported in connection with bioactive endophytes or biocontrol endophytes (Abdou et al., 2021). Therefore, there is plenty of room for research and development of novel microorganisms for pest management and for increasing plants' environmental tolerance to cope with climate change effects (del Pozo et al., 2019).

3.4 *Farmer Adaptation to Cope with Adverse Climate*

Practicing adverse climate adaptation at farm level can effectively reduce the climate change impacts and simultaneously increase agricultural productivity (Khanal et al., 2018; Roco et al., 2017). However, still the adaptation frequency of specific and global impacts is low (Roco et al., 2017; Mees et al., 2014) and the situation becomes worse when farmers are reluctant to adapt the practices recommended by experts. So, it is important not only to generate adaptation technologies but also to understand the factors associated with the ability to adapt, as well as to develop effective strategies for transferring such technologies to farmers (Roco et al., 2014). The implementation of adaptation decision depends on farmers' perceptions about climate change and farmers' perceptions are affected by the available information and socio-demographic characteristics. Farmers, who accept the concept of climate change, work on adaptation responses within their farms, such as increasing their water-storage capacity in contrast to those that reject the phenomenon (Raymond and Spoehr, 2013).

Therefore, considering the understanding the farmers' behavior is an important factor for developing adaptation technologies. Understanding farmers' behavior towards adaptation practices requires a broad view of different approaches that include public policy, economics, geography, sociology and psychology, as well as different layers of analyses. Strong governance structures with effective (achieving the desired goals) and efficient (optimal allocation of public resources) policy incentives can favor adaptation among famers. Suitable adaptation techniques along with favorable public policy could create a favorable environment (del Pozo et al., 2019).

4. Conclusion and Prospects

Adequate amount of fruit consumption is needed for building a sound health. Fig fruit that bears enormous medicinal values is also delicious and nutritious. People consume fig fruit in fresh, dry or different forms of value-added products (jam, jelly, pickle, tea, etc.) and the demand for fig in this group of people is enormous. On the other hand, wild fig is a key food source in many forest ecosystems. However, like other agricultural crops, fig cultivation is also being affected by climate change even more severely because the fig-cultivation zone already has pre-conditions (high atmospheric temperature and water scarcity in summer). Further climate change impact in this region (Mediterranean climate) would be harsh for fig production. To minimize the climate change impacts, combined efforts of plant scientists, climate change scientists, agronomists, policy makers and political leaders are an immediate and crying need. Different levels of goals with working strategy (immediate, short-term and long-term) are recommended to combat climate change disasters. Immediate action might prevent further degradation of environment with farmers' adaptation of best (available) agronomic practices. Short-term action might heal the environmental damage, natural water reservation and irrigation infrastructure development, improve agronomic technology and encourage drought and heat stress tolerance. Long-term strategy would be breeding for higher water-use efficiency, functional symbiosis for stress tolerance and maintaining environment health.

References

Abdel-Aty, A.M., Hamed, M.B., Salama, W.H., Ali, M.M., Fahmy, A.S. and Mohamed, S.A. (2019). *Ficus carica, Ficus sycomorus* and *Euphorbia tirucalli* latex extracts: Phytochemical screening, antioxidant and cytotoxic properties. *Biocatalysis and Agricultural Biotechnology*, 20: 101199. https://doi.org/10.1016/j.bcab.2019.101199.

Abdou, R., Mojally, M. and Attia, G.H. (2021). Investigation of bioactivities of endophytes of *Ficus carica* L. *Fam Moraceae. Bulletin of the National Research Centre*, 45(1): 1–7.

Araus, J.L., Slafer, G.A., Royo, C. and Serret, M.D. (2008). Breeding for yield potential and stress adaptation in cereals. *Critical Reviews in Plant Science*, 27(6): 377–412.

Atkinson, C.J., Brennan, R.M. and Jones, H.G. (2013). Declining chilling and its impact on temperate perennial crops. *Environmental and Experimental Botany*, 91: 48–62. Doi: https://doi.org/10.1016/j.envexpbot.2013.02.004.

Babikova, Z., Gilbert, L., Bruce, T.J., Birkett, M., Caulfield, J.C., Woodcock, C., Pickett, J.A. and Johnson, D. (2013). Underground signals carried through common mycelial networks warn neighbouring plants of aphid attack. *Ecology Letters*, 16(7): 835–843.

Barolo, M.I., Mostacero, N.R. and López, S.N. (2014). *Ficus carica* L. (Moraceae): An ancient source of food and health. *Food Chemistry*, 164: 119–127.

Bescansa, P., Imaz, M., Virto, I., Enrique, A. and Hoogmoed, W. (2006). Soil water retention as affected by tillage and residue management in semiarid Spain. *Soil and Tillage Research*, 87(1): 19–27.

Bindi, M. and Howden, S. (2004). Challenges and opportunities for cropping systems in a changing climate. *In*: *Proceedings of the 4th International Crop Science Congress*. Brisbane, Australia, 2004, CDROM.

Bisbis, M.B., Gruda, N. and Blanke, M. (2018). Potential impacts of climate change on vegetable production and product quality—A review. *Journal of Cleaner Production*, 170: 1602–1620. Doi:https://doi.org/10.1016/j.jclepro.2017.09.224.

Bita, C.E. and Gerats, T. (2013). Plant tolerance to high temperature in a changing environment: scientific fundamentals and production of heat stress-tolerant crops. *Frontiers in Plant Science*, 4: 273–273. Doi:10.3389/fpls.2013.00273.

Brachmann, A. and Parniske, M. (2006). The most widespread symbiosis on earth. *PLoS Biology*, 4(7): e239.

Brennan, J., Condon, A., Van Ginkel, M. and Reynolds, M. (2007). An economic assessment of the use of physiological selection for stomatal aperture-related traits in the CIMMYT wheat breeding programme. *The Journal of Agricultural Science*, 145(3): 187.

Camargo, A.V. and Lobos, G.A. (2016). Latin America: A development pole for phenomics. *Frontiers in Plant Science*, 7: 1729.

Challinor, A.J., Watson, J., Lobell, D.B., Howden, S.M., Smith, D.R. and Chhetri, N. (2014). A meta-analysis of crop yield under climate change and adaptation. *Nature Climate Change*, 4(4): 287–291. Doi:10.1038/nclimate2153.

Collier, R.R., Fellows, J., Adams, S., Semenov, M. and Thomas, B. (2008). Vulnerability of horticultural crop production to extreme weather events. *Aspects of Applied Biology*, 88: 1–14.

Cramer, W., Guiot, J., Fader, M., Garrabou, J., Gattuso, J.-P., Iglesias, A., Lange, M.A., Lionello, P., Llasat, M.C., Paz, S., Peñuelas, J., Snoussi, M., Toreti, A., Tsimplis, M.N. and Xoplaki, E. (2018). Climate change and interconnected risks to sustainable development in the Mediterranean. *Nature Climate Change*, 8(11): 972–980. Doi:10.1038/s41558-018-0299-2.

Cutajar, S. and Mifsud, D. (2017). *Good Agricultural Practice (GAP) for Fig Tree Cultivation*. Malta: Plant Protection Directorate, Last access: November 2021; https://www.um.edu.mt/library/oar/handle/123456789/64205.

Dáder, B., Fereres, A., Moreno, A. and Trębicki, P. (2016). Elevated CO_2 impacts bell pepper growth with consequences to Myzus persicae life history, feeding behaviour and virus transmission ability. *Scientific Reports*, 6: 19120–19120. Doi:10.1038/srep19120.

de Amorin, A., Borba, H.R., Carauta, J.P., Lopes, D.S. and Kaplan, M.A. (1999). Anthelmintic activity of the latex of Ficus species. *Journal of Ethnopharmacology*, 64(3): 255–258.

del Pozo, A., Brunel-Saldias, N., Engler, A., Ortega-Farias, S., Acevedo-Opazo, C., Lobos, G.A., Jara-Rojas, R. and Molina-Montenegro, M.A. (2019). Climate change impacts and adaptation strategies of agriculture in Mediterranean-Climate Regions (MCRs). *Sustainability*, 11(10): 2769.

Dhondt, S., Wuyts, N. and Inzé, D. (2013). Cell to whole-plant phenotyping: The best is yet to come. *Trends in Plant Science*, 18(8): 428–439.

Diab, A.A., Teulat-Merah, B., This, D., Ozturk, N.Z., Benscher, D. and Sorrells, M.E. (2004). Identification of drought-inducible genes and differentially expressed sequence tags in barley. *Theoretical and Applied Genetics*, 109(7): 1417–1425.

Doaa, A.D., El-Berry, I., Mustafa, N., Hagagg, F. and Fatma, S. (2015). Detecting drought tolerance of fig (*Ficus carica,* L.) cultivars depending on vegetative growth and peroxidase activity. *International Journal of ChemTech Research*, 8: 1520–1532.

Dong, J.-l., Li, X., Nazim, G. and Duan, Z.-Q. (2018a). Interactive effects of elevated carbon dioxide and nitrogen availability on fruit quality of cucumber (*Cucumis sativus* L.). *Journal of Integrative Agriculture*, 17(11): 2438–2446. doi:https://doi.org/10.1016/S2095-3119(18)62005-2.

Dong, J., Gruda, N., Lam, S.K., Li, X. and Duan, Z. (2018b). Effects of elevated CO_2 on nutritional quality of vegetables: a review. *Frontiers in Plant Science*, 9: 924. Doi:10.3389/fpls.2018.00924.

Dong, J., Xu, Q., Gruda, N., Chu, W., Li, X. and Duan, Z. (2018c). Elevated and super-elevated CO_2 differ in their interactive effects with nitrogen availability on fruit yield and quality of cucumber. *Journal of the Science of Food and Agriculture*, 98(12): 4509–4516. Doi:10.1002/jsfa.8976.

El Yaacoubi, A., Malagi, G., Oukabli, A., Hafidi, M. and Legave, J-M. (2014). Global warming impact on floral phenology of fruit trees species in Mediterranean region. *Scientia Horticulturae*, 180: 243–253.

Fahad, S., Bajwa, A.A., Nazir, U., Anjum, S.A., Farooq, A., Zohaib, A., Sadia, S., Nasim, W., Adkins, S., Saud, S., Ihsan, M.Z., Alharby, H., Wu, C., Wang, D. and Huang, J. (2017). Crop production under drought and heat stress: plant responses and management options. *Frontiers in Plant Science*, 8: 1147. Doi:10.3389/fpls.2017.01147.

FAOSTAT Agricultural Data. (2012). http://.faostat.fao.org/site/570/DesktopDefault.aspx accessed 26.10 2015.

Fischer, R. (2007). Understanding the physiological basis of yield potential in wheat. *The Journal of Agricultural Science*, 145(2): 99.

Flaishman, M.A., Rodov, V. and Stover, E. (2007). The fig: botany, horticulture, and breeding. *In*: Janick, J. (ed.). *Horticultural Reviews*. https://doi.org/10.1002/9780470380147.ch2.

Gholami, M., Rahemi, M. and Rastegar, S. (2012). Use of rapid screening methods for detecting drought tolerant cultivars of fig (*Ficus carica* L.). *Scientia Horticulturae*, 143: 7–14.

Hasanuzzaman, M., Nahar, K., Alam, M.M., Roychowdhury, R. and Fujita, M. (2013). Physiological, biochemical, and molecular mechanisms of heat stress tolerance in plants. *International Journal of Molecular Science*, 14(5): 9643–9684. doi:10.3390/ijms14059643.

Hatfield, J.L. and Prueger, J.H. (2015). Temperature extremes: Effect on plant growth and development. *Weather and Climate Extremes*, 10: 4–10. Doi:https://doi.org/10.1016/j.wace.2015.08.001.

Hijmans, R.J. (2003). The effect of climate change on global potato production. *American Journal of Potato Research*, 80(4): 271–279.

Houle, D., Govindaraju, D.R. and Omholt, S. (2010). Phenomics: The next challenge. *Nature Reviews Genetics*, 11(12): 855–866.

Hu, H. and Xiong, L. (2014). Genetic engineering and breeding of drought-resistant crops. *Annual Review of Plant Biology*, 65: 715–741. Doi:10.1146/annurev-arplant-050213-040000.

Hullé, M., Cœur d'Acier, A., Bankhead-Dronnet, S. and Harrington, R. (2010). Aphids in the face of global changes. *Comptes Rendus Biologies*, 333(6): 497–503. Doi:https://doi.org/10.1016/j.crvi.2010.03.005.

IPCC. (2013). Summary for policymakers. *In*: Stocker, T.F., Qin, D., Plattner, G.-K., Tignor, M., Allen, S.K., Boschung, J., Nauels, A., Xia, Y., Bex, V. and Midgley, P.M. (eds.). *Climate Change 2013: The Physical Science Basis*. Contribution of Working Group I to the Fifth Assessment Report of the Intergovernmental Panel on Climate Change. Cambridge University Press, Cambridge, United Kingdom and New York, NY, USA.

Jacob, D., Petersen, J., Eggert, B., Alias, A., Christensen, O.B., Bouwer, L.M., Braun, A., Colette, A., Déqué, M., Georgievski, G., Georgopoulou, E., Gobiet, A., Menut, L., Nikulin, G.,

Haensler, A., Hempelmann, N., Jones, C., Keuler, K., Kovats, S., Kröner, N., Kotlarski, S., Kriegsmann, A., Martin, E., van Meijgaard, E., Moseley, C., Pfeifer, S., Preuschmann, S., Radermacher, C., Radtke, K., Rechid, D., Rounsevell, M., Samuelsson, P., Somot, S., Soussana, J.-F., Teichmann, C., Valentini, R., Vautard, R., Weber, B. and Yiou, P. (2014). EURO-CORDEX: New high-resolution climate change projections for European impact research. *Regional Environmental Change*, 14(2): 563–578. Doi:10.1007/s10113-013-0499-2.

Jafari, M., Rahemi, M. and Zare, H. (2015). Increasing the tolerance of fig tree to drought stress by trunk thinning. pp. 189–194. *In: V International Symposium on Fig 1173.*

Karimi, S., Hojati, S., Eshghi, S., Moghaddam, R.N. and Jandoust, S. (2012). Magnetic exposure improves tolerance of fig 'Sabz'explants to drought stress induced *in vitro*. *Scientia Horticulturae*, 137: 95–99.

Khanal, U., Wilson, C., Lee, B. and Hoang, V.-N. (2018). Do climate change adaptation practices improve technical efficiency of smallholder farmers? Evidence from Nepal. *Climatic Change*, 147(3): 507–521.

Kitajima, S., Aoki, W., Shibata, D., Nakajima, D., Sakurai, N., Yazaki, K., Munakata, R., Taira, T., Kobayashi, M., Aburaya, S. and Savadogo, E.H. (2018). Comparative multi-omics analysis reveals diverse latex-based defense strategies against pests among latex-producing organs of the fig tree (*Ficus carica*). *Planta*, 247(6): 1423–1438.

Kumar, S. and Singh, A. (2014). Biopesticides for integrated crop management: Environmental and regulatory aspects. *Journal of Biofertilizer Biopesticide*, 5: e121.

Lampurlanés, J., Plaza-Bonilla, D., Álvaro-Fuentes, J. and Cantero-Martínez, C. (2016). Long-term analysis of soil water conservation and crop yield under different tillage systems in Mediterranean rainfed conditions. *Field Crops Research*, 189: 59–67.

Lansky, E.P., Paavilainen, H.M., Pawlus, A.D. and Newman, R.A. (2008). Ficus spp. (fig): Ethnobotany and potential as anticancer and anti-inflammatory agents. *Journal of Ethnopharmacology*, 119(2): 195–213.

Lazreg-Aref, H., Mars, M., Fekih, A., Aouni, M. and Said, K. (2012). Chemical composition and antibacterial activity of a hexane extract of Tunisian caprifig latex from the unripe fruit of *Ficus carica*. *Pharmaceutical Biology*, 50(4): 407–412.

Leung, C.W., Epel, E.S., Ritchie, L.D., Crawford, P.B. and Laraia, B.A. (2014). Food insecurity is inversely associated with diet quality of lower-income adults. *Journal of the Academy of Nutrition and Dietetics*, 114(12): 1943–1953, e1942.

Li, S.Z. (1578). *Outline of Materia Medica*. vol. 31 (Section on Fruits), People's Health Publisher, Beijing.

Litton, M.M. and Beavers, A.W. (2021). The relationship between food security status and fruit and vegetable intake during the covid-19 pandemic. *Nutrients*, 13(3): 712.

Lobato-Gómez, M., Hewitt, S., Capell, T., Christou, P., Dhingra, A. and Girón-Calva, P.S. (2021). Transgenic and genome-edited fruits: Background, constraints, benefits, and commercial opportunities. *Horticulture Research*, 8(1): 1–16.

Lobos, G.A. and Hancock, J.F. (2015). Breeding blueberries for a changing global environment: A review. *Frontiers in Plant Science*, 6: 782.

Loizzo, M.R., Bonesi, M., Pugliese, A., Menichini, F. and Tundis, R. (2014). Chemical composition and bioactivity of dried fruits and honey of *Ficus carica* cultivars Dottato, San Francesco and Citrullara. *Journal of the Science of Food and Agriculture*, 94(11): 2179–2186.

Mavlonov, G., Ubaidullaeva, K.A., Rakhmanov, M., Abdurakhmonov, I.Y. and Abdukarimov, A. (2008). Chitin-binding antifungal protein from *Ficus carica* latex. *Chemistry of Natural Compounds*, 44(2): 216–219.

Mees, H.L., Dijk, J., van Soest, D., Driessen, P.P., van Rijswick, M.H. and Runhaar, H. (2014) A method for the deliberate and deliberative selection of policy instrument mixes for climate change adaptation. *Ecology and Society*, 19(2): 58.

Molina-Montenegro, M.A., Oses, R., Torres-Díaz, C., Atala, C., Núñez, M.A. and Armas, C. (2015). Fungal endophytes associated with roots of nurse cushion species have positive effects on native and invasive beneficiary plants in an alpine ecosystem. *Perspectives in Plant Ecology, Evolution and Systematics*, 17(3): 218–226.

Moniruzzaman, M., Yaakob, Z. and Taha, R. (2015). *In vitro* production of fig (*Ficus carica* L.) plantlets. pp. 231–236. *In: V International Symposium on Fig*, 1173.

Moniruzzaman, M., Yaakob, Z., Khatun, R. and Awang, N. (2017). Mealybug (Pseudococcidae) infestation and organic control in fig (*Ficus carica*) orchards of Malaysia. *Biology and Environment: Proceedings of the Royal Irish Academy*, 117B(1): 25–32.

Moniruzzaman, M., Anuar, N., Yaakob, Z., Islam, A.K.M.A. and Al-Khayri, J.M. (2020). Performance evaluation of seventeen common fig (*Ficus carica* L.) cultivars introduced to a tropical climate. *Horticulture, Environment, and Biotechnology*, 61(5): 795–806. Doi:10.1007/s13580-020-00259-1.

Moniruzzaman, M., Yaakob, Z. and Anuar, N. (2021). Factors affecting *in vitro* regeneration of *Ficus carica* L. and genetic fidelity studies using molecular marker. *Journal of Plant Biochemistry and Biotechnology*, 30(2): 304–316. Doi:10.1007/s13562-020-00590-9.

Oliveira, A.P., Silva, L.S.R., Andrade, P.B., Valentão, P.C., Silva, B.M., Gonçalves, R.F., Pereira, J.A. and Guedes de Pinho, P. (2010). Further insight into the latex metabolite profile of *Ficus carica*. *Journal of Agricultural and Food Chemistry*, 58(20): 10855–10863.

Oyebode, O., Gordon-Dseagu, V., Walker, A. and Mindell, J.S. (2014). Fruit and vegetable consumption and all-cause, cancer and CVD mortality: Analysis of Health Survey for England data. *Journal of Epidemiol Community Health*, 68(9): 856–862.

Palmeira, L., Pereira, C., Dias, M.I., Abreu, R.M.V., Corrêa, R.C.G., Pires, T.C.S.P., Alves, M.J., Barros, L. and Ferreira, I.C.F.R. (2019). Nutritional, chemical and bioactive profiles of different parts of a Portuguese common fig (*Ficus carica* L.) variety. *Food Research International*, 126: 108572. Doi:https://doi.org/10.1016/j.foodres.2019.108572.

Polat, A.A. and Caliskan, O. (2017). Effect of different environments on fruit characteristics of table fig (*Ficus carica* L.) cultivars. *Modern Agricultural Science and Technology*, 3: 11–14.

Raafat, K. and Wurglics, M. (2019). Phytochemical analysis of *Ficus carica* L. active compounds possessing anticonvulsant activity. *Journal of Traditional and Complementary Medicine*, 9(4): 263–270. Doi:https://doi.org/10.1016/j.jtcme.2018.01.007.

Raskovic, B., Lazic, J. and Polovic, N. (2016). Characterisation of general proteolytic, milk clotting and antifungal activity of *Ficus carica* latex during fruit ripening. *Journal of the Science of Food and Agriculture*, 96(2): 576–582.

Raymond, C.M. and Spoehr, J. (2013). The acceptability of climate change in agricultural communities: Comparing responses across variability and change. *Journal of Environmental Management*, 115: 69–77.

Reynolds, M., Manes, Y., Izanloo, A. and Langridge, P. (2009). Phenotyping approaches for physiological breeding and gene discovery in wheat. *Annals of Applied Biology*, 155(3): 309–320.

Reynolds, M. and Langridge, P. (2016). Physiological breeding. *Current Opinion in Plant Biology*, 31: 162–171.

Richards, R.A., Rebetzke, G.J., Condon, A.G. and van Herwaarden, A.F. (2002). Breeding opportunities for increasing the efficiency of water use and crop yield in temperate cereals. *Crop Science*, 42(1): 111–121. Doi:10.2135/cropsci2002.1110.

Roco, L., Engler, A., Bravo-Ureta, B. and Jara-Rojas, R. (2014). Farm level adaptation decisions to face climatic change and variability: Evidence from Central Chile. *Environmental Science & Policy*, 44: 86–96.

Roco, L., Bravo-Ureta, B., Engler, A. and Jara-Rojas, R. (2017). The impact of climatic change adaptation on agricultural productivity in Central Chile: A stochastic production frontier approach. *Sustainability*, 9(9): 1648.

Rubnov, S., Kashman, Y., Rabinowitz, R., Schlesinger, M. and Mechoulam, R. (2001). Suppressors of cancer cell proliferation from fig (*Ficus carica*) resin: isolation and structure elucidation. *Journal of Natural Products*, 64(7): 993–996.

Ruggiero, A., Punzo, P., Landi, S., Costa, A., Van Oosten, M.J. and Grillo, S. (2017). Improving plant water use efficiency through molecular genetics. *Horticulturae*, 3(2): 31.

Saadi, S., Todorovic, M., Tanasijevic, L., Pereira, L.S., Pizzigalli, C. and Lionello, P. (2015). Climate change and Mediterranean agriculture: Impacts on winter wheat and tomato crop evapotranspiration, irrigation requirements and yield. *Agricultural Water Management*, 147: 103–115. Doi:https://doi.org/10.1016/j.agwat.2014.05.008.

Santos, J.A., Costa, R. and Fraga, H. (2017). Climate change impacts on thermal growing conditions of main fruit species in Portugal. *Climatic Change*, 140(2): 273–286. Doi:10.1007/s10584-016-1835-6.

Seligman, H.K., Laraia, B.A. and Kushel, M.B. (2010). Food insecurity is associated with chronic disease among low-income NHANES participants. *The Journal of Nutrition*, 140(2): 304–310.

Sowers, J., Vengosh, A. and Weinthal, E. (2011). Climate change, water resources, and the politics of adaptation in the Middle East and North Africa. *Climatic Change*, 104(3): 599–627.

Stover, E., Aradhya, M., Ferguson, L. and Crisosto, C.H. (2007). The fig: overview of an ancient fruit. *HortScience*, 42(5): 1083–1087.

Tapia, R., Botti, C., Carrasco, O., Prat, L. and Franck, N. (2003). Effect of four irrigation rates on growth of six fig tree varieties. pp. 113–118. *In: International Society for Horticultural Science (ISHS)*. Leuven, Belgium. Doi:10.17660/ActaHortic.2003.605.17.

Tondelli, A., Francia, E., Barabaschi, D., Aprile, A., Skinner, J.S., Stockinger, E.J., Stanca, A.M. and Pecchioni, N. (2006). Mapping regulatory genes as candidates for cold and drought stress tolerance in barley. *Theoretical and Applied Genetics*, 112(3): 445–454.

Trappe, J.M. (2005). AB Frank and mycorrhizae: The challenge to evolutionary and ecologic theory. *Mycorrhiza*, 15(4): 277–281.

Tuberosa, R., Turner, N.C. and Cakir, M. (2014). Two decades of InterDrought conferences: are we bridging the genotype-to-phenotype gap? *Journal of Experimental Botany*, 65(21): 6137–6139. Doi:10.1093/jxb/eru407.

Türk, F.H. and Aksoy, U. (2011). *Comparison of Organic, Biodynamic and Conventional Fig Farms under Rain-fed Conditions in Turkey*.

Van Damme, V., Berkvens, N., Moerkens, R., Berckmoes, E., Wittemans, L., De Vis, R., Casteels, H., Tirry, L. and De Clercq, P. (2015). Overwintering potential of the invasive leafminer Tuta absoluta (Meyrick) (*Lepidoptera: Gelechiidae*) as a pest in greenhouse tomato production in Western Europe. *Journal of Pest Science*, 88(3): 533–541. Doi:10.1007/s10340-014-0636-9.

Van de Perre, E., Jacxsens, L., Liu, C., Devlieghere, F. and De Meulenaer, B. (2015). Climate impact on Alternaria moulds and their mycotoxins in fresh produce: The case of the tomato chain. *Food Research International*, 68: 41–46. Doi:https://doi.org/10.1016/j.foodres.2014.10.014.

Varshney, R.K., Thudi, M., Nayak, S.N., Gaur, P.M., Kashiwagi, J., Krishnamurthy, L., Jaganathan, D., Koppolu, J., Bohra, A. and Tripathi, S. (2014). Genetic dissection of drought tolerance in chickpea (*Cicer arietinum* L.). *Theoretical and Applied Genetics*, 127(2): 445–462.

Venuprasad, R., Lafitte, H.R. and Atlin, G.N. (2007). Response to direct selection for grain yield under drought stress in rice. *Crop Science*, 47(1): 285–293.

Wang, J., Wang, X., Jiang, S., Lin, P., Zhang, J., Lu, Y., Wang, Q., Xiong, Z., Wu, Y. and Ren, J. (2008). Cytotoxicity of fig fruit latex against human cancer cells. *Food and Chemical Toxicology*, 46(3): 1025–1033.

Wang, P.Y., Fang, J.C., Gao, Z.H., Zhang, C. and Xie, S.Y. (2016). Higher intake of fruits, vegetables or their fiber reduces the risk of type 2 diabetes: A meta-analysis. *Journal of Diabetes Investigation*, 7(1): 56–69.

Wang, T., Zhang, H. and Zhu, H. (2019). CRISPR technology is revolutionizing the improvement of tomato and other fruit crops. *Horticulture Research*, 6(1): 1–13.

Wang, X., Ouyang, Y., Liu, J., Zhu, M., Zhao, G., Bao, W. and Hu, F.B. (2014). Fruit and vegetable consumption and mortality from all causes, cardiovascular disease, and cancer: Systematic review and dose-response meta-analysis of prospective cohort studies. *BMJ*, 349.

Woodland, D.W. (1997). *Contemporary Plant Systematics*. 2nd ed., Andrews University Press, Berrien Springs, MI.

Yang, W., Guo, Z., Huang, C., Duan, L., Chen, G., Jiang, N., Fang, W., Feng, H., Xie, W. and Lian, X. (2014). Combining high-throughput phenotyping and genome-wide association studies to reveal natural genetic variation in rice. *Nature Communications*, 5(1): 1–9.

Zalidis, G., Stamatiadis, S., Takavakoglou, V., Eskridge, K. and Misopolinos, N. (2002). Impacts of agricultural practices on soil and water quality in the Mediterranean region and proposed assessment methodology. *Agriculture, Ecosystems & Environment*, 88(2): 137–146. Doi:https://doi.org/10.1016/S0167-8809(01)00249-3.

Zhang, H.-C., Ma, Y.-M., Liu, R. and Zhou, F. (2012a). Endophytic fungus *Aspergillus tamarii* from *Ficus carica* L., a new source of indolyl diketopiperazines. *Biochemical Systematics and Ecology*, 45: 31–33.

Zhang, H.C., Ma, Y.M. and Liu, R. (2012b). Antimicrobial additives from endophytic fungus *Fusarium solani* of *Ficus carica*. pp. 783–786. *In*: *Applied Mechanics and Materials*. Trans Tech. Publ.

5

Cactus Pear (*Opuntia* spp.) A Multipurpose Crop with High Tolerance to Adverse Climate

Mouaad Amine Mazri,[1,*] *Ilham Belkoura,*[2] *Reda Meziani,*[3] *Riad Balaghi*[4] and *Meriyem Koufan*[5]

1. Introduction

In recent years, climate change has aroused significant international interest as well as attention of scientists, stakeholders, politicians, decision makers and the general public. In fact, climate change has become a major concern because of its biotic and abiotic impacts that threaten global agricultural production and food security. Changes in atmospheric temperatures and rainfall patterns are the two aspects of climate change that affect plant phenology, biodiversity, reproduction, distribution and

[1] Agro-Biotechnology Research Unit, Regional Center of Agricultural Research of Marrakech, National Institute of Agricultural Research, Avenue Ennasr, BP 415 Rabat Principale, 10090 Rabat, Morocco.
[2] *In Vitro* Culture Laboratory, Department of Basic Sciences, National School of Agriculture, BP S/40, 50001 Meknes, Morocco.
[3] Oasis Systems Research Unit, Regional Center of Agricultural Research of Errachidia, National Institute of Agricultural Research, Avenue Ennasr, BP 415 Rabat Principale, 10090 Rabat, Morocco.
[4] National Institute of Agricultural Research, Avenue Ennasr, BP 415 Rabat Principale, 10090 Rabat, Morocco.
[5] Natural Resources and Local Products Research Unit, Regional Center of Agricultural Research of Agadir, National Institute of Agricultural Research, Avenue Ennasr, BP 415 Rabat Principale, 10090 Rabat, Morocco.
* Corresponding author: mouaadamine.mazri@inra.ma

may increase pest and pathogen invasion. Some consequences of climate change, such as water scarcity, pronounced drought, soil degradation and accelerated desertification have negative impacts on natural ecosystems and agricultural production. Thus, efforts are now being devoted to cultivating plant genus and species characterized by high adaptability and tolerance to biotic and abiotic stresses caused by climate change.

Opuntia spp. is a plant genus that contains around 300 species (Arba, 2020). It belongs to the family Cactaceae and is commonly known as cactus pear, prickly pear and bunny ears (Mazri, 2018). Cactus pear is a fruit tree that has fleshy stems with clusters of glochids or spines which function in photosynthesis. The root system is extensive. The leaves are either small or absent. The solitary flower has many petals and sepals (Glimn-Lacy and Kaufman, 2006). The fruit, which is a berry known as tuna, Indian fig and nopal fruit, consists of peel, pulp and numerous small seeds. It has an ovoid-spherical shape and is rich in minerals, antioxidants and fibers. At maturity, the fruit differs in color and size (Arba, 2020; El Kharrassi et al., 2015, 2016; Mazri, 2018; Patel, 2015). Cactus pear is a Crassulacean Acid Metabolism (CAM) plant. It is characterized by stomatal opening at night and closing during daytime, and therefore has a lower transpiration than C_3 and C_4 plants and better water use-efficiency. This results in a notable adaptation to environments characterized by limited water availability, high temperatures and drought conditions (Inglese et al., 2017).

Cactus pear is a native of tropical and subtropical Americas, but is now cultivated in many countries and under various climatic conditions (Mazri, 2018; Sáenz, 2013a). It is found in deserts, in arid and tropical regions and in areas with very low temperatures (Kumar et al., 2018). The most important producer countries of cactus pear fruits are Mexico, Tunisia, Argentina, Algeria and Italy (Jiménez-Aguilar et al., 2014, 2015; Timpanaro and Foti, 2014).

Cactus pear can be cultivated for various purposes. It may be used for human consumption, for pharmaceutical applications and to manufacture cosmetic products (El Kharrassi et al., 2018; Mazri, 2018; Sáenz, 2013b; Tilahun and Welegerima, 2018; Yahia and Sáenz, 2018). Cactus pear can be used also to feed livestock, especially under drought conditions since its inclusion in ruminant diets reduces their water consumption (Borges et al., 2019; Costa et al., 2009; Magalhães et al., 2019). In addition, cactus pear cultivation has many ecological benefits. The specific physiological and anatomical traits of this genus makes it highly adaptable to harsh climatic and environmental conditions. In fact, cactus pear can be used to restore degraded lands, to preserve biodiversity, to improve the physical properties of soil, to prevent soil erosion, to preserve water resources and to combat desertification (Nefzaoui and Ben Salem, 2002; Nefzaoui et al., 2014). Among all cultivated cactus pear species, *O. ficus indica* is the most economically important one and is used for both fruit and cladode production.

Under different climate change scenarios, cactus pear can be considered as a highly valuable plant genus. In fact, cactus pear shows high tolerance to most abiotic stresses caused by climate change, such as increased temperatures, high atmospheric CO_2 concentrations, soil degradation, droughts and water deficit. This chapter presents an overview on cactus pear cultivation, description of the climate change impacts and how to mitigate these impacts. It also describes the climatic requirements of cactus pear and its natural adaptation to changes in climatic and environmental conditions. It reports the findings of prediction and simulation studies under future climate change scenarios. It discusses the major biotic stresses affecting this plant genus and which could be increased by climate change and finally reports some mitigation strategies against these stresses.

2. An Overview of the General Impacts of Climate Change on Agricultural Production and Food Security

Nowadays, climate change is considered as a major threat to worldwide agriculture and food security. Changes in rainfall frequencies, water availability and atmospheric temperatures, as well as the increase of atmospheric CO_2, sea levels, and the frequencies of drought, storms, flooding, cold and heat waves are all consequences of climate change (Uprety et al., 2019d). All these factors have a direct impact on agricultural production and food security since they affect food availability, access, stability and utilization (Balaghi et al., 2010).

2.1 *Rise in CO_2 Concentrations*

The three major greenhouse gases responsible for climate change are nitrous oxide (N_2O), methane (CH_4) and carbon dioxide (CO_2). Among these three gases, CO_2 is the main factor responsible for global warming (Sthapit and Scherr, 2012a). In fact, the rise in CO_2 concentrations increases global temperatures (Uprety et al., 2019b). Moreover, the increase in atmospheric CO_2 levels causes changes in plant morphology, anatomy and physiology, alters their nutritional quality, affects their defense mechanisms against pathogens and pests, and promotes weed dissemination (Uprety et al., 2019a, c).

2.2 *Increased Temperatures*

Plant nutritional quality is affected by increased temperatures (Uprety et al., 2019a). In addition, high temperatures increase the risk of drought, pest invasion, pathogen dissemination and biodiversity loss. Besides, warm winters can cause an acceleration in the loss of soil carbon, which affects soil fertility and soil water availability (Pryor et al., 2014). In areas with high temperatures, further augmentations will affect fruit quality and yield, damage fruits and, when combined with high rainfall and humidity, it will provide the perfect conditions for disease and pathogen proliferation (Dinesh and Reddy, 2012).

2.3 *Rainfall, Water Availability and Soil Erosion*

The fruit yield of many crop species is affected by water availability. Thus, irrigation is an important factor to maintain high productivity as well as sustainability of fruit and vegetable production. Unfortunately, in many regions of the world, extreme weather events, high evapotranspiration and restricted rainfall limit the range of crop species that can be cultivated (Olsen and Bindi, 2002). In fact, a reduction in water resources makes it difficult to cultivate crop species that require high quantities of water. On the other hand, more intense spring rainfall may increase runoff, flood and erosion rates, leading to crop losses (Pryor et al., 2014).

2.4 *Drought and Heat Waves*

Drought is an abiotic stress that causes hydrological imbalances and negatively affects plant growth and crop production (Uprety et al., 2019d). It is increased by high temperatures and low precipitation. Besides, drought is one of the major factors threatening world food security (Li et al., 2009). Fruit crops may be disturbed by early-spring heat waves followed by frost events. Moreover, heat waves during reproduction events, such as flowering and pollination, significantly decrease crop yield (Pryor et al., 2014).

2.5 *Pest and Disease Invasion*

Climate change can be a causing factor of wide and rapid dissemination of pests and diseases. In fact, an increase in temperature associated with reduction in rainfall frequencies may facilitate pest invasions (Abrha et al., 2018). Furthermore, droughts and high temperatures may accelerate pathogen evolution, increase their transmission rates and stimulate their dissemination and invasion of new areas (Pautasso et al., 2012). Plants may also become more sensitive to many diseases as a result of warm temperatures associated with spring precipitation (Pautasso et al., 2012).

2.6 *Plant Morphology, Phenology, Abundance and Distribution*

The distribution and abundance of plant species may be shifted as a consequence of climate change (Hegland et al., 2009). As a result, the natural and native habitat of many plant species will change. Plants will tend to occupy new areas where temperatures are suitable for their metabolism. Moreover, there might be changes in their morphology, anatomy as well as phenological events, such as shooting, flowering and fruiting (Root et al., 2003). Changes in plant morphology and anatomy can result in change in plant-pathogen interactions and increase the risk of infections (Uprety et al., 2019c).

2.7 *Ecological Instability and Biodiversity Loss*

High temperatures, rise of sea levels, rise of greenhouse gas emissions and deforestation are all causes of species loss and extinction, flora destruction and natural habitat degradation. Moreover, high temperatures affect plant pollination by influencing the abundance, distribution and activity of pollinating insects, resulting in poor fruit set and quality (Bhuiyan et al., 2018; Hegland et al., 2009).

All these facts enforce cultivation of plant species characterized by resilience to climate change impacts. Cactus pear (*Opuntia* spp.) is a plant genus that shows high tolerance to drought, water scarcity and high temperatures. In addition, cactus pear could be cultivated for multiple purposes: it can be used for human consumption, to feed cattle, in cosmetic and pharmaceutical industries, to combat desertification and soil erosion, and as fence. Furthermore, cactus pear has a good phenological adaptability to different climatic and geographic conditions, and is characterized by high genetic variability that could be used to create more tolerant or resistant genotypes to climate change impacts.

3. Climatic Requirements and Natural Adaptation of Cactus Pear

Cactus pear is native to tropical and subtropical regions of the American continent (Sáenz, 2013a). The best climatic conditions for its growth are as follows: mean temperatures ranging between 11.2–27.1°C, mean annual rainfall ranging between 116.7–1805 mm and an altitude range of 0–2675 m (Badii and Flores, 2001). The optimum day/night temperature for CO_2 uptake is 25/15°C, whereas temperatures above 30°C may affect fruit growth, development and post-harvest storage (Kumar et al., 2018). Generally, cactus pear species are well adapted to various types of soils (gravely, rocky, heavy and alkaline) and environments (Kumar et al., 2018).

Today, species of the genus *Opuntia* are cultivated in many regions of the world and under different climate conditions (Mazri, 2018). They can be found in deserts below sea level, in high altitudes, in regions of Canada where the temperature is very low (–40°C), and in tropical regions characterized by temperatures above 5°C all through the year

(Kumar et al., 2018). Cactus pear plants are also found in countries with the most extensive dry lands since they are well adapted to extreme conditions of water deficit, high temperatures and poor soils (Kumar et al., 2018). They have acquired physiological and structural adaptations to arid environments under which they can produce high green biomass quantities (Mulas and Mulas, 2004; Nefzaoui and Ben Salem, 2002). Indeed, cactus pear cladodes are covered by wax and thick epidermis to reduce evapotranspiration while an impermeable layer to water covers their fine roots. The stomata close almost throughout the day and open at night, resulting in reduced water loss. Under extreme conditions of drought and water deficit, shoots decrease transpiration while roots reduce their permeability to water (Nefzaoui and Ben Salem, 2002).

It is well known that climate change can alter the phenological stages of fruit tree species. For example, increased temperatures, insufficient chilling requirement and changes in rainfall pattern affect bud dormancy and differentiation as well as fruit set and quality (Dinesh and Reddy, 2012; Sthapit and Scherr, 2012b). To deal with such situations, it becomes necessary to use fruit trees characterized by high phenological adaptability and tolerance to climate change. Cactus pear plants have shown variations in periods of occurrence of their phenological events, depending on species and regions of cultivation (El Kharrassi et al., 2015). In the Northern Hemisphere, floral-bud development starts in March, while in the Southern hemisphere, it starts in September. Regarding anthesis, it occurs between May and July in USA and Europe, February and August in Mexico, April and November in Ecuador, October and February in Brazil, and August and January in Argentina (Reyes-Agüero et al., 2006). Fruiting periods range from May to June in Morocco, June to July in Spain, July to August in Italy, August to September in USA, throughout the year in Mexico (depending on Mexican regions), and between June and November in Ecuador (El Kharrassi et al., 2015; Reyes-Agüero et al., 2006). These different periods of time during which the phenological events of cactus pear occur indicate the high adaptability of the species of this genus to various geographical regions and climatic conditions.

4. Cactus Pear Cultivation for Enhanced Resilience to Climate Change

Under different climate change scenarios, and in order to ensure future food security, it is necessary to prioritize the cultivation of plant species resilient to climate change impacts. Such plants should be well adapted to extreme conditions of drought, able to improve soil fertility and structure, to prevent erosion and desertification, to use water resources efficiently, to preserve biodiversity, to help in land rehabilitation, and to sustain economy. Cactus pear seems to be a perfect plant genus for cultivation to mitigate climate change impacts. In fact, in addition to its high phenological adaptability to various climatic and environmental conditions, cactus pear is characterized by high tolerance to abiotic stresses associated with climate change.

4.1 *Cactus Pear: A CAM Plant to Mitigate Increasing Atmospheric CO_2*

Cactus pear is a crassulacean acid metabolism (CAM) plant and thus it is able to exhibit high productivity under water stress conditions (Iqbal et al., 2020; Nefzaoui and Ben Salem, 2002). In fact, CAM plants are characterized by stomatal opening at night, resulting in gaseous exchanges at low temperatures and therefore a reduction in plant water loss in comparison with C_3 and C_4 plants. During the day, when stomata are closed, CAM plants synthesize starch, decarboxylase maltase and fix CO_2 by Rubisco. At night, their pH is decreased as a result of carboxylation, which consists of malic acid production and accumulation, and starch breakdown (Mulas and Mulas, 2004). In addition, CAM plants are able to store high amounts of water and tolerate advanced cellular dehydration (Nobel, 2009). All these characteristics of CAM plants make them suitable for cultivation in regions where C_3 and C_4 plants can hardly survive (Nefzaoui and Ben Salem, 2002).

It was reported that cactus pear species can increase their productivity under increased greenhouse effect. In fact, an increase in atmospheric CO_2 concentration will result in an increase in cactus pear biomass production. Furthermore, the enhancement of atmospheric CO_2 concentration could be attenuated by the high potential productivity of cactus pear (Nobel, 1994; Potgieter, 2007). Cactus pear plants were also reported to survive even after losing up to 90% of their hydrated water content, while C_3 and C_4 plants may suffer irreversible damages after losing 30% of their water content (Nefzaoui, 2009).

4.2 *Degraded Land Rehabilitation and Biodiversity Preservation*

Cactus pear can be used to restore degraded lands (Middleton, 2002). In fact, the root system of *Opuntia* spp. plants can grow and develop in degraded soils (Nefzaoui and Ben Salem, 2002). Accordingly, contour planting of cactus pear was used in some North African countries as a valuable and cheap tool for marginal land rehabilitation (Nefzaoui and El Mourid, 2009). For example, in Algeria and Tunisia, contour planting of cactus pear was done to restore and stabilize badlands (Le Houérou, 1996). In Algeria, *O. ficus indica* was planted to rehabilitate degraded rangelands of drylands (Neffar et al., 2013). As a result, *O. ficus indica* has facilitated the development, abundance and colonization of herbaceous species by improving micro-environmental conditions (Neffar et al., 2013). In addition, plants of *O. ficus indica* were reported to preserve animal biodiversity by creating refuges and shelters for some birds and mammals, and by being a feed source for them (Le Houérou, 1996). Le Houérou (1996) also reported that planting cacti is a cheap, quick and easy way to rehabilitate degraded lands with an increase in their productivity by a factor of up to 10 times. In Tunisia, *Opuntia* plantations were used to preserve natural resources and to prevent deterioration of ecologically weak environments (Nefzaoui and Ben Salem, 2002). In Morocco, planting cactus pear resulted in regeneration and preservation of the native plant species of Rhamna region (Mazhar et al., 2002). In addition, the yield in dry matter of annual vegetation was higher in lands covered by cactus pear than those planted with eucalyptus and grazing lands (Mazhar et al., 2002).

4.3 Soil Conservation and Improvement, and Erosion Prevention

Achieving global food security requires better utilization of land to increase crop production. Soil conservation and erosion prevention have become major challenges to ensure agricultural sustainability and resilience to climate change. Accordingly, a better management of land is now a priority in many countries. In fact, many practices such as contouring, rotations, hedge plantations, conservation tillage, chemical control and cover crops have been in use for many years. In addition to these practices, selecting and cultivating plant species able to conserve soils, to improve their properties and to prevent erosion is now a global priority.

Planting cactus pear on hillsides was reported to control soil erosion since the root system of cactus pear plants is able to maintain the soil surface of hilly zones (Mulas and Mulas, 2004). When used as hedge, cactus pear plays important roles in improving the physical properties of soil and controlling erosion. In fact, cactus pear hedges improve topsoil structural stability, reduce erosion risks and increase nitrogen and organic matter of soils by up to 200% (Nefzaoui et al., 2014). In the same vein, it was reported that the characteristics of cactus pear root system make this plant genus able to resist to wind and rain erosion and, as a result, prevent soil erosion (Nefzaoui and Ben Salem, 2002). In North African countries, cactus pear plants have been widely used as hedges to protect orchards and groves, and as evidence of land ownership. These cactus hedges also play major roles in preventing and controlling erosion since they constitute physical obstacles to runoff (Le Houérou, 1996). In addition, such obstacles are cheap and efficient to prevent topsoil loss and to control and protect against wind erosion that can cause damage to agricultural soil (Nefzaoui, 2009). Other researchers also reported that cactus pear hedges significantly improve soil physical properties and organic matter content, with an increase in organic matter and nitrogen of up to 200%

(Monjauze and Le Houérou, 1965) and improve topsoil structural stability (Le Houérou, 1996). Le Houérou (1996) revealed that cactus pear plants improve soil fertility by improving soil structure, enriching top soils with organic matter, cycling trace element and improving aggregate stability. In agroforestry systems, cactus pear showed high ability of soil conservation than many other plant species (Nefzaoui et al., 2014).

According to Neffar et al. (2013), *O. ficus indica* could indirectly improve soil properties by mitigating harsh environmental conditions, which stimulate the growth and development of grasses and other plant species. *O. ficus indica* is also able to maintain herbaceous vegetation that will improve soil qualities. Neffar et al. (2011) reported that cactus pear plantations improve organic matters between 2.47–4.97%, significantly decrease active limestone, from 12% in control to 6% in cactus pear-planted lands, and protect against erosion due to its expanded root system. In Rhamna (Morocco), cactus pear plantations improved soil quality by making it more valuable and productive, and made slopes resistant to wind and water erosion (Mazhar et al., 2002).

4.4 *Water Conservation and Water Use Efficiency*

Water resource conservation is one of the most important issues to address in order to develop a sustainable agriculture resilient to climate change. Thus, plant cultivation and agricultural production should prioritize better management and use of water resources. Cactus pear is a plant genus well adapted to arid regions. It is characterized by specific physiological and anatomical traits that allow it to use water efficiently and then to be cultivated with little irrigation (Mazri, 2018).

Since cactus pear is a CAM plant, the stomata are generally open at night. This results in a total water loss of only 20–35% in comparison to the total water lost by C_3 and C_4 plants (Nobel, 2001). This is due to low night temperatures that reduce the internal water vapor content in CAM plants (Nefzaoui et al., 2014). Furthermore, the specific characteristics of cactus pear make it a plant of high water-retention ability and high water-use efficiency. The water conservation efficiency in cactus pear plants is also a result of mucilage, a closed oxalic acid metabolism and a hydrophilic mucus that combines with the absorbed water and slows its evaporation (Nefzaoui and Ben Salem, 2002). Plants of the genus *Opuntia* are able to store high quantities of water, tolerate advanced cellular dehydration and survive even after losing 80–90% of their water content. In addition, they are able to produce five to 10 times more biomass per unit of water consumed than C_3 and C_4 plants (Nefzaoui et al., 2014; Nobel, 2009). The high water-use efficiency of cactus pear plants makes them highly suitable for cultivation in arid regions, where C_3 and C_4 plants can hardly survive without irrigation (Nefzaoui, 2009; Nobel, 2009).

Cactus pear spines, cladodes and roots also play important roles in water conservation and water-use efficiency. One of the many functions of cactus pear spines is condensing water from the air. The cladodes contain a waxy and thick outer surface cuticle that limits evapotranspiration during drought season and prevents water loss, while shoot transpiration is decreased and root surface is restricted under extreme drought conditions, resulting in energy saving as well as a decrease in the permeability to water (Mazri, 2018; Mulas and Mulas, 2004; Nefzaoui and Ben Salem, 2002; Sudzuki Hills, 1995).

Cactus pear is a good source of water for livestock. In fact, plants of this genus are able to store high quantities of water in their cladodes (Nefzaoui and Bensalem, 2002). In arid regions, it is difficult for livestock to find water sources and to meet their requirements in water, which may affect their body weight (Mulas and Mulas, 2004). Accordingly, using cactus pear to feed cattle is a good solution in such situations. Indeed, cactus pear cladodes contain 90% of water (Mulas and Mulas, 2004). Feeding livestock with cactus pear cladodes is a valuable solution to address water scarcity in dry regions. Under such a diet, the cattle require no more or little water supplementation (Nefzaoui and Ben Salem, 2002).

Using cactus pear as a hedge was reported to improve water storage capacity in soil (Le Houérou, 1996; Nefzaoui, 2009). Neffar et al. (2011) reported that, in Algeria, plants of *O. ficus indica* have significantly increased soil humidity, from 1.83 to 2.85%, in comparison with control (i.e., soils without cactus pear plantations).

4.5 Drought Tolerance and Desertification Prevention

Opuntia spp. is a drought-resistant plant genus, characterized by high tolerance to elevated temperatures (Mulas and Mulas, 2004). Cactus pear plants have an enzymatic system which is able to function under extreme temperatures of deserts (Mulas and Mulas, 2004). In addition, the cladodes are characterized by a waxy and thick outer surface cuticle that limits water loss and evapotranspiration under drought conditions (Mazri, 2018). Besides, the stomata are closed for most of the day and even at night, when the temperature is high, reduced photosynthesis occurs (Mulas and Mulas, 2004; Nefzaoui and Ben Salem, 2002). The photosynthetic system of *O. ficus indica* cladodes makes this plant species able to produce organic carbon even when soil water capacity is very low, under 5%, and as a result, to store energy that will be used to stimulate plant growth and development until water is available again (Mulas and Mulas, 2004). The root system of cactus pear plays as well an important role in drought tolerance since the fine roots are covered by a layer that prevents water loss in dry soils (Nefzaoui and Ben Salem, 2002).

Under extreme conditions of droughts, cactus pear produces higher amounts of dry matter than C_3 and C_4 plants and thus cactus pear cladodes are used to feed livestock (Nefzaoui and Ben Salem, 2002). In addition, due to the specific characteristics of cactus pear, for example, its specific photosynthetic system, the CAM metabolism, the succulence trait and its high tolerance to arid conditions, the production of dry matter per unit of consumed water in *Opuntia* species is higher than that of many other plant species (Nefzaoui and Ben Salem, 2002). On the other hand, cactus pear plants are characterized by asynchronous organ development. This means that even when cactus pear plants are cultivated under extreme conditions, there are always some parts that are not affected. More interestingly, cactus pear plants can survive under high temperatures reaching 70°C (Nefzaoui, 2009; Nobel, 2009). Furthermore, cactus pear species are characterized by acceptable productivity even during several years of extreme drought conditions.

In some North African countries, plantations of *O. ficus indica* were used to direct and slow sand movements, prevent desertification, establish evergreen fodder to feed livestock during drought seasons and rehabilitate vegetation cover (Nefzaoui and Ben Salem, 2002). In Peru, an oasis was successfully created in the desert of Ica, using cactus pear plants (Nefzaoui and Ben Salem, 2002). Planting two rows of cactus pear cladodes in land terraces was reported to prevent water runoff and to enhance terrace stability, while combining cactus pear plants with palm leaves or cement barriers was efficient to control sand movements and wind erosion, and to rehabilitate vegetation (Nefzaoui and Ben Salem, 2002). Mulas and Mulas (2004) reported that, in some West Asian and North African countries, planting cactus pear resulted in a successful control of desertification and the establishment of a new fodder source to feed livestock.

In sum, while high temperatures, consecutive drought seasons and scarcity of water are some consequences of climate change that adversely affect global agricultural production and food security, the above literature review demonstrates that cactus pear plants are well adapted to these harsh conditions under which they are able to maintain high biomass productivity. Accordingly, *Opuntia* spp. could be considered as a highly valuable plant genus to mitigate climate change. Besides, plants of cactus pear can be cultivated for different purposes. Indeed, in addition to the above-mentioned benefits, the fruit of many cactus pear species is delicious, highly nutritious and greatly appreciated for human consumption. Furthermore, cactus pear can be cultivated for medicinal and cosmetic purposes. The most important uses of cactus pear are detailed in the next section.

5. Cactus Pear Utilization

Cactus pear is a fruit crop that can be cultivated for multiple purposes. The delicious and nutritious fruit is highly appreciated for fresh consumption and for juice production. In some countries, cactus pear cladodes are used in salad. However, their most common use is for animal nutrition, especially under drought conditions. Besides, *Opuntia* species have been used for cosmetic and pharmaceutical purposes (Mazri, 2018).

5.1 Cactus Pear Cultivation for Human Consumption

Cactus pear fruit is a delicious and highly nutritious ovoid-spherical berry that can be green, purple, red, yellow or orange at maturity (Mazri, 2021). It consists of peel, pulp and seeds and can be consumed as fresh fruit or processed in the form of juice (Mazri, 2018). Due to its high nutritional values and health benefits, cactus pear fruits that belong to different species and geographic areas were characterized by several researchers. It was found that the physical characteristics and phytochemical properties vary depending on the species, stage of maturity and geographic origin.

Chahdoura et al. (2019) evaluated the bioactivity and phytochemical composition of pulps and skins of two cactus pear species, *O. macrorhiza* and *O. microdasys*, harvested from Monastir (Tunisia). Regarding the edible fraction (fruit pulp), it was found that in both species the major isoform of vitamin E is α-tocopherol, the major fatty acid is lauric acid, the major organic acid is citric acid, and both species have antioxidant activity. Besides, the major volatile compound in *O. macrorhiza* was ethyl acetate while it was camphor in *O. microdasys*. On the other hand, fructose was the major sugar in *O. macrorhiza* while it was sucrose in *O. microdasys*. Katanić et al. (2019) compared the phenolic and flavonoid compounds of extracts of *O. dillenii* fruits collected from two Moroccan localities, Nador and Essaouira. It was found that total phenolic and total flavonoid contents vary, depending on the geographic origin, extraction method and fruit fraction (skin, seeds or juice). Besides, the fruit extracts showed important antioxidant, antibacterial and antifungal activities, which highlight their nutritional and pharmaceutical values. García-Cayuela et al. (2019) compared the physical characteristics and phytochemical composition of some Mexican and Spanish *O. ficus indica* cultivars and found variations depending on the genotype. While no significant differences were found in terms of whole fruit weight among the six genotypes, there were differences in terms of titratable acidity, pH, soluble solids as well as betalain and phenolic compounds. Betanin and indicaxanthin were the major betalains, whereas the observed phenolics corresponded mostly to flavonoid (isorhamnetin, quercetin and kaempferol), glycosides and piscidic acid. Idir et al. (2018) determined pulp and nectar (35 and 45%

pulp) characteristics of *O. ficus indica* from Algeria. It was found that fruit pulp is a good source of vitamin C, sugars, phenolic compounds, betalains, β-carotene and minerals. The pulp nectars are nutraceutical and are a good source of carbohydrates, vitamin C and minerals. El Kharrassi et al. (2016) determined the characteristics of fruit and juice of 30 accessions of *O. ficus indica* and *O. megacantha* from Morocco. Fruit length and weight, skin weight and water content in the skin varied significantly among the evaluated accessions. Regarding the fruit juice characteristics, there were significant differences in terms of pH, soluble solids, titratable acidity, total carotenoids and vitamin C, while no significant difference was observed in reducing sugar contents. These authors recommended the use of Moroccan cactus pear fruits for juice production. Mena et al. (2018) evaluated the bioactive compounds of different botanical parts (fruit pulp, fruit skin, young and adult cladodes) of six *O. ficus indica* cultivars collected from Spain. Up to 41 phytochemical compounds were detected and significant differences in their concentrations were reported, depending on the genotype and botanical part.

All these examples of *Opuntia* spp. fruit characterization show their high nutritional and health values, and highlight the benefits of cultivating cactus pear for human consumption. Besides, cladodes of cactus pear can also be consumed as vegetable in salads (Du Toit et al., 2021; Yahia and Sáenz, 2018).

5.2 Cactus Pear Cultivation to Feed Livestock

Climate change consequences, such as increased temperatures, poor soils and scarcity of water negatively affect forage production. In such conditions, cactus pear represents a valuable source of fodder (De Wit and Fouché, 2021). As previously reported, cactus pear is tolerant to these climate change stresses under which it can grow and produce biomass. Besides, cactus pear cladodes have a high nutritional value and are desirable for their high water content and dietary fibers (Yahia and Sáenz, 2018). Consequently, they have been widely used to feed livestock, especially in semi-arid regions (Paula et al., 2018). Generally, the chemical composition of cactus pear cladodes depends on many factors, such as their age, the species and cultural conditions (Yahia and Sáenz, 2018).

Magalhães et al. (2019) evaluated the effects of including cactus pear in the diet of castrated male Santa Ines sheep (55 ± 0.84 kg body weight) on intake, digestibility and rumen parameters. The inclusion of cactus pear in the diet decreased ether extract intake, increased mineral matter intake and crude protein digestibility, and significantly decreased water consumption. Borges et al. (2019) evaluated the effects of including cactus pear in diets of F1 Holstein/Zebu cows and reported that cactus pear (*O. ficus indica* cv. Gigante) consumption does not affect milk production while it has many benefits. For example, it decreased dry matter intake, reduced water consumption by 44.52% and improved the digestibility of nutrients. Costa et al. (2009) substituted corn meal with cactus pear (*O. ficus indica*) in dairy goat diet, then evaluated its effect on milk production, feed intake and water consumption. It was found that cactus pear does not affect milk production but decreases milk fat and increases dry and fresh matter intakes. Besides, cactus pear significantly decreased water consumption.

In addition to its utilization for nutritional purposes, cactus pear has many other benefits when used as forage. The effect of cactus pear (*O. robusta*) cladode-based diet on the behavior of Landrance × Duroc castrated pigs of 115 ± 4 d age and 64.7 ± 5.2 kg weight was evaluated by Mendez-Llorente et al. (2018). It was found that cactus pear diet improves pig behavior. The pigs fed with *O. robusta* cladodes spent significantly more time lying down, more time eating and less time fighting than those used as control (not fed with *O. robusta* cladodes), while increased cladode concentration in the diet increased chewing time. The effect of adding cactus pear in diets on greenhouse gas emission by ruminants was also evaluated. Along this line, it was found that the inclusion of cactus pear in the diet of Holstein steers resulted in a decrease in greenhouse gas production (Elghandour et al., 2018). More specifically, replacing corn grain by cactus pear showed a decrease in carbon dioxide and methane production. Elghandour et al. (2018) concluded that substituting corn grain with cactus pear and the addition of *Moringa oleifera* in ruminant diets represented an effective strategy to reduce their greenhouse gas emission. Santos et al. (2018) demonstrated the beneficial effects of *O. ficus indica* essential oils against gastrointestinal nematodes of sheep. Indeed, essential oils from cladode peel showed an egg hatch inhibition percentage of up to 90%, and a larval migration inhibition percentage of up to 77.26%. These authors concluded that the essential oils from cladode peel of cactus pear can be considered as an ecofriendly antiparasiticide to control parasite infections in sheep. In a different study, the chemical composition and antioxidant activity of milk of lactating donkeys was evaluated after feeding them with cladodes of *O. ficus indica* (Valentini et al., 2017). While there was no difference between groups fed with cactus pear cladodes and control regarding the chemical composition of milk, an increase in the antioxidant activity of milk was reported in those fed with *O. ficus indica* cladodes, suggesting the use of cactus pear to feed animals especially in the regions where water is scarce (Valentini et al., 2017).

5.3 Other Uses of Cactus Pear

In addition to the above-mentioned utilizations, cactus pear can be also used for pharmaceutical and cosmetic purposes. In fact, it has been shown that cactus pear plants are rich in bioactive compounds, which confers to them antimicrobial, anti-inflammatory, antitumor, antiulcer, antidiabetic, antioxidant, anticancer, hypolipidemic, hypocholesterolemic, hepatoprotective, hypoglycemic, hypolipidemic and neuroprotective properties (Benayad et al., 2014; Katanić et al., 2019; Majeed et al., 2021; Melgar et al., 2017; Tilahun and Welegerima, 2018). Besides, many cosmetic products can be manufactured from cactus pear (Nazhand et al., 2021; Sáenz, 2013b).

6. Genetic Diversity and Resource Conservation of Cactus Pear

6.1 Genetic Diversity of Cactus Pear

Cactus pear is characterized by high genetic variability (Mazri, 2021). In fact, the genus *Opuntia* is composed by around 300 species (Arba, 2020). The most economically important species is *O. ficus indica*. However, many other species are cultivated for fruit production, like *O. amyclaea, O. hyptiacantha, O. megacantha, O. robusta, O. streptacantha* and *O. xoconostle* (Sáenz, 2013a; Yahia and Sáenz, 2018).

Cactus pear characterization and its genetic diversity within and among species have been assessed by researchers from different countries (e.g., El Kharrassi et al., 2015, 2017; Ganopoulos et al., 2015; Valadez-Moctezuma et al., 2015; Zarroug et al., 2015), and a high level of genetic variability was reported. The high genetic diversity of cactus pear is very valuable since it can be exploited in breeding programs to increase adaptation to climate change.

6.2 Cactus Pear Propagation

Cactus pear plants are commonly propagated vegetatively using cladodes: segments or whole cladodes are detached from the mother plant, dried under shade for two weeks then planted in soil. A plant growth regulator could be used to stimulate rooting. After plantation, new shoots are initiated from areoles (Mazri, 2018). Cactus pear can also be propagated through seeds or through *in vitro* techniques (Alam-Eldein et al., 2021; Mazri, 2018).

6.3 Genetic Resource Conservation

The main approach used for conservation of cactus pear genetic resources is *ex situ* conservation (Mazri, 2018). It consists of conserving cactus pear species, genotypes and accessions away from their natural habitat. The main advantage of this approach over *in situ* conservation (which consists of conserving the genetic resources in their native habitat) is that many factors may threaten the natural habitat of cactus pear; for example, expansion of urban areas and natural threats associated with climate change (Mondragón Jacoboa and Chessa, 2017).

Many collection orchards have been established worldwide to conserve cactus pear genetic resources. For example, Italy has a germplasm collection based in Sardinia that contains 2,200 accessions belonging to different species. The accessions are originated from different countries and include wild and local genotypes as well as hybrids and selected lines (Chessa, 2010; Mondragón Jacoboa and Chessa, 2017). Mexico has established many germplasm collection sites which contain hundreds of genotypes that can be used for different purposes (Mondragón Jacoboa and Chessa, 2017). In Morocco, many collection orchards have been created by the National Institute of Agricultural Research to conserve the genetic resources of cactus pear (Fig. 1). The orchards are located in experimental stations and contain genotypes and ecotypes collected from different regions of Morocco. Other countries, such as Tunisia, South Africa, Argentina, USA and Brazil, were also reported to have *ex situ* conservation collections of cactus pear (Mondragón Jacoboa and Chessa, 2017; Mazri, 2018).

Fig. 1. *Ex situ* conservation of cactus pear.

7. Intelligent Prediction

Owen et al. (2016) predicted the productivity of *O. ficus indica* under various climate change scenarios by using the Nobel environmental productivity Index (EPI) methodology, but with some refinements to include the characteristics of soil water retention, contrasting day and night temperatures as well as CO_2 uptake under drought conditions. Outputs from Representative Concentration Pathway (RCP) scenarios from the *IPCC's 5th Assessment Report* were used to simulate productivity under future climate scenarios, and macro-scale land-use constraints were applied to estimate productivity potential on low-grade lands.

These authors reported that *O. ficus indica* is highly resilient to climate change even under the worst-case scenarios. In addition, *O. ficus indica* outperforms C_3 and C_4 bioenergy plants in terms of low-grade land productivity, and is able to meet future bioenergy demands.

Abrha et al. (2018) evaluated the impacts of climate change and the invasion of cochineal (*Dactylopius coccus* Costa) on the future distribution of *O. ficus indica*. Various scenarios were predicted, at near, mid and end-century using R-programing language, two emissions (RCP 4.5 and 8.5) and one General Circulation Model (GCM), which was GCM-5: Climate Community System Model Version 4 (CCSM4). According to these authors, climate change alone will not have a significant impact on cactus pear distribution. However, the invasion of the cochineal and its combined effect with that of climate change will significantly reduce *O. ficus indica* distribution. The findings of this study indicated that the cochineal invasion, which is mainly affected by cactus pear presence, will increase in mid- and end-century.

8. Significant Impacts of Climate Change on Cactus Pear

Though cactus pear species show high tolerance to the most critical climate change impacts, such as drought, high atmospheric CO_2 concentrations, water deficit and high temperatures, it is important to note that biotic factors, such as pest invasion and spread, which can be accelerated because of global warming and climate change, are the most serious threats to cactus pear.

Pests of the genus *Dactylopius*, which are phytophagous hemipterous insects known as 'cochineal scale insects' or 'cochineal insects' (Chávez-Moreno et al., 2009), constitute a serious and increasing threat to cactus pear. In fact, many *Opuntia* species are hosts to *Dactylopius*. The genus *Dactylopius* is native to the Americas and includes nine to 11 species, since there is no consensus yet over two species (Chávez-Moreno et al., 2009; Mazzeo et al., 2019). The females of *Dactylopius*, which are sessile, live in colonies and produce a white cotton-type wax that covers their bodies and that may spread to cover the whole cladode (Fig. 2). This results in fruit drop, drying out and necrosis of cladodes and finally plant death (Mazzeo et al., 2019). Originally, the females of the genus *Dactylopius* have been used as a biological agent to control the invasion of cactus pear plants. In fact, in many countries, *Opuntia* species were considered as invasive weeds. Besides, the females of *Dactylopius* have been also used to produce natural red dyes (Mazzeo et al., 2019).

The invasion of the cochineal (*Dactylopius coccus* Costa) was simulated by Abrha et al. (2018). The life cycle of *D. coccus* Costa depends on climatic variables. Areas with temperatures ranging between –5 to 40°C, and mean annual rainfall below 700 mm are suitable for its invasion (Abrha et al.,

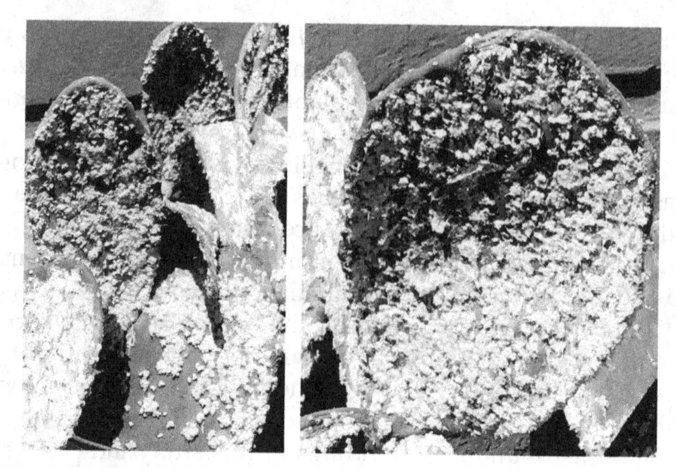

Fig. 2. Symptoms of *Dactylopius opuntiae* attacks on cactus pear cladodes.

2018). In addition to climatic variables, it was found that *D. coccus* Costa invasion is affected by cactus pear presence. According to these authors, cochineal invasion will increase in mid- and end-century with up to 32.3% as compared to the present, will invade up to 94% of cactus pear resources and as a result will have a significant impact on cactus pear distribution. Hence, cactus pear is seriously threatened by *D. coccus* Costa (Abrha et al., 2018). Efficient management interventions, for example, integrated strategies to combat this pest, must be taken into consideration to control its invasion and to mitigate its damage.

Regarding *D. opuntiae,* the most destructive pest of *Opuntia* spp., climate change has played an important role in its dissemination in the Mediterranean regions. Indeed, temperature increase is a major factor that affects insect life cycles and development, and consequently, the host plants. While Mediterranean regions are predicted to be hotter and drier, cultivating more resilient plants, such as cactus pear seems to be a good alternative to replace other crops. In fact, plants of the genus *Opuntia* are tolerant to high temperatures and drought, and can be used to combat desertification. Nevertheless, the harmfulness of *D. opuntiae* is also predicted to be more important, and invasion risks will become more serious in countries not yet invaded by this pest, but where cactus pear is cultivated; for example, in Italy (Mazzeo et al., 2019). In Morocco, *D. opuntiae* has caused massive damage to cactus pear plantations since 2015, especially in the regions of Doukkala and Rhamna. This resulted in high agronomical, environmental and socio-economic losses (Mazri, 2018). In such situations, and with the risks of more spread of *D. opuntiae* under climate change, it is necessary to develop efficient measures to control this destructive pest.

9. Mitigation Strategies against the Biotic Stresses of Climate Change

Due to the severe damage caused by the cochineal scale insects to cactus pear plantations and with more risk of infestation in the next few years due to climate change, it is important to develop efficient and integrated strategies to control these pests and reduce their damage.

Biological strategies might play an important role in the fight against the cochineal. Along this line, the use of natural enemies was suggested by many researchers. For example, Vanegas-Rico et al. (2010) reported the existence of seven natural predators of *D. opuntiae* in Tlalnepantla, Morelos (Mexico). These natural predators are *Chilocorus cacti, Hyperaspis trifurcata, Laetilia coccidivora, Leucopis bellula, Sympherobius angustus, Sympherobius barberi* and *Salpingogaster cochenillivorus.* Hernández-González and Cruz-Rodríguez (2018) evaluated the effect of *Chilocorus cacti* L. (Coleoptera: Coccinellidae) on the survival of *D. opuntiae*. It was found that adults and larvae of *C. cacti* prefer to consume the nymphs of *D. opuntiae*. This is due to the wax produced by adult females of *D. opuntiae* and that constitutes an obstacle to *C. cacti*. Besides, a higher consumption of adult females without wax was noted in comparison to those with wax. Generally, *C. cacti* showed a significant negative impact on *D. opuntiae* survival. In fact, the survival rate of *D. opuntiae*, 60 days after the introduction of *C. cacti*, was 57.59%, while it was 90.63% in control (colonies without *C. cacti*). *Hyperaspis trifurcata* is also a natural predator of *D. opuntiae* (Vanegas-Rico et al., 2016). Individuals of this species were reported to consume around 5,000 *D. opuntiae* nymphs during their life cycles (Vanegas-Rico et al., 2016). In Morocco, Bouharroud et al. (2018) evaluated the effect of the adults of *Cryptolaemus montrouzieri* Mulsant (Coleoptera: Coccinellidae) on the females of *D. opuntiae*. According to these authors, the highest reduction in the population of *D. opuntiae* was observed between days 77 and 92 after *C. montrouzieri* release, and it reached 95%. By the end of experiment (105 days after *C. montrouzieri* release), the reduction in *D. opuntiae* population was 66%. In a different study, it was reported that the coccinellid *Hyperaspis campestris* is also a natural predator of *D. opuntiae* (Bouharroud et al., 2019). Laboratory studies showed that adults of *H. campestris* consume an average of 10 larvae per day in the case of first larval instars, and 7.5 larvae per day in the case of second larval instars (Bouharroud et al., 2019).

In addition to natural predators, many natural products were recommended by researchers to control *D. opuntiae*. Vigueras et al. (2009) evaluated the effects of extracts of some medicinal and aromatic plants mixed with different emulsifier against nymphs in stages I and II as well as against adult females of *D. opuntiae* under laboratory conditions. The highest mortality rate (99.2%) was observed after 72 h in stage II nymphs treated with extracts from *Tagetes florida* L., mixed with Tween 20®. In addition, it was found that the botanical extracts from the plant species evaluated affect the reproductive parameters of *D. opuntiae* by reducing oviposition of adult females. Borges et al. (2013a) evaluated different alternative products against *D. opuntiae* under greenhouse conditions and found that *Neem*-based product results in the highest reduction in *D. opuntiae* infestation after 40 days of treatment. D-limonene was also suggested to control the females of *D. opuntiae* (Bouharroud et al., 2018). The highest mortality rate caused by this product under field conditions was 99% and was observed at the concentration of 150 ppm, six days after treatment (Bouharroud et al., 2018).

In some countries, selection and cultivation of genotypes resistant to *D. opuntiae* have been in use to control the spread of this pest. In Brazil, the clones 'Orelha de Elefante Africana' and 'Orelha de Elefante Mexicana', both belonging to the genus *Opuntia*, were reported to be immune to *D. opuntiae* (Borges et al., 2013b; Torres and Giorgi, 2018). In fact, there was no attachment or development of the pest on the cladodes of these clones during the experiment (Borges et al., 2013b). In Morocco, eight cactus pear genotypes were selected to reconstitute groves destroyed by the cochineal scale insect. These genotypes, called 'Marjana', 'Belara', 'Karama', 'Ghalia', 'Angad', 'Cherratia', 'Melk Zhar' and 'Aakria', are all characterized by high resistance to *D. opuntiae*, high fruit quality and could be used to feed livestock as well (Sbaghi et al., 2018). Rapid and mass propagation of these resistant genotypes through *in vitro* culture techniques, particularly adventitious organogenesis (Fig. 3), might be a good approach to rehabilitate the destroyed groves.

Regarding chemical pesticides, they might be effective to control *D. opuntiae*. However, their use is hampered by some factors, such as

Fig. 3. Cactus pear propagation through adventitious organogenesis.

their expensiveness, the negative impact on environment and the specific national regulations since in some countries, efficient products are not authorized for use (Mazzeo et al., 2019).

In addition to biological and chemical means to control *D. opuntiae*, a mechanical method could also be envisaged. It consists of removing infested cactus pear plants and then burning them in order to avoid infestation spread. This mechanical approach has been used in Morocco since the detection of *D. opuntiae* (Mazri, 2018).

10. Conclusion and Prospects

Under different climate change scenarios, global agricultural production and food security have become a major and universal concern. Cultivating plant species characterized by high resilience to climate change is a global priority. Cactus pear (*Opuntia* spp.) is a plant genus well adapted to extreme conditions of drought. It shows high tolerance to increased temperatures and elevated atmospheric CO_2 concentrations, and is able to improve soil structure and fertility. Cactus pear also helps to prevent erosion and desertification and to rehabilitate degraded lands. It contributes to maintaining biodiversity and can survive and facilitate high biomass production under adverse conditions. All these characteristics make this plant genus a perfect candidate for enhanced resilience to climate change.

In addition, cactus pear can be used for human consumption since its fruit is delicious and highly nutritious. It can be used also for cosmetic and pharmaceutical purposes since the plant is rich in bioactive compounds. However, the most important use of cactus pear under climate change is as fodder to feed livestock, especially in arid and semi-arid regions since the cladodes are characterized by high water contents. Accordingly, including cactus pear cladodes in livestock diets results in a significant decrease in water consumption.

Unfortunately, cactus pear is threatened by pests of the genus *Dactylopius* (the cochineal insects), which produce a white cotton-type wax that covers the cladodes and results in plant death. Climate change has played an important role in the invasion of the cochineal insects in the Mediterranean regions. Prediction studies showed that the cochineal invasion will increase in mid- and end-century, and that its combined effect with that of climate change will significantly reduce cactus pear distribution. Moreover, the cochineal insects may invade areas not yet invaded but where cactus pear is cultivated. Therefore, integrated strategies must be undertaken in order to control this dangerous pest.

Conservation of cactus pear genetic resources can be done *ex situ* as it is the case in many countries. In this approach, the best genotypes are collected from different regions and planted away from their native habitat.

This method may be preferable to *in situ* conservation since it protects cactus pear plants from many factors associated with climate change and that may threaten their natural habitat. For example, the cochineal invasion. On the other hand, cactus pear is characterized by a high level of genetic diversity that could be exploited in breeding programs in order to increase adaptation to climate change. For example, such genetic diversity may be used to create genotypes, resistant to the cochineal insects and characterized by a high potential of fruit and cladode production.

References

Abrha, H., Birhane, E., Zenebe, A., Hagos, H., Girma, A., Aynekulu, E. and Alemie, A. (2018). Modeling the impacts of climate change and cochineal (*Dactylopius coccus* Costa) invasion on the future distribution of cactus pear (*Opuntia ficus-indica* (L.) Mill.) in northern Ethiopia. *Journal of the Professional Association for Cactus Development*, 20: 128–150.

Alam-Eldein, S.M., Omar, A.E.D.K., Ennab, H.A. and Omar, A.A. (2021). Cultivation and cultural practices of *Opuntia* spp. pp. 121–158. *In*: Ramadan, M.F., Ayoub, T.E.M. and Rohn, S. (eds.). *Opuntia Spp.: Chemistry, Bioactivity and Industrial Applications*. Springer, Cham, Switzerland.

Arba, M. (2020). The potential of cactus pear (*Opuntia ficus-indica* (L.) Mill.) as food and forage crop. pp. 335–357. *In*: Hirich, A., Choukr-Allah, R. and Ragab, R. (eds.). *Emerging Research in Alternative Crops*. Springer, Cham, Switzerland.

Badii, M.H. and Flores, A.E. (2001). Prickly pear cacti pests and their control in Mexico. *Florida Entomologist*, 84: 503–505.

Balaghi, R., Badjeck, M.C., Bakari, D., De Pauw, E., De Wit, A., Defourny, P., Donato, S., Gommes, R., Jlibene, M., Ravelo, A.C., Sivakumar, M.V.K., Telahigue, N. and Tychon, B. (2010). Managing climatic risks for enhanced food security: Key information capabilities. *Procedia Environmental Sciences*, 1: 313–323.

Benayad, Z., Martinez-Villaluenga, C., Frias, J., Gomez-Cordoves, C. and Es-Safi, N.E. (2014). Phenolic composition, antioxidant and anti-inflammatory activities of extracts from Moroccan *Opuntia ficus-indica* flowers obtained by different extraction methods. *Industrial Crops and Products*, 62: 412–420.

Bhuiyan, M.A., Jabeen, M., Zaman, K., Khan, A., Ahmad, J. and Hishan, S.S. (2018). The impact of climate change and energy resources on biodiversity loss: Evidence from a panel of selected Asian countries. *Renewable Energy*, 117: 324–340.

Borges, L.R., Santos, D.C., Gomes, E.W.F., Cavalcanti, V.A.L.B., Silva, I.M.M., Falcão, H.M. and da Silva, D.M.P. (2013a). Use of biodegradable products for the control of Dactylopius opuntiae (*Hemiptera: Dactylopiidae*) in cactus pear. *Acta Horticulturae*, 995: 379–386.

Borges, L.R., Santos, D.C., Cavalcanti, V.A.L.B., Gomes, E.W.F., Falcão, H.M. and da Silva, D.M.P. (2013b). Selection of cactus pear clones regarding resistance to carmine cochineal Dactylopius opuntiae (*Dactylopiidae*). *Acta Horticulturae*, 995: 359–365.

Borges, L.D.A., Rocha Júnior, V.R., Monção, F.P., Soares, C., Ruas, J.R.M., Silva, F.V.E., Rigueira, J.P.S., Costa, N.M., Oliveira, L.L.S. and Rabelo, W.O. (2019). Nutritional and productive parameters of Holstein/Zebu cows fed diets containing cactus pear. *Asian-Australasian Journal of Animal Sciences* (in press). Doi: https://doi.org/10.5713/ajas.18.0584.

Bouharroud, R., Sbaghi, M., Boujghagh, M. and El Bouhssini, M. (2018). Biological control of the prickly pear cochineal *Dactylopius opuntiae* Cockerell (Hemiptera: Dactylopiidae). *EPPO Bulletin*, 48: 300–306.

Bouharroud, R., El Aalaoui, M., Boujghagh, M., Hilali, L., El Bouhssini, M. and Sbaghi, M. (2019). New record and predatory activity of *Hyperaspis campestris* (Herbst 1783) (Coleoptera: Coccinellidae) on *Dactylopius opuntiae* (Hemiptera: Dactylopiidae) in Morocco. *Entomological News*, 128: 156–160.

Chahdoura, H., Barreira, J.C.M., Barros, L., Dias, M.I., Calhelha, R.C., Flamini, G., Soković, M., Achour, L. and Ferreira, I.C.F.R. (2019). Bioactivity, hydrophilic, lipophilic and volatile compounds in pulps and skins of *Opuntia macrorhiza* and *Opuntia microdasys* fruits. *LWT – Food Science and Technology* 105: 57–65.

Chávez-Moreno, C.K., Tecante, A. and Casas, A. (2009). The Opuntia (Cactaceae) and Dactylopius (Hemiptera:Dactylopiidae) in Mexico: a historical perspective of use, interaction and distribution. *Biodiversity and Conservation*, 18: 3337–3355.

Chessa, I. (2010). Cactus pear genetic resources conservation, evaluation and uses. pp. 43–53. *In:* Nefzaoui, A., Inglese, P. and Belay, T. (eds.). *Cactusnet Newsletter: Improved Utilization of Cactus Pear for Food, Feed, Soil and Water Conservation and Other Products in Africa*. The Proceedings of International Workshop Held in Mekelle, 19–21 October 2009, Mekelle, Ethiopia.

Costa, R.G., Filho, E.M.B., de Medeiros, A.N., Givisiez, P.E.N., Queiroga, R.C.R.E. and Melo, A.A.S. (2009). Effects of increasing levels of cactus pear (*Opuntia ficus-indica* L. Miller) in the diet of dairy goats and its contribution as a source of water. *Small Ruminant Research*, 82: 62–65.

De Wit, M. and Fouché, H. (2021). South African perspective on *Opuntia* spp.: cultivation, human and livestock food and industrial applications. pp. 13–48. *In*: Ramadan, M.F., Ayoub, T.E.M. and Rohn, S. (eds.). *Opuntia Spp.: Chemistry, Bioactivity and Industrial Applications*. Springer, Cham, Switzerland.

Dinesh, M.R. and Reddy, B.M.C. (2012). Physiological basis of growth and fruit yield characteristics of tropical and subtropical fruits to temperature. pp. 45–73. *In*: Sthapit, B., Rao, V.R. and Sthapit, S. (eds.). *Tropical Fruit Tree Species and Climate Change*. Biodiversity International, New Delhi, India.

Du Toit, A., Mpemba, O., De Wit, M., Venter, S.L. and Hugo, A. (2021). The effect of size, cultivar and season on the edible qualities of nopalitos from South African cactus pear cultivars. *South African Journal of Botany*, 142: 459–466.

Elghandour, M.M.Y., Rodríguez-Ocampo, I., Parra-Garcia, A., Salem, A.Z.M., Greiner, R., Márquez-Molina, O., Barros-Rodríguez, M. and Barbabosa-Pilego, A. (2018). Biogas production from prickly pear cactus containing diets supplemented with *Moringa oleifera* leaf extract for a cleaner environmental livestock production. *Journal of Cleaner Production*, 185: 5470553.

El Kharrassi, Y., Mazri, M.A., Mabrouk, A., Nasser, B. and El Mzouri, E.H. (2015). Flowering and fruiting phenology, and physico-chemical characteristics of two-year-old plants of six species of *Opuntia* from eight regions of Morocco. *The Journal of Horticultural Science and Biotechnology*, 90: 682–688.

El Kharrassi, Y., Mazri, M.A., Benyahia, H., Benaouda, H., Nasser, B. and El Mzouri, E.H. (2016). Fruit and juice characteristics of 30 accessions of two cactus pear species (*Opuntia ficus indica* and *Opuntia megacantha*) from different regions of Morocco. *LWT – Food Science and Technology*, 65: 610–617.

El Kharrassi, Y., Mazri, M.A., Sedra, M.H., Mabrouk, A., Nasser, B. and El Mzouri, E.H. (2017). Characterization of genetic diversity of cactus species (*Opuntia* spp.) in Morocco by morphological traits and molecular markers. *Current Agriculture Research Journal*, 5: 146–159.

El Kharrassi, Y., Maata, N., Mazri, M.A., El Kamouni, S., Talbi, M., El Kebbaj, R., Moustaid, K., Essamadi, A.K., Andreoletti, P., El Mzouri, E.H., Cherkaoui-Malki, M. and Nasser, B. (2018). Chemical and phytochemical characterizations of argan oil (*Argania spinosa* L. skeels), olive oil (*Olea europaea* L. cv. Moroccan picholine), cactus pear (*Opuntia megacantha* salm-dyck) seed oil and cactus cladode essential oil. *Journal of Food Measurement and Characterization*, 12: 747–754.

Ganopoulos, I., Kalivas, A., Kavroulakis, N., Xanthopoulou, A., Mastrogianni, A., Koubouris, G. and Madesis, P. (2015). Genetic diversity of barbary fig (*Opuntia ficus-indica*) collection in Greece with ISSR molecular markers. *Plant Gene*, 2: 29–33.

García-Cayuela, T., Gómez-Maqueo, A., Guajardo-Flores, D., Welti-Chanes, J. and Cano, M.P. (2019). Characterization and quantification of individual betalain and phenolic compounds in Mexican and Spanish prickly pear (*Opuntia ficus-indica* L. Mill) tissues: A comparative study. *Journal of Food Composition and Analysis*, 76: 1–13.

Glimn-Lacy, J. and Kaufman, P.B. (2006). *Botany Illustrated: Introduction to Plants, Major Groups, Flowering Plant Families*. Springer, New York, USA, 146 pp.

Hegland, S.J., Nielsen, A., Lázaro, A., Bjerknes, A. and Totland, Ø. (2009). How does climate warming affect plant-pollinator interactions? *Ecology Letters*, 12: 184–195.

Hernández-González, I.A. and Cruz-Rodríguez, J.A. (2018). *Chilocorus cacti* (Coleoptera: Coccinellidae) as a biological control agent of the wild cochineal (Hemiptera: Dactylopiidae) of prickly pear cactus. *Environmental Entomology*, 47: 334–339.

Idir, L., Cherrared, Z. and Amir, Y. (2018). Characterization and transformation of the *Opuntia ficus indica* fruits. *Journal of Food Measurement and Characterization*, 12: 2349–2357.

Inglese, P., Liguori, G. and de la Barrera, E. (2017). Ecophysiology and reproductive biology of cultivated cacti. pp. 29–41. *In*: Inglese, P., Mondragon, C., Nefzaoui, A. and Sáenz, C. (eds.). *Crop Ecology, Cultivation and Uses of Cactus Pear*. FAO-ICARDA, Rome, Italy.

Iqbal, M.A., Hamid, A., Imtiaz, H., Rizwan, M., Imran, M., Sheikh, U.A.A. and Saira, I. (2020). Cactus pear: A weed of dry-lands for supplementing food security under changing climate. *Planta Daninha*, 38: e020191761.

Jiménez-Aguilar, D.M., Mújica-Paz, H. and Welti-Chanes, J. (2014). Phytochemical characterization of prickly pear (*Opuntia* spp.) and of its nutritional and functional properties: A review. *Current Nutrition & Food Science*, 10: 57–69.

Jiménez-Aguilar, D.M., López-Martínez, J.M., Hernández-Brenes, C., Gutiérrez-Uribe, J.A. and Welti-Chanes, J. (2015). Dietary fiber, phytochemical composition and antioxidant activity of Mexican commercial varieties of cactus pear. *Journal of Food Composition and Analysis*, 41: 66–73.

Katanić, J., Yousfi, F., Caruso, M.C., Matić, S., Monti, D.M., Loukili, E.H., Boroja, T., Mihailović, V., Galgano, F., Imbimbo, P., Petruk, G., Bouhrim, M., Bnouham, M. and Ramdani, M. (2019). Characterization of bioactivity and phytochemical composition with toxicity studies of different *Opuntia dillenii* extracts from Morocco. *Food Bioscience*, 30: 100410.

Kumar, K., Singh, D. and Singh, R.S. (2018). *Cactus Pear: Cultivation and Uses*. ICAR-Central Institute for Arid Horticulture, Bikaner, India, 38 pp.

Le Houérou, H.N. (1996). The role of cacti (*Opuntia* spp.) in erosion control, land reclamation, rehabilitation and agricultural development in the Mediterranean Basin. *Journal of Arid Environments*, 33: 135–159.

Li, Y., Ye, W., Wang, M. and Yan, X. (2009). Climate change and drought: A risk assessment of crop-yield impacts. *Climate Research*, 39: 31–46.

Magalhães, A.L.R., Sousa, D.R., Júnior, J.R.S.N., Gois, G.C., Campos, F.S., dos Santos, K.C., do Nascimento, D.B. and de Oliveira, L.P. (2019). Intake, digestibility and rumen parameters in sheep fed with common bean residue and cactus pear. *Biological Rhythm Research* (in press). https://doi.org/10.1080/09291016.2019.1592351.

Majeed, S., Zafar, M., Ahmad, M., Ozdemir, F.A., Kilic, O., Hamza, M., Sultana, S., Yaseen, G., Lubna and Raza, J. (2021). Ethnobotany, medicinal utilization and systematics of *Opuntia* species from deserts of Pakistan. pp. 49–80. *In*: Ramadan, M.F., Ayoub, T.E.M. and Rohn, S. (eds.). *Opuntia spp.: Chemistry, Bioactivity and Industrial Applications*. Springer, Cham, Switzerland.

Mazhar, M., Arif, A., Chriyâa, A., El Mzouri, L. and Derkaoui, M. (2002). Cactus protects soil and livestock in Rhamna region. *Acta Horticulturae*, 581: 329–332.

Mazri, M.A. (2018). Cactus pear (*Opuntia* spp.) breeding. pp. 307–341. *In*: Al-Khayri, J., Jain, S. and Johnson, D. (eds.). *Advances in Plant Breeding Strategies: Fruits*. Springer, Cham, Switzerland.

Mazri, M.A. (2021). Cactus pear (*Opuntia* spp.) species and cultivars. pp. 83–107. *In*: Ramadan, M.F., Ayoub, T.E.M. and Rohn, S. (eds.). *Opuntia spp.: Chemistry, Bioactivity and Industrial Applications*. Springer, Cham, Switzerland.

Mazzeo, G., Nucifora, S., Russo, A. and Suma, P. (2019). *Dactylopius opuntiae*, a new prickly pear cactus pest in the Mediterranean: An overview. *Entomologia Experimentalis et Applicata*, 167: 59–72.

Melgar, B., Dias, M.I., Ciric, A., Sokovic, M., Garcia-Castello, E.M., Rodriguez-Lopez, A.D., Barros, L. and Ferreira, I. (2017). By-product recovery of *Opuntia* spp. peels: Betalainic and phenolic profiles and bioactive properties. *Industrial Crops and Products*, 107: 353–359.

Mena, P., Tassotti, M., Andreu, L., Nuncio-Jáuregui, N., Legua, P., Del Rio, D. and Hernández, F. (2018). Phytochemical characterization of different prickly pear (*Opuntia ficus-indica* (L.) Mill.) cultivars and botanical parts: UHPLC-ESI-MS^n metabolomics profiles and their chemometric analysis. *Food Research International*, 108: 301–308.

Mendez-Llorente, F., Aguilera-Soto, J.I., Hernández-Briano, P., Carrillo-Muro, O., Medina-Flores, C.A., Rincón-Delgado, R.M. and López-Carlos, M.A. (2018). Behavioral characteristics of pigs fed with *Opuntia robusta*. *International Journal of Applied Research in Veterinary Medicine*, 16: 163–167.

Middleton, K. (2002). Opportunities and risks: A cactus pear in Madagascar. *Acta Horticulturae*, 581: 63–73.

Mondragón Jacoboa, C. and Chessa, I. (2017). Nopal (*Opuntia* spp.) genetic resources. pp. 43–49. *In*: Inglese, P., Mondragon, C., Nefzaoui, A. and Sáenz, C. (eds.). *Crop Ecology, Cultivation and Uses of Cactus Pear*. FAO-ICARDA, Rome, Italy.

Monjauze, A. and Le Houérou, H.N. (1965). *Le rôle des Opuntia dans l'économie agricole nord africaine. Bulletin de l'Ecole Nationale Supérieure d'Agronomie de Tunis*, 8/9: 85–164.

Mulas, M. and Mulas, G. (2004). *The Strategic Use of Atriplex and Opuntia to Combat Desertification*. Desertification Research Group, University of Sassari, Sassari, Italy, 101 pp.

Nazhand, A., Durazzo, A., Lucarini, M., Raffo, A., Souto, E.B., Lombardi-Boccia, G., Lupotto, E. and Santini, A. (2021). *Opuntia* spp. in cosmetics and pharmaceuticals. pp. 953–959. *In*: Ramadan, M.F., Ayoub, T.E.M. and Rohn, S. (eds.). *Opuntia spp.: Chemistry, Bioactivity and Industrial Applications*. Springer, Cham, Switzerland.

Neffar, S., Beddiar, A., Redjel, N. and Boulkheloua, J. (2011). *Effets de l'âge des plantations de figuier de Barbarie (Opuntia ficus indica f. inermis) sur les propriétés du sol et la végétation à Tébessa (zone semi-aride de l'est algérien). Ecologia Mediterranea*, 37: 5–15.

Neffar, S., Chenchouni, H., Beddiar, A. and Redjel, N. (2013). Rehabilitation of degraded rangeland in drylands by prickly pear (*Opuntia ficus-indica* L.) plantations: Effect on soil and spontaneous vegetation. *Ecologia Balkanica*, 5: 63–76.

Nefzaoui, A. and Ben Salem, H. (2002). Cacti: efficient tool for rangeland rehabilitation, drought mitigation and to combat desertification. *Acta Horticulturae*, 581: 295–315.

Nefzaoui, A. (2009). Cactus: A crop to meet the challenges of climate change in dry areas. *Annals of Arid Zone*, 48: 1–18.

Nefzaoui, A. and El Mourid, M. (2009). Cacti: A key-stone crop for the development of marginal lands and to combat desertification. *Acta Horticulturae*, 811: 365–372.

Nefzaoui, A., Louhaichi, M. and Ben Salem, H. (2014). Cactus as a tool to mitigate drought and to combat desertification. *Journal of Arid Land Studies*, 24: 121–124.

Nobel, P.S. (1994). *Remarkable Agaves and Cacti*. Oxford University Press, New York, USA, 180 pp.

Nobel, P.S. (2001). Ecophysiology of *Opuntia ficus-indica*. pp. 13–20. *In*: Mondragon-Jacobo, C. and Perez-Gonzalez, S. (eds.). *Cactus (Opuntia spp.) as Forage*. FAO, Rome, Italy.

Nobel, P.S. (2009). Desert Wisdom Agaves and Cacti: CO_2, Water, Climate Change. *iUniverse*, New York, USA, 196 pp.

Olsen, J.E. and Bindi, M. (2002). Consequences of climate change for European agricultural productivity, land use and policy. *European Journal of Agronomy*, 16: 239–262.

Owen, N.A., Fahy, K.F. and Griffiths, H. (2016). Crassulacean acid metabolism (CAM) offers sustainable bioenergy production and resilience to climate change. *GCB Bioenergy*, 8: 737–749.

Patel, S. (2015). Opuntia fruits as a source of inexpensive functional food. pp. 15–30. *In*: Patel, S. (ed.). *Emerging Bioresources with Nutraceutical and Pharmaceutical Prospects*. Springer, Cham, Switzerland.

Paula, T., Véras, A.S.C., Guido, S.I., Chagas, J.C.C., Conceição, M.G., Gomes, R.N., Nascimento, H.F.A. and Ferreira, M.A. (2018). Concentrate levels associated with a new genotype of cactus (*Opuntia stricta* [Haw]). *Cladodes in the Diet of Lactating Dairy Cows in a Semi-arid Region*, 156: 1251–1258.

Pautasso, M., Döring, T.F., Garbelotto, M., Pellis, L. and Jeger, M.J. (2012). Impacts of climate change on plant diseases—opinions and trends. *European Journal of Plant Pathology*, 133: 295–313.

Potgieter, J.P. (2007). *The Influence of Environmental Factors on Spineless Cactus Pear (Opuntia spp.) Fruit Yield in Limpopo Province, South Africa*. M.S.A. thesis, University of the Free State, Bloemfontein, South Africa.

Pryor, S.C., Scavia, D., Downer, C., Gaden, M., Iverson, L., Nordstrom, R., Patz, J. and Robertson, G.P. (2014). Midwest. pp. 418–440. *In*: Melillo, J.M., Richmond, T. and Yohe, G.W. (eds.). *Climate Change Impacts in the United States: The Third National Climate Assessment*. US Global Change Research Program, Washington, USA.

Reyes-Agüero, J.A., Aguirre, R.J.R. and Valiente-Banuet, A. (2006). Reproductive biology of *Opuntia*: A review. *Journal of Arid Environments*, 64: 549–585.

Root, T.L., Price, J.T., Hall, K.R., Schneider, S.H., Rosenzweig, C. and Pounds, J.A. (2003). Fingerprints of global warming on wild animals and plants. *Nature*, 421: 57–60.

Santos, C., Campestrini, L.H., Vieira, D.L., Pritsch, I., Yamassaki, F.T., Zawadzki-Baggio, S.F., Maurer, J.B.B. and Molento, M.B. (2018). Chemical characterization of *Opuntia ficus-indica* (L.) Mill. hydroalcoholic extract and its efficiency against gastrointestinal nematodes of sheep. *Veterinary Sciences*, 5: 80.

Sáenz, C. (2013a). *Opuntias* as a natural resource. pp. 1–5. *In*: Sáenz, C., Berger, H., Rodríguez-Félix, A., Galletti, L., García, J.C., Sepúlveda, E., Varnero, M.T., de Cortázar, V.G., García, R.C., Arias, E., Mondragón, C., Higuera, I. and Rosell, C. (eds.). *Agro-industrial Utilization of Cactus Pear*. FAO, Rome, Italy.

Sáenz, C. (2013b). Industrial production of non-food products. pp. 89–101. *In*: Sáenz, C., Berger, H., Rodríguez-Félix, A., Galletti, L., García, J.C., Sepúlveda, E., Varnero, M.T., de Cortázar, V.G., García, R.C., Arias, E., Mondragón, C., Higuera, I. and Rosell, C. (eds.). *Agro-industrial Utilization of Cactus Pear*. FAO, Rome, Italy.

Sbaghi, M., Bouharroud, R., Boujghagh, M. and El Bouhssini, M. (2018). *Huit Nouvelles Variétés de Cactus Résistantes à la Cochenille*. INRA-edition, Rabat, Morocco, 20 pp.

Sthapit, S.R. and Scherr, S.J. (2012a). Tropical fruit trees and climate change. pp. 11–23. *In*: Sthapit, B., Rao, V.R. and Sthapit, S. (eds.). *Tropical Fruit Tree Species and Climate Change*. Biodiversity International, New Delhi, India.

Sthapit, S.R. and Scherr, S.J. (2012b). Tropical fruit trees and opportunities for adaptation and mitigation. pp. 141–150. *In*: Sthapit, B., Rao, V.R. and Sthapit, S. (eds.). *Tropical Fruit Tree Species and Climate Change*. Biodiversity International, New Delhi, India.

Sudzuki Hills, F. (1995). Anatomy and morphology. pp. 28–35. *In*: Barbera, G., Inglese, P. and Pimienta Barrios, E. (eds.). *Agro-ecology, Cultivation and Uses of Cactus Pear*. FAO, Rome, Italy.

Tilahun, Y. and Welegerima, G. (2018). Pharmacological potential of cactus pear (*Opuntia ficus Indica*): A review. *Journal of Pharmacognosy and Phytochemistry*, 7: 1360–1363.

Timpanaro, G. and Foti, V.T. (2014). The structural characteristics, economic performance and prospects for the Italian cactus pear industry. *Journal of the Professional Association for Cactus Development*, 16: 32–50.

Torres, J.B. and Giorgi, J.A. (2018). Management of the false carmine cochineal *Dactylopius opuntiae* (Cockerell): Perspective from Pernambuco state, Brazil. *Phytoparasitica*, 46: 331–340.

Uprety, D.C., Reddy, V.R. and Mura, J.D. (2019a). Introduction. pp. 1–5. *In*: Uprety, D.C., Reddy, V.R. and Mura, J.D. (eds.). *Climate Change and Agriculture*. Springer, Singapore, Singapore.

Uprety, D.C., Reddy, V.R. and Mura, J.D. (2019b). Greenhouse gases: A historical perspective. pp. 31–41. *In*: Uprety, D.C., Reddy, V.R. and Mura, J.D. (eds.). *Climate Change and Agriculture*. Springer, Singapore, Singapore.

Uprety, D.C., Reddy, V.R. and Mura, J.D. (2019c). Crop responses. pp. 53–58. *In*: Uprety, D.C., Reddy, V.R. and Mura, J.D. (eds.). *Climate Change and Agriculture*. Springer, Singapore, Singapore.

Uprety, D.C., Reddy, V.R. and Mura, J.D. (2019d). Climate-resilient agriculture. pp. 59–66. *In*: Uprety, D.C., Reddy, V.R. and Mura, J.D. (eds.). *Climate Change and Agriculture*. Springer, Singapore, Singapore.

Valadez-Moctezuma, E., Samah, S. and Luna-Paez, A. (2015). Genetic diversity *of Opuntia* spp. varieties assessed by classical marker tools (RAPD and ISSR). *Plant Systematics and Evolution*, 301: 737–747.

Valentini, V., Allegra, A., Adduci, F., Labella, C., Paolino, R. and Cosentino, C. (2017). Effect of cactus pear (*Opuntia ficus-indica* (L.) Miller) on the antioxidant capacity of donkey milk. *International Journal of Dairy Technology*, 71: 579–584.

Vanegas-Rico, J.M., Lomeli-Flores, J.R., Rodrígues-Leyva, E., Mora-Aguilera, G. and Valdez, J.M. (2010). *Enemigos naturales de* Dactylopius opuntiae *(Cockerell) en Opuntia ficus-indica* (L.) *Miller en el centro de México. Acta Zoológica Mexicana*, 26: 415–433.

Vanegas-Rico, J.M., Rodríguez-Leyva, E., Lomeli-Flores, J.R., González-Hernández, H., Pérez-Panduro, A. and Mora-Aguilera G. (2016). Biology and life history of *Hyperaspis trifurcata* feeding on *Dactylopius opuntiae*. *BioControl*, 61: 691–701.

Vigueras, A.L., Cibrián-Tovar, J. and Pelayo-Ortiz, C. (2009). Use of botanical extracts to control wild cochineal (Dactylopius opuntiae Cockerell) on cactus pear. *Acta Horticulturae*, 811: 229–234.

Yahia, E.M. and Sáenz, C. (2018). Cactus pear fruit and cladodes. pp. 941–956. *In*: Yahia, E.M. (ed.). *Fruit and Vegetable Phytochemicals: Chemistry and Human Health*. Wiley, Chichester, UK.

Zarroug, M.B., Hannachi, A.S., Souid, S., Zourgui, L., Barbato, M. and Chessa, I. (2015). Molecular research on the genetic diversity of cactus (*Opuntia* spp.) using the SSR method. *Acta Horticulturae*, 1067: 53–58.

6
Cultivation of Date Palm for Enhanced Resilience to Climate Change

Adel Ahmed Abul-Soad,[1,*] *Nashwa Hassan Mohamed,*[1]
Ricardo Salomón-Torres[2] and *Jameel M Al-Khayri*[3]

1. Introduction

Prevailing climate conditions determine the suitability of a region for the cultivation of certain fruit species. According to climatic factors, such as temperature, rainfall, humidity and light, the climatic zone of fruit is classified as 'tropical, sub-tropical, temperate and arid zones' (Borchert, 2009). Moreover, some plant species, such as date palm, can be cultivated in two or more climatic zones without affecting fruit production. Date palm, which belongs to arid zone, can be cultivated in temperate climate as in Pakistan and recently in Thailand, where climate is arid to semi-arid with significant variations of spatiotemporal climate (Farooqi et al., 2005).

Climate change is defined as significant, long-term changes in the global climate elements including the sun, the earth, ocean, wind, rain, snow, forest, desert, savanna and people's activities (IPCC, 2007). However,

[1] Horticulture Research Institute, Agricultural Research Center, 9 Cairo University St., Orman 12619, Giza, Egypt.
[2] Unidad Académica San Luis Rio Colorado, Universidad Estatal de Sonora. Km. 6.5 carretera a Sonoyta, SLRC, Sonora 83500, México.
[3] Department of Agricultural Biotechnology, College of Agriculture and Food Sciences, King Faisal University, Al-Ahsa 31982, Saudi Arabia.
Emails: nanosh28@yahoo.com; ricardo.salomon@ues.mx; jkhayri@kfu.edu.sa
* Corresponding author: adel.aboelsoaud@arc.sci.eg

changes in temperature, atmospheric carbon dioxide (CO_2), the frequency and intensity of extreme weather conditions could have a significant impact on crop yield and disrupt food availability and quality (USGCRP, 2014). Date palm (*Phoenix dactylifera* L.) is one of the economically important fruit crops subjected to the impact of climate change which is expected to influence the fruit's distribution and productivity. Therefore, it is essential to develop cultivars with marketable date characteristics, tolerant to pests and diseases and resilient to suboptimal cultivation conditions. Various organizations, such as The International Center for Agricultural Research in the Dry Areas (ICARDA), are working in reduction of poverty in arid areas, where the date palm is an inhabitant, enhance environmental health in the face of global challenges, such as climate change by reducing atmospheric CO_2 percentage and where palm trees have high capacity of absorbing CO_2 from the atmosphere, then convert it into sugars and other important compounds that lead to air cleaning. Moreover, each single tree of date palm can absorb 400 L/tree/day of CO_2. Date palm is considering one of the most important desert fruit crops as it can be active photo-synthetically and is productive for at least 70 years with minimum cultivation requirement elements. So, date palm is a suitable fruit species for desert reclamation (Suliman, 2019). In addition, Food and Agriculture Organization (FAO) promotes on-farm diversification to increase the resilience to climate change and the preservation of biodiversity through the adaptation of biodiversity-supportive practices. Above all, date palm research centers in major date palm producing countries, i.e., Egypt, Saudi Arabia and Pakistan are making efforts to identify adaptive landraces to propagate as new productive cultivars. The impact of climate change has resulted in prolonged winters and summers, with severe changes in temperature, causing the migration of plant species. The main component of this effect has been the scarcity of water, resulting in severe droughts, which have compromised agricultural production in arid and semi-arid countries, such as Saudi Arabia (FAO, 2020).

Three international commitments were made to build climate resilience and to orient a big budget of industrial countries towards low-emission of GHGs. where the emissions of six gases (carbon dioxide, methane, nitrous oxide, hydrofluorocarbons, perfluorocarbons, and hexafluoride sulfur) are responsible for global warming. The first was the International Kyoto Protocol that was set out and adopted in Kyoto, 1997; second commitment of the Doha, in Qatar, 2012; and third, in Paris, France, 2015 with the adoption of the Paris Agreement. Through these agreements, all governments worldwide agreed to limit global warming below 2°C and to pursue efforts to limit global mean temperature (GMT) to 1.5°C in order to increase adaptation to the adverse impacts of climate change (Samuel and Smith, 2004; Poulopoulos, 2016).

This chapter describes the climatic conditions including temperature, precipitation, wind and humidity of date palm cultivation and sorting the reported impacts of climate change and extreme weather on the date palm cultivation. It also presents important emerging technologies and agricultural practices to enhance resilience. This information is believed to encourage passionate farmers to cultivate date palm outside their traditional cultivation zones, such as the practice in Thailand, India, Australia, Peru, Mexico, Nigeria, South Africa and Namibia.

2. Distribution

Date palm is one of the oldest cultivated fruit trees in arid zones, with economic importance in Middle Eastern countries, such as Saudi Arabia, Egypt, Iran, Iraq, as well as some European countries, like Turkey and Spain. Egypt is the origin of date-palm tree. It was known in Egypt since 4,000 years and appeared on paintings of date palm trees on the walls of ancient Egyptian temples. Date palm is a multipurpose tree and is considered one of the main income sources for local farmers. Many cultivars, that are significantly different in their vegetative growth and yield characteristics, are spread geographically within all over the Nile Valley and the Western Desert Oasis. The distribution of date palm culture in Egypt follows a geographic pattern, including locations for the successful production of soft, semi-dry and dry types of dates. Moreover, date palm is playing an important role in Egyptian agriculture and reclamation program. Date palm is cultivated in arid and semi-arid regions, which are characterized by five-month long hot summers of 25–35°C as temperature average, low rainfall below 200 mm/year and low relative humidity, that is, below 60% during the fruit development and ripening period of five to seven months. Also, date palm was established in America decades after its discovery and was dispersed throughout the American continent, along with the Spanish conquest at the beginning of the 16th century, in countries such as United States, Mexico, El Salvador, Panama, Colombia, Venezuela, Brazil, Peru, Chile, Uruguay, Cuba, the Dominican Republic and other Caribbean islands (Shabani et al., 2012; Rivera et al., 2013). Its successful production could only be achieved in the arid regions of Peru, Chile, the southwest of the United States and the northwest of Mexico. The date palms planted in Baja California, Mexico and Peru, originating from the Spanish conquest, are adapted to the climatic conditions of the desert oases near the missions, maintaining important production for a period of 200 years. However, it is possible to cultivate palms outside their natural range in more temperate climates.

Distribution of date palm is expected to change as a result of climate change. Therefore, it is essential to identify the regions that will take to cultivating date palms in the future as well as identifying the regions that may be negatively affected. For that, researchers are trying to predict the future of date palm distribution under climate change, using many climatic models. Moreover, it is important to consider the effect of both climatic and non-climatic parameters when predicting the future of species distribution (Shabani et al., 2014b; Allbed et al., 2017). Some studies indicate that large areas of Algeria and Saudi Arabia will become climatically unsuitable and unable to cultivate date palm crop to the same extent in the future. The results are based on the current observations of decrease in production of date palm Tree in Middle Eastern countries during 1990–2000 (Jain, 2011; Zaid, 2012).

2.1 *Temperature Range*

Date palm is considered one of the important yields in the agrarian economy of several countries in arid regions of the world. During the last two decades, there has been a significant increase in the production area of date palms, especially in the Middle East and North Africa which are the pre-dominant date-palm-growing regions in the world. Moreover, the Food and Agriculture Organization of the UN estimates that more than 100 million date-palm trees have an annual world production of nearly 8 million tons (FAOSTAT, 2010). It is essential for a date-palm grower to consider the climate when selecting a particular cultivar that would be best suited to the cultivation area for obtaining high-quality fruit production. So, growing the optimal date-palm cultivar will ensure economic success and avoid future losses and discouragement. Furthermore, two aspects are used to judge the suitability of a cultivar for a particular climate area—the first one is according to the accumulated heat hours and air's relative humidity during five to seven months of fruit development on the tree; the second aspect is the fruit's moisture content at edible stage. There are three cultivar groups: soft, semi-dry and dry date palm cultivars that require approximately 4200 h, 3600 h and 2800 h temperature above 18°C, respectively. Temperature level varies according to varieties and to local climatic condition, where temperatures in the shade are less than 18°C do not produce fruits. Fruits are obtained only when the temperature is 25°C in the shade (Johnson and Hodel, 2007).

The best growing and the highest quality production of dates located at regions characterize with long hot, dry summer and early fall (Abdul-Baki et al., 2007). Thus as a result to climate change some areas that are now climatically unsuitable may be suitable for date palm cultivation in the future. This fact is important for future choices in the location of date-palm farms and associated industries. Climate change may have direct impact on the economy by affecting agricultural output. For instance, the total annual income from date palms in Middle Eastern countries has reduced from 1990 to 2000 because of plant diseases and water shortages

(Zaid and Arias Jiménez, 2002). Temperature range during winter season affected flowering. For example, Saudi Arabia, as a result of reduction in available water resource, the yield of crop was affected during the 2010 season when many farmers noticed unusual early blooming of date palm as a direct effect of climate change (Darfaoui and Assiri, 2009; Zatari, 2011).

2.2 *Health and Medicinal Benefits*

Date palm fruit is considered an ideal and main food in many countries, such as the Arab Gulf region because of its content of essential nutrients, like calcium, iron, fluorine, and selenium. Carbohydrates constitute a major component with 70% of sucrose, glucose and fructose. The fruit is also a good source of fiber and many important vitamins, like thiamin, vitamin C and vitamin A. Moreover, besides a wide range of saturated and unsaturated fatty acids, dates contain antioxidant and antimutagenic properties. It is noticeable that these concentrations differ according to the production region and date variety (Erskine et al., 2004; Ismail et al., 2006; Allaith, 2008; Ali et al., 2009; Saafi et al., 2009).

On the medical side, date can be used as a practical supplement for iron deficiency. It is suitable for people suffering from hypertension and cardiac disorders after diarrhea, vomiting or diuretic medications because of its high potassium and low sodium contents (Ali et al., 2009). Furthermore, date can be used for treating fever, stomach disorders, memory disturbances, nervous disorders, as well as an aphrodisiac and to boost the immunity. Also, it protects against many chronic diseases, like cancer and heart diseases (Allaith, 2008; Vyawahare et al., 2009; Al-Mssallem et al., 2019).

The most important micronutrient deficiencies people suffer from are lack of iron, vitamin A and zinc (Diaz et al., 2003; Kosek et al., 2005). According to studies eating 100 g of fresh dates provides about half of the daily recommended micronutrients (Shabani et al., 2015). Climate change can contribute adversely to world hunger. Moreover, food systems will be at risk and unstable because of short-term variability in supply. Moreover, climate change will increase food insecurity in areas currently facing hunger and malnutrition (Wheeler and Braun, 2013).

2.3 *Versatile Uses*

Humans can get many products from date palm, for instance, the fruit as the main product, whether it fresh or dried, is used in cereal, pudding, bread, pressed cakes, cookies, candy bars, ice cream, juice, vinegar, wine, beer, sugar, syrup, honey, chutney, pickle, paste, dip and food flavoring (Glasner et al., 2002). Furthermore, the trunk can be used as timber, wood, or fuel. The fiber from the trunk and leaves can be made into bags, baskets, camel saddles, cords, crates, fans, food covers, furniture, mats, paper, ropes, trays and twine. Shades, roofs, separating walls and enclosures can be made from dried bundles of leaves. At the environmental level, date-palm groves play a central role in the desert's ecological system, in controlling desertification and land reclamation (Gotch et al., 2006).

2.4 *International Trade Status*

During the last three decades, the world's total production areas were Middle East and north Africa, Australia, India, Mexico, southern Africa, South America, Pakistan and the USA. According to Saker and Moursy (2003), the date palm production is one of the oldest economic activities and a strategic sector of most Arab countries. There is a remarkable increase in date trade in international markets because of increasing awareness about the health benefits of dates, as well as its use as a source of sugar and medicinal alcohol in many parts of the world (Mansouri et al., 2005; Al-Farsi et al., 2007). In 2016, the Arab region was the world leader in date cultivation with almost 75% of global area under date palm, with around 77% of world production and approximately 69% of world total export of dates. According to the FAO, KSA, Oman and United Arab Emirates (UAE) had the highest harvested areas in 2016, with 145,516 ha; 24,120 ha and 93,561 ha in the three countries, respectively. At the same time, Egypt was the largest producer of dates in the world with approximately 1,084,529 metric tons of dates. Though Egypt has increased date cultivation by more than 100% since 1993 and currently has an estimated 15,582,000 date palm trees, it accounts for less than 3% of world export of dates. While Iran produces 947,809 metric tons annually, the majority of Iran's dates are exported to Asian countries. The biggest importers are India (16%) and Malaysia (11%), followed by Russia (9.9%). In comparison, Saudi Arabia is at the third position in date production with 836,983 metric tons produced per year. It exports about 8.8% of the world's dates and totals to around $94.3 million. The primary importers are Jordan (19%), Yemen (17%), and Kuwait (15%). Furthermore, Iraq produced 675,440 metric tons every year and was responsible for 7.3% of global date export. In 2014, Iraq exported $77.5 million worth of dates and 79% of it went to India that is considered the world's largest importer of dates, followed by UAE and Morocco, with import dates valued at $119 m and $96 m, respectively (Amber, 2017; Boubaker et al., 2018).

Climate change has a direct and indirect impact on agricultural production as it changes the distribution of economically important crops through changes in their physiological ability to adapt to water shortage and other climate elements (Jain, 2011). For example, the annual income from date palms in the Middle Eastern countries fell during 1990–2000 (Zaid, 2012).

2.5 Production in Oasis

The oasis agro-ecosystem suffers due to non-availability of suitable irrigation water. However, even with sufficient water, its use under the usually hot, dry climate is often not sustainable, leading to soil salinization as a consequence of unsuitable irrigation and drainage systems. The oasis agro-ecosystem is a standard model for a cropping system of date palms, fruit trees and annual crops. Date palms, fruit trees and annual crops approximately intercept 20, 20 and 40% of daily net radiation, respectively. Highly adapted cultivars of date palm, fruit trees and annual crops are managed through refined social practices and institutions.

Most of the unique oasis agro-ecosystems are found in MENA countries. These oases cover about one million hectares, where the most important crop is date palm—Al-Qatif and Al-Ahsa in Saudi Arabia; Al-Ain, UAE; Buraimi, Maghta, and Bahla in Oman; Bahraiya, Farafra, and Siwa in Egypt; Ghadames and Kufra in Libya; Ouargla, Taut, and Timimoun in Algeria; Tozeur and Tamerza in Tunisia; and Tafilalt and Ourzazat in Morocco are the most famous oases in MENA (Fig. 1). The importance of cultivars in each oasis is related to the climatic conditions, the number of trees per each cultivar and the quality of the fruit. As such, the oasis agro-ecosystem has very complex ecological, social, and economic infrastructure.

Nearly 14–17 different date palm cultivars and about of 107 plant species have been recorded in three small oases in the northern mountains of Oman. Date palms and annual crops are cultivated in oases that are watered either by springs or by '*aflaj*' tunnel systems dug in the ground or carved into the rock to use underground water. These systems are needed for oases located below plateaus or the foot of cliffs (Abdullah, 2011).

(A) (B)

Fig. 1. Examples of different oasis agro-ecosystems: (A) Siwa, Egypt and (B) Ourzazat, Morocco (*Source*: Google Earth).

Moreover, climate change affects the seasonal distribution of rainfall in Algeria. In a study at Biskra Oasis, continued for 23 years, since 1990–2013, the season for rainfall shifted from autumn to spring. During the ten years of the last century, extensive rainfall occurred in the autumn (37%) and winter (39%), while by the end of the first decade of the current century, most of the rainfall occurred in winter (33%) and spring (37%). This new distribution of precipitation affected pollination and the date's harvesting time (Tarai and Belhamara, 2014). In a particular case of Mexico, there are two very different areas of production—the San Luis Rio Colorado Valley (32°18′19″ N, 114°56′43″ W), with offshoots imported from Yuma, Arizona and the other is spread in the Mexicali valley (32°22′27″ N, 115°07′13″ W), with offshoots imported from southern California. While production of 97% in Mexico is concentrated in these two valleys, the production of Medjool cultivar stands out with 94% of the national production (Ortiz-Uribe et al., 2019). These valleys are characterized by low rainfall and high temperatures during the harvesting of this fruit. At the beginning of the 2020 season, an atypical cold lasting till the end of March, caused the male palms to produce their pollen until mid-April. Similarly, female palms delayed the appearance of their inflorescence, thus delaying the pollination process for up to a month-and-a-half. This meant that the harvest lasted until the end of October 2020, which commonly ended in the beginning of the same month. This could also have been affected by atypical increase in humidity this summer.

Recent studies carried out in these valleys have identified the pollen cultivar that allows early ripening of the fruit compared to the use of other varieties of pollen (Salomon-Torres et al., 2017; Salomón-Torres et al., 2018). Likewise, research is currently being undertaken on the use of various growth biostimulators as they encourage greater growth in the fruit and early ripening. These measures could help counteract the lag effect on the date harvest.

3. Cultivation and Water Requirements

Date palm belongs to areas with long dry summers and mild winters. It grows well in desert and oasis with high temperature and underground water close to the surface. Fruit production is dependent on certain heat requirement according to the date varieties, be they dry, semi-dry or soft. Dry varieties are found in dry areas, while semi-dry and soft varieties are located in humid and semi-dry areas (Erskine et al., 2004). Date palms are suitable to grow in different soil types, light, but deep soil gives the best production. Moreover, it is tolerant to high levels of salinity when growth and fruit productivity may be affected by high values of EC (Nazir, 2010). Date palm is one of the crops that requires a limited amount of irrigation water and is thus considered drought-tolerant. It is a major crop in most hot arid and semi-arid regions of the world (Abd-Elgawad et al., 2019). Date palms are particularly well adapted to saline soil and excessive rise in the water-table, but data are needed to evaluate the effects precisely. The date-palm tree requires irrigation water at 16,000–20,000 m^3/ha/year for optimal fruit production, depending on the soil type and climate conditions (Tarai and Belhamara, 2014).

Water scarcity has not yet been a climate-change problem in the date-producing region of Mexico. However, some preventive measures have been taken to conserve water, such as cementing of irrigation canals and some parts of the Colorado river (which comes from the USA and feeds both valleys) in order to avoid water seepage. Likewise, according to data from the Ministry of Agriculture, a higher yield has been reported in plantations where drip irrigation systems have been employed than the flood irrigation systems (Maged et al., 2020; SIAP, 2020). With this, the government hopes to encourage the use of this system to save irrigation water.

Arid areas, where long periods of dryness last three to four years and are coupled with stormy wind-bearing sand particles, blemishes develop on the growing fruits. As a mitigation measure, the fruit bunches are covered with a container woven from date-palm leaves to protect the fruit against drying and hot-air-bearing sand particles (Fig. 2). Another consequence is the drop in groundwater level below 8 m as checked out in many wells of date-palm oases in northwest of Balochistan province on the border between Iran and Pakistan (Abul-Soad et al., 2015). Date-palm plantation in some areas, as in 'Yach Mach' village, requires irrigation water to be pumped from aquifers. Also, farmers dig a circular furrow

Fig. 2. Bunches of date fruit are covered in windy arid land of 'Yach Mach' village, Nicondi, Balochistan in Pakistan. Date palm is irrigated through pumped underground water in a basin dug around the tree to facilitate efficient use of water.

around the tree trunk to collect the water at the absorbing root zone of 120 cm depth.

4. Climate Change Constraints

4.1 Abiotic Stresses

By all odds, increase in world temperatures leading to climate change affect significantly the crop at different growth stages (Mathur et al., 2014; Lamaoui et al., 2018). Salinity, drought and heat stress commonly account for a decline in photosynthetic pigment substances, especially chlorophyll. Under field conditions, abiotic stresses like drought and salinity happen simultaneously. Thus many stresses impact and cannot be determined based on a single factor (Rastogi et al., 2019). The high temperature reduces the synthesis of photosynthetic pigments, damages the chloroplast membranes, lowers chlorophyll and other pigments besides increasing the content of abscisic acid which controls stomata closure and compatible osmolyte induces the expression of adaptation-response genes (Zandalinas et al., 2016). In southern Iraq, where various abiotic stress factors, such as salinity, drought and heat predominate, date palm is drought-and-salt tolerant because of its ability to adjust growth and developmental processes which reduce the magnitude of injury caused by the environmental extremes. However, recent water shortages, which increase salinity in both groundwater and soil, have threatened the continued productivity of the crop (Shareef, 2019; Hussein et al., 2020).

Salinity causes changes in plant anatomy, physiology and growth. These changes are counteracted by mechanisms that mitigate the effects of stresses, including osmotic and ion toxicity stress. Moreover, salinity inhibits water uptake and increases the concentration of toxic ions, such as Na^+, accumulation of Reactive Oxygen Species (ROS) and contributes to imbalances in nutrient uptake, thereby impacting negatively plant physiology, reducing photosynthetic capacity and altering cellular metabolism. All of the above lead to reduced growth rate, increased rate of senescence and lower yield of crop. Many early studies revealed that date palm responds to salt through different strategies (Munns and Tester, 2008; Alhammadi and Kurup, 2012; Hussain et al., 2012), salt stress triggering the production of osmolytes and compatible solutes, including proline. Other studies have reported that date palm ameliorate the effects of reactive oxygen species by increasing the expression of anti-oxidant enzymes and increasing the concentrations of antioxidant metabolites (Djibril et al., 2005; Yaish, 2015; Ait-El-Mokhtar et al., 2019; Al Kharusi et al., 2019a). Also, Hussein and Al-Khayri (2021) in their study found that date palm forms a new protein, under the influence of stress conditions, as a reaction to adapt to the influence of excessive environmental stress.

In natural conditions, heat and drought occur at the same time with heat enhancing the severity of drought stress. Heat stress results in the expression of heat shock proteins (Saidi et al., 2011). Plants treated with different stresses either avoid the stressor (stress avoidance) or gain the ability to maintain plant function in the presence of the stressor (stress tolerance). Many plants use both methods to face environmental challenges. Moreover, date palms use complex mechanisms to maintain high water potential in drought conditions. There are many anatomical characteristics that contribute to tolerance conditions of date palms as hyper-arid. For example, date palms maintain thick, waxy cuticle and pinnate, compound leaves covered with many spines, which insulate the tip-growing point. The deep root system in date palms traps water in various types of soil. These traits reduce evaporation and the date palm microbiomes and abiotic stress maximize the water uptake and contribute to *P. dactylifera*-tolerance for drought stress. Moreover, long periods of drought negatively impact date palm by reducing growth, fruit quality and yield (Elshibli et al., 2016).

Noticeably date palm photosynthesis is not affected by drought and heat stresses even with reduction in concentration of antioxidants in leaves (ascorbate and glutathione) (Arab et al., 2016). This reduction may recompense increase in the activity of antioxidant enzymes, such as glutathione reductase. Moreover, a change in fatty acid composition under drought was noticed but not due to heat. Thus date palms have independent response pathways to drought and heat stress. Furthermore, date palm seedlings accumulate proline, not only in response to drought and salinity stress but also in response to extreme temperatures, that has lead researchers to use proline production as a marker in date-palm breeding programs for improving drought and salt tolerance (Yaish, 2015; Arab et al., 2016).

4.1.1 Temperature Fluctuations

Photosynthesis pigments are a fundamental indicator of the effect of various abiotic stress factors on plants. Periods of stress that occur commonly during the two hottest months of the summer season indicated that high temperature is the primary factor that determines the growth of plants. Recovery of photosynthesis following low-temperature alleviation for the most part determines the plant flexibility in adapting to water deficiency and salinity (Hussein et al., 2020).

4.1.1.1 Disturbed Flowering and Fruit Set Failure

The temperature fluctuations, be it in winter through warm temperatures or during the hot and dry waves of high temperatures, disturb the normal growth of the tree and affect the fruit production too. It was observed in Egypt, particularly in the date-palm oasis at Al-Bahariya, that warm

Fig. 3. Malformed late flowering due to long warm winter intermitted with heat waves causing fruit set failure (seen in Al-Wahat Al-Bahariya oasis, Egypt in mid-May 2018).

winter encouraged the early emergence of flower bunches. This would inevitably lead to an early crop, but long winter reduces the rate of fruitset. In addition, the late flowering spathes were deformed and not well pollinated, leading to formation of parthenocarpic fruit (Fig. 3).

4.1.1.2 Sudden Drying of Fruit Bunches of Dates

Climate change occurring all around the world caused this phenomenon in Egypt and Iran. Some cultivars are sensitive to this phenomenon. There are many reasons for over-drying of maturing fruit bunches of dates on the tree. One of the reasons is the heat wave as it happened a few years ago at Siwa oasis in Egypt, leading to 'frihy' dry cultivar. The heat weaves of about 40–45°C caused non-commercial shrinking of fruit. Covering the fruit bunch with the pinnae of date-palm fronds is a practice in desert plantations to protect the fruit bunch from the sandy heat waves (Fig. 4). Heavy bunches, pest infestation and other factors may also lead to bunch drying. In general, any reason causes weakness in the joint of fruit bunch with the heart tissues of the tree, resulting in imbalance in the water passage to the developing fruit and subsequently to the dryness. In order to mitigate this phenomenon irrigation should be controlled.

An exceptional fruit set failure was noticed on the Barhee cultivar trees in Egypt during the season of 2019. An open field observation indicated that most of these farms were in the northern part of Egypt, where the temperature is relatively low along with a long winter season. The fruit set failed, although the Barhee trees were pollinated by various regular practices at different places. This supports the view regarding the adverse impact of low temperature at pollination time, subsequently reducing

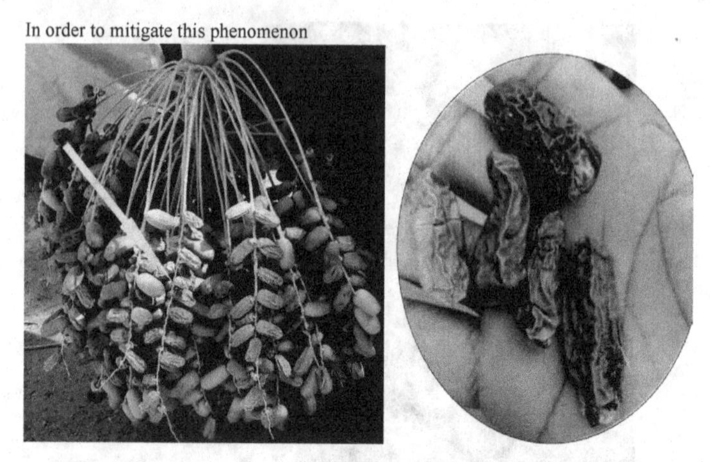
In order to mitigate this phenomenon

Fig. 4. Dry cultivar 'frihi' at Siwa oasis in Egypt in season 2015; a couple of weeks' hot waves led to inferior fruit.

the size of the fruit set. Such exceptional fruit set failure was attributed mostly to the change in climate during the sensitive growth stages of a particular cultivar that requires special precautions to be taken during the pollination process.

On the other hand, due to the long cold winter season, trees of most cultivars produce small and late unfertile spathes which reduced the overall fruit yield (Fig. 5). It is well known that a date-palm tree produces fruit spathes in three batches/rows. The upper one comes last and misses the proper timing for pollination. Covering Barhee cultivar trees and other similar cultivars with perforated paper bags after pollination provides

Fig. 5. Notice the small late spathes come out after pollination due to a long cold winter season and subsequently cause failure in the fruit set in Barhee cultivar trees, in the northern region of Egypt.

some warmth and increases the ambient temperature suitable for the pollination process. Besides, bunch cover with brown paper bags protects pollen grain to be swept away from the stigma of the female florets after pollination by wind or rain.

4.1.2 Flood Water

Another effect of climate change in the oases, such as at Baja California, has been the rainy season accompanied with hurricanes in summer, when date-palm crops have been wiped out in the oases and up to 5,000 palms detached from the ground (Elinformanate, 2019). In such areas, date fruits dry naturally and/or artificially because of the rains. In the month of August, a flood-hit Pakistan was witness to date-palm trees being threatened at Khairpur, Sindh. As date palm is the cornerstone of livelihood for the locals, the entire life came to a standstill. Flood waters covered 1–3 meters of tree trunks in some areas of Khairpur (Fig. 6). Some of the young trees (offshoots) of new plantations died, while the attached offshoots of adult trees suffered a less severe impact. However, the adult date-palm tree has the ability to continue growing even if its base is entirely covered with water. All activities of harvesting and dates getting cured under the heat of the sun were affected. However, large air pockets in the tissues of the roots apparently played a role in the respiratory system of the date-palm tree. The climate change indirectly impacted the pest-and-disease incidences in the fruit species. Furthermore, the high moisture stress led to increase in fruit blemishes on the date-palm fruit (Fig. 7).

Fig. 6. Flooded date-palm orchard at Khairpur, Sindh in Pakistan during the summer monsoon rains in 2010.

Fig. 7. Fruit checking for quality due to cultivation in moist areas near to Karachi more than 60% (RH) crop affected.

4.1.3 Forest Fires

Date-palm plantations located in the oases of Baja California are considered a cultural heritage of Mexico as they were derived from the Spanish conquest. But over time their production has decreased greatly since the dates they produce are not commercially attractive and actually are grown only for local consumption. Besides, the oases that are far from urban centers and are commonly affected by natural catastrophes. A direct effect of climate change in these regions has been the high temperatures and droughts in summer which have led to continuous forest fires (Fig. 8). In 2019, the forest fire destroyed more than 70 ha of date palms in the oases and even the adjacent farms (Fig. 9) (BCSnoticias, 2019).

Fig. 8. View of the forest fire recorded in the oases of Baja California and which happens almost every year.

Fig. 9. Damage caused by forest fire to date palms located in the oases of Baja California península in Mexico.

4.2 Biotic Stresses

Date palm is attacked by a wide range of insect pests and phytophagous mites, causing serious losses in the yield and even death of the tree in some areas. During the last two decades, there has been a significant increase in the cultivated area of date palm that provides an ideal ecological housing for biotic stresses, including insect pests and diseases (FAOSTAT, 2012). Enhanced date-palm cultivation in many countries besides global warming, unbounded use of chemical insecticides and extensive international trade is likely to have an impact on the species diversity and density of pest in the date-palm agro-ecosystem. So, development of sustainable pest management strategies in date palm is vital to meet the existing and emerging pest challenges.

Furthermore, climate change and large-scale movement of palm species for farming and ornamental gardening has also increased the attacks by invasive species, thereby multiplying the problems of crop protection. For instance, during 1992 and 2012, the Maghreb region of north Africa and the Gulf countries showed a significant increase in new date palm plantations. These young plantations were exposed to attacks by several insect pests, especially Red Palm Weevil (*Rynchophorus ferrugineaus*) RPW, which prefers to invade date palms less than 20 years old (Abraham et al., 1998). Moreover, there were 54 species of insect pests and mites which were found on date palms. Out of the listed 112 species of mites and insects, 22 species attacked stored dates worldwide. Ten species of arthropod pests were considered major, while 45 predators and parasitoids correlated with the insect pest complex of date palm. El-Shafie in 2012.

Red palm weevil (*Rynchophorus ferrugineaus*) insect is native to Southeast Asia and is considered the most harmful pest of palms belonging to the Arecaceae family. The attack by Red Palm Weevil started three to four decades ago, causing a real threat to date palm plantations. In San Ignazio, Mexico, the RPW insect nested in palms and despite the fact that it was the first time that this had appeared, it put the oasis of San Ignaezio at risk of disappearing as well as the other ecosystems to which it could migrate (Radiokashana, 2019). For the recovery of palms in the oases, an adequate crop management was recommended, which included cultural tasks, such as irrigation, pruning, weeding and fertilization as well as the placement of traps for an early identification of the presence of this insect and of other invasive pests.

Recently the presence of the monk parakeet (*Myiopsittam onachus*) has been observed. It is an invasive species that has adapted to the climate of the cities of northwestern Mexico, where date palms grow (Tinajero and Rodríguez-Estrella, 2015). This has become a threat since this species likes to eat the palm leaves and the fruit (Fig. 10). To date it has only approached the small plantations near the city of Mexicali, where producers have chosen to bring predators of this bird—cats—which have been able to control this pest. However, the local authorities are analyzing how to exterminate or reduce these colonies of birds before they infest the entire date palm plantation. The presence of pests and diseases in these valleys has been very scarce. However, the presence of the round-tailed ground squirrel (*Xerospermophilus tereticaudus*) and gophers (*Thomomys bottae*) has been identified (Ortiz-Uribe et al., 2019). Producers, who have faced this problem, have solved it by introducing cats in their plantations and by placing metal sheets around the trunk of the palm to prevent these rodents from climbing the palm and damaging the fruit.

Measures taken to control pests are as follows:

1. Control program of date mite, *Oligonychus afrasiaticus.* The fruit bunch is protected against *Oligonychus afrasiaticus* in Pakistan with covers made from pinnae of fronds (Fig. 11) or alternatively by using mineral oil solution 1.5 L/100 L water, or 1 kg/wet sulfur in June.
2. Control of palm-frond borer, *Phonapate prontalis.* The fruit bunch is covered to protect from infestation with wasps or alternatively by using one or more of the following treatments: light traps, broken leaf and stalk pruning and burning it, horticultural management (dryness or greenish), and Cidial or Pasodin 300 cm/100 L water in May.

Management of palm pests usually starts with the application of insecticides, but gradually progresses to Integrated Pest Management (IPM) programs because of the negative aspects of insecticide-based control programs. Integrated Pest Management (IPM) is an overall process to control pests, that includes the collection of information on the pest's biological side and bind them with the environmental factors to identify

Fig. 10. Tunnels created by *Phonapate prontalis* and broken fronds seen in Siwa oasis, Egypt.

Fig. 11. Infestation with *Oligonychus afrasiaticus* in north of Baluchistan, Pakistan.

available control methods to manage the pest in more economically and efficient way that is more environmentally friendly and less risky to the human health. Agricultural countries need to improve their information on pest and disease management. When date palms were cultivated far from their original location attacked by new pests, farmers turned to countries that faced the same pest previously for help and advice in order to control the insect pest. Date palms are fiercely attacked by scale insects (Hemiptera: Coccidea) which is controlled by using organophosphates but intensified use endorses IPM programs. For example, according to Soroker et al. (2005) and Blumberg (2008), the IPM endeavor in date palm were aimed at controlling fruit pests, scale insects and tissue borers like (RPW) in Israel. Biological control for pests attacking date palm in Iraq, reduced the infestation levels of, targeting stem borers, lesser date moth and dubas bug resulted in 90.5%, 80% and 96.7%, respectively from 2009 to 2017 using sustainable IPM, Methodology included solar light, hand collection of borers, azadirachtin sprays against dubas and biological control of the lesser date moth employing treatments with *Bacillus thuringiensis* (Al-Jboory, 2007; Waqas et al., 2015).

5. Approaches to Enhance Climate Change Resilience

Different strategies used now to enhance sustainable crop production and decrease the climate change effect are—high-throughput Single Nucleotide Polymorphism (SNP) genotyping, genomic selection and trait mapping. These tools are essential not only for an in-depth understanding of trait variations but also for the transformative engineering required to accelerate the plant breeding efforts (Anup and Bijalwan, 2015; Nutan et al., 2020a). Moreover, the international orientation is to develop suitable mechanisms of adaptation and mitigation against changing climatic parameters. Effective adaptation to climate change requires Information Communication Technologies (ICT), which not only reduce the time but also systemically transform the information through networked governance. The effect of climate change can be assessed by ICTs based on updating data to be used for monitoring, measuring climate change effects and controlling the interactions with the environment. The practical applications of ICT are widely used, such as:

a) High-value remote monitoring equipment, like satellites.
b) Networks of remote sensors.
c) Global positioning and Geographic Information System (GIS) applications.
d) Communications services, such as the internet, mobile networks and SMS.
e) Handheld devices, such as mobile phones and a personal digital assistant (PDAs).

5.1 Crop Management to Mitigate Climate Change

Climate change has a negative effect on crop production. Plant researchers are focused on finding solutions to minimize these negative effects. The risk of climate change includes the periodic and severe droughts, floods and fires and the complex biological impacts on the productivity and stability of livelihoods that depend on natural resources and human health. Moreover, there are a combination of social, economic, political and physical factors that determine the amount of damage and also predict crop resistance. Dealing with climate-related risks, the starting point for adaptation measures is understanding of current sensitivity to climate variability and extremes. Thus, it has to differentiate between short- and medium-term adaptations and the longer-term natural, biological adaptation of ecosystems for livelihood systems and human settlement (Geoff et al., 2006).

There are a lot of biotic and abiotic factors that affect the management of a healthy date-palm crop. Global climate change could be an intimidation to plantations of date palm; thus, climate models predict the shrinkage in suitable regions for date palm growth, especially in the Middle East. Moreover, water is a limiting factor in growth, though excess water can also reduce yield. While application of insufficient water slows plants growth, excess irrigation depletes the reservoirs' groundwater in arid regions. Also, salinity has a negative effect on date palm growth (Shabani et al., 2015; Abdul-Sattar and Hama, 2016).

5.2 *GIS and Remote Sensing Systems*

Geographic Information System (GIS) is a computer-based tool for mapping and analyzing feature events on Earth, while Remote Sensing (RS) is the art and science of making measurements for objects far from the Earth using sensors on airplanes or satellites. Remote sensing (RS) and geographic information system (GIS) are useful technical tools or applications to monitor, investigate and map date palm distribution, health, diversity, density and changes on the cultivated area due to the impact of climate change all around the world.

As mentioned before, habitats of date palm trees are characterized by hot and dry conditions with an annual rainfall of about 100–150 mm and are suited to Mediterranean climate with salty and alkaline soils. Climatic parameters, land degradation, desertification, water shortage and plant diseases caused by climate change are the factors which are responsible for the future of date palm production (Chao and Krueger, 2007; Shabani et al., 2013). For that, distribution of date palm will change as an upshot of climate change. So, it is essential to identify which regions will take the opportunity of cultivating date palms in future and which may be negatively affected. Therefore, many climatic models are being developed to predict the future of date palm under the effect of climate change. Many studies have been made using the CLIMEX software "which is an eco-climatic modelling package that predicts the potential distribution of species using climatic information, biological data and the natural geographic distribution of that species." It will help to develop a global model of the climate response of *P. dactylifera* based on its native and cultivation distribution. Because of the suitability of this software to geographically determine the climatic parameters that describe species' response to climate, the results were run with the A2 SRES (Special Report on Emissions Scenarios) for 2030, 2050, 2070 and 2100. According to studies, the CLIMEX results showed that by 2100 many areas with a suitable climate for date palm are expected to be climatically unsuitable in North Africa, while South America, such as south-eastern Bolivia and northern Venezuela, will become climatically more suitable. Also the climate suitability of Saudi Arabia (Figs. 12, 13), Iraq and western Iran will decrease by 2070. Cold and dry stress will play important roles in date palm distribution in the future (Shabani et al.,

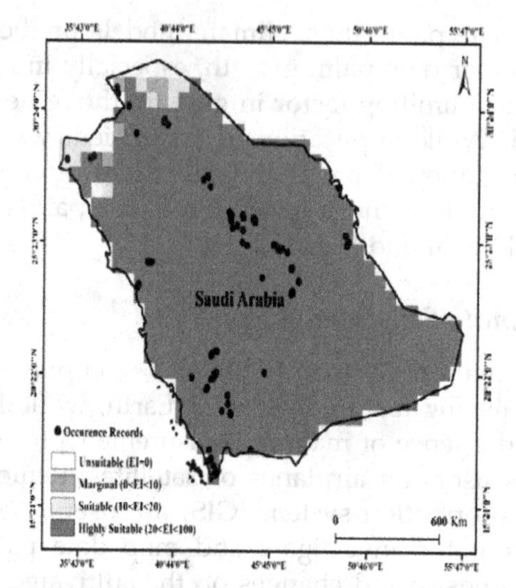

Fig. 12. Current and modelled potential distribution of *P. dactylifera* in Saudi Arabia. EI, eco-climatic index; *P. dactylifera* (Allbed et al., 2017).

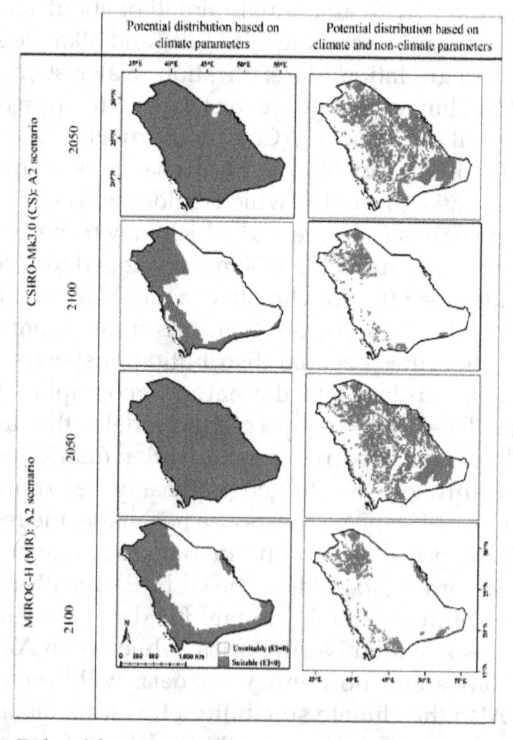

Fig. 13. The EI for *P. dactylifera* for 2050 and 2100 under CS and MR GCMs running with the A2 emission scenario. Colour online: CS, CSIRO-Mk3·0; EI, eco-climatic index; MR, MIROC-H (Allbed et al., 2017).

2012; Shabani et al., 2016; Allbed et al., 2017). Subsequently, by the next 60 years, in some developing countries, most of the regions that are suffering from micronutrient deficiencies would become highly conducive for date palm cultivation.

Remote sensing provides a bird's eye-view of the plantation and a way of counting the trees automatically. Counting trees manually takes a lot of effort, cost and has a large percentage of error, specially in non-systemic fields with overlapping rows, and trees of different sizes and ages. For example, most plantations have resorted to estimate the figures by multiplying the total area with the number of palms per hectare, which obviously is not accurate due to heterogeneity of the land surface and its features (river, land, or forest). Thus, the number of palm trees was calculated from the satellite image programmatically. Taking advantage of the accuracy of the spatial resolution of the satellite image and the ability of software recognition and characteristics of the palm tree, a systematic top view can be distinguished from the satellite image, besides the manner of cultivation, vertical growth and stability for long periods of time (Fig. 14). In a general sense, automatic tree counting involves image processing techniques, such as object segmentation (Khai et al., 2017; Emad and Al-Helaly, 2018), for which researchers have devised many automatic and semi-automatic approaches. Single trees are detected from high spatial resolution images of different types of sensors. However, it is not an easy or simple process to detect individual trees using remote sensing. There are many limitations, like the presence of small ground vegetation and tree canopy overlapping (Maillard and Gomes, 2016). Nevertheless, sustainability of date palm requires continuous monitoring of spatio-temporal variations in these trees over time. So modern geo-spatial technologies provide a fast, cheaper and accurate mechanism utilizing high spatial-resolution remote sensing data integrated with image processing technique and GIS functionalities in automatically extracting palm trees. This method was used to count date palm trees in Al-Ahssa region of Saudi Arabia and it successfully extracted date palm tree locations with an overall accuracy of 93% according to Shawky and Chockalingam (2020) (Fig. 15). On the other hand, tree counting is an important and necessary practice for yield estimation and monitoring, replanting and layout planning. For instance, in monitoring oil palm plantation, remote sensing has been used in various applications, including land-cover classification, automatic tree counting, change detection, age estimation, above ground biomass (AGB), carbon and yield estimation; also, pest and disease detection (Khai et al., 2017).

Moreover, Geographic Information System (GIS) was used to map the distribution of red palm weevil (*Rhynchophorus ferrugineus*) which is considered an important palm pest and to determine all the locations

Fig. 14. A: The top view of palm. B: Palm trees upper view (Emad and Al-Helaly, 2018).

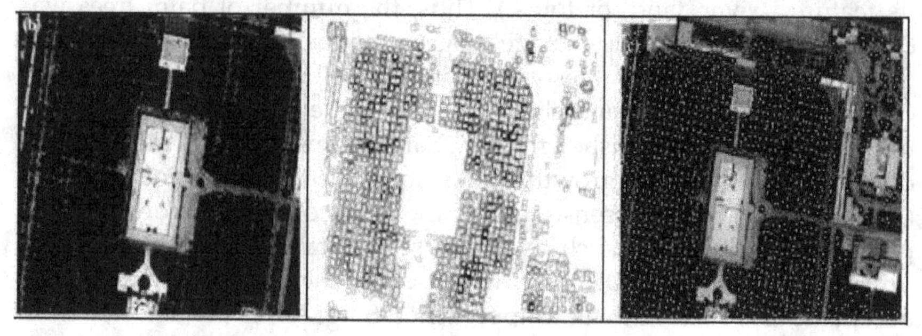

Fig. 15. Illustrates identification of date palm trees: represents tree crowns (Shawky and Chockalingam, 2020).

with the highest concentrations of this pest in the Caribbean region (Roda et al., 2011). Furthermore, in Oman, GIS was used to study infestations by the Dubas bug (DB), *Ommatissus lybicus* Bergevin, in date palm from 2006 to 2015. Results showed that the distribution pattern varied considerably with time and space. There was a relationship between date palm tree density and *O. lybicus* infestation. Dubas bug is found to be the main pest attacking date palm crops. Therefore, remote sensing and geographic information systems were used to determine the number of date palm trees in traditional agriculture locations by using high spatial resolution satellite imagery. However, only a few studies have counted date palm trees *P. dactylifera*. Many studies have counted oil palm trees and coconut palm trees. Moreover, other studies have counted and classified different fruit trees, such as various citrus tree varieties, olive trees and mango *Mangifera indica* Linnaeus. Furthermore, this technology has been adapted to count and classify different forest trees, such as *Cupressus sempervirence*, and pine forest tree (Al-Yahyai and Khan, 2015; Agustin et al., 2016; Khalifa et al., 2017; Rashid et al., 2018).

Satellite and digital imagery offer quick and precise information to observe vegetation changes at different periods (Howari, 2003). Moreover, mapping the vegetation through change detection depends on vegetation indices (VI), such as Soil Adjusted Vegetation Index (SAVI) and Normalized Difference Vegetation Index (NDVI). These indices are based on reflectance in the visible region of the electromagnetic spectrum. Then VI is correlated to the vegetation biomass to detect any decline in the plant pigment or condition. Also, response of the date palm to salinity that needs a long time to detect can be measured. In a study by Al Hammadi and Glenn (2008), two types of satellite sensors—(TM) and (ETM+)—were used to study the health of date palm trees grown in a salt-affected environment. A serious decrease in date palm trees' health was associated with increased salinity levels. Also Al Hammadi, 2010 and Al Hammadi and Kurup, 2012 used a high resolution satellite sensor, QuickBird 60 × 60 cm, to assess the growth rate of 18 cultivars of date palm trees grown in three salinity levels (5, 10, 15 dS m^{-1}) during the years 2003 and 2008. Results showed that QuickBird sensor was able to detect significant variations in the growth of date palm cultivars at the three salinity levels. Furthermore, the health of the date palm trees was highest in the year 2003 but subsequently it started to decrease over time. Rationally, using high-resolution sensors for future studies is recommended.

5.3 *Adaptation and Natural Resources Conservation*

Agriculture and the future of global food security are the most important issues in climate change negotiations. Agriculture, rural livelihoods, sustainable management of natural resources and food security are strongly related with the development and climate change challenges of the twenty-first century. We can achieve successful adaptation and mitigation responses in agriculture within the ecologic, economic and social sustainability goals of the World Food Summit, the Millennium Development Goals and the UNFCCC. There is insufficiency between the climate and socio-economic systems on which greenhouse gas emissions depend. So, we are obligated to confront a degree of climate change and its negative impacts, regardless of the mitigation strategy used. Hence, the sooner the mitigation activities begin, the lower would be the probable effects (FAO, 2012).

Adaptation to climate change is described as management of ecological, economic and social systems in response to observed or expected changes in climatic motivations and their effects. Also we can define adaptation as a series of access and actions that man takes to face climate change. However, it is a long-term strategy for the coming generations. So, adaptation is needed to minimize the impact of climate change. For instance, remote sensing techniques have been used for environmental monitoring purposes and decision support guides. There are two main practices of adaptation—autonomous and planned adaptation. Autonomous adaptation is mainly affected by ecological changes in natural systems whereas planned adaptation is the result of a thoughtful policy decision, creation of awareness and applying tools of adaptation (IPCC, 2007). Moreover, there is the necessity to compare the cost and benefits of adaptation strategies for creating equilibrium in the society (Sathaye et al., 2006).

So adaptation strategies are those that enable better management of climate change, the effects of which will serve to exacerbate anthropogenic-related stresses in natural communities. Twelve common adaptation strategies have been identified and used to begin development of more specific adaptation strategies. Some common adaptation strategies are:

a) Create special protections (for keystone species, corridors, processes, habitats).
b) Reduce anthropogenic stresses (e.g., pollution, development, over-harvest).
c) Ensure genetic diversity (species or ecosystems).
d) Restore altered ecosystems to original function.
e) Create or maintain resources.
f) Relocate species.
g) Anticipate and prepare for shifting wildlife movement patterns.
h) Maintain key ecosystem services.
i) Monitor and plan to support adaptive management goals.
j) Ensure legislative and regulatory flexibility to climate change.
k) Increase awareness and knowledge of stakeholders.
l) Promote sustainable use of resources (subsistence, recreational and commercial).

These categories can be applied to a wide variety of situations, habitats and species. In a study on Sudanese and Moroccan date palm cultivars, Sakina (2009) showed that cultivars and phenotypes possess specific direct or interaction effects due to water availability on a range of morphological and physiological traits. Soft and dry phenotypes responded differently to different levels of water stress, while the dry phenotype was more sensitive and conservative. Also, results indicated that date palm has high fixation capacity to photosynthetic CO_2 supply with interaction effect to water availability, which can be considered as advantageous when coping with stresses that may arise with climate change. Although a large amount of diversity exists among date palm germplasm, the role of biological nature of the tree, isolation by distance and environmental effects on structuring date palm genome was highly influenced by human impacts. Identity of date palm cultivars as developed and manipulated by date-palm growers, in the absence of scientific breeding programs, may continue to depend mainly on tree morphology and fruit characters. The pattern of genetic differentiation may cover specific morphological and physiological traits that contribute to adaptive mechanisms in each phenotype. These traits can be considered for further studies related to drought adaptation in date palm.

The main objective of Egypt's National Strategy for Adaptation to Climate Change and Disaster Risk Reduction is to enhance the flexibility of the Egyptian society to the risks and disasters caused by climate change and its effect on different sectors and activities. So, the strategy provides a summary of the efforts of monitoring, evaluation and follow-up at two levels: (a) efforts on the ground related to climate change phenomena, including the efforts made to handle the issue of uncertainty and, (b) monitoring, evaluation and follow-up of the application of the strategy for measuring and identifying the required future steps and measures. Furthermore, it is necessary to emphasize the dynamic nature of this strategy, that is, it needs to be updated, depending on the changes and developments that may occur in the projections and potential impact. It is also necessary to emphasize that the strategy presents itself as a guide to all state institutions and the different sectors during the development of their own operational action plans (UNDP, 2011).

6. Conclusion and Prospects

As a result of climate change, some cultivated areas may become unsuitable for cropping and some tropical grassland may become more and more arid. In sub-Saharan Africa alone, projections predict a loss of 10–20 million hectares of land available for double cropping and 5–10 million hectares for triple cropping as a result of climate change. At the regional level, the biggest losses in cropland due to climate change are likely to be in Africa. Date palm trees have a significant impact on improving the environment for living beings, particularly in oases and desert's local communities. Date palm trees are characterized by their potent ability to grow in desert, arid and semi-arid environments. This kind of tree plays a major role in preserving the environment and combating desertification. Nevertheless, dates are highly valued as a source of food, medicine and in other industries. Nonetheless, date palm growth, productivity, health and distribution are affected by climate change. This necessitates further studies to monitor and predict the future situation of this important crop. Distinctly technical advances in bioinformatics, which facilitate collecting large data about date palm and impact of climate change, play an important role in the management and prediction. Understanding the likely potential distribution of this crop under current and future climate scenarios will enable effective preparation of appropriate strategies to manage the environmental changes.

References

Abd-Elgawad, H., Saleh, A.M., Al Jaouni, S., Selim, S., Hassan, M.O., Wadaan, M.A.M., Shuikan, A.M., Mohamed, H.S. and Hozzein, W.N. (2019). Utilization of actinobacteria to enhance the production and quality of date palm (*Phoenix dactylifera* L.) fruits in a semi-arid environment. *Sci. Total Environ.*, 665: 690–697.

Abdul-Baki, A., Aslan, S., Linderman, R., Cobb, S. and Davis, A. (2007). *Soil, Water and Nutritional Management of Date Orchards in the Coachella Valley and Bard.*

Abdullah, A.J. (2011). *Soils, Plant Growth and Crop Production – Biodiversity of Date.* USDA-ARS, 803 Iowa Avenue, Morris, MN 56267 USA. http://greenplanet.eolss.net/EolssLogn/mss/C19/E1-05/E1-05-16.

Abdul-Sattar, A.A. and Hama, N.N. (2016). Integrated management for major date palm pests in Iraq. *Emirates Journal of Food and Agriculture*, 28(1): 24–33. Doi: 10.9755/ejfa.-01-032 http://www.ejfa.me/.

Abraham, V.A., Al Shuaibi, M., Faleiro, J.R., Abozuhairah, R.A. and Vidyasagar, P.S.P.V. (1998). An integrated approach for the management of red palm weevil *Rhynchophorus ferrugineus* Oliv. – A key pest of date palm in the Middle-East. Sultan Qaboos University. *Journal of Scientific Research (Agricultural Science)*, 3: 77–83.

Abul-Soad, A.A., Mahdi, S.M. and Markhand, G.S. (2015). Date palm status and perspective in Pakistan. pp. 153–205. *In*: Al-Khayri, J.M., Jain, S.M. and Johnson, D.V. (eds.). *Date Palm Genetic Resources and Utilization*. vol. 2, Asia and Europe, Springer, Netherlands.

Agustin, S., Devi, P.A.R., Sutaji, D. and Fahriani, N. (2016). Oil palm age classification on satellite imagery using fractal-based combination. *J. Theor. Appl. Inf. Technol.*, 89: 18.

Ait-El-Mokhtar, M., Ben Laouane, R., Anli, M., Boutasknit, A., Wahbi, S. and Meddich, A. (2019). Use of mycorrhizal fungi in improving tolerance of the date palm (*Phoenix dactylifera* L.) seedlings to salt stress. *Sci. Hortic.*, 253: 429–438. Doi: 10.1016/j.scienta.2019.04.066.

Al-Farsi, M., Morris, A. and Baron, M. (2007). Functional properties of Omani dates (*Phoenix dactylifera* L.). *Acta Hort.*, 736: 479–487.

Al Hammadi, M.S. and Glenn, E.P. (2008). Detecting date palm trees health and vegetation greenness change on the eastern coast of the United Arab Emirates Using SAVI. *International Journal of Remote Sensing*, 29: 1745–1765.

Al Hammadi, M.S. (2010). Using QuickBird Satellite Images to Study the Salinity Effect on Date Palm Field. *International Congress Geotunis*, 29 November–3 December 2010, Tunis.

Al Hammadi, M.S. and Kurup, S.S. (2012). Impact of salinity stress on date palm (*Phoenix dactylifera* L.)—A review. Dr. Peeyush Sharma (ed.). *Crop Production Technologies*. ISBN: 978-953-307-787-1, InTech. Available from: http://www.intechopen.com/books/crop-production-technologies/impact-of-salinitystress-on-date-palm-phoenix-dactylifera-l-a-review.

Ali, A., Al-Kindi, M. and Al-Said, F. (2009). Chemical composition and glycemic index of three varieties of Omani dates. *International Journal of Food Science and Nutrition*, 60(S4): 51–62.

Al-Jboory, I.J. (2007). Survey and identifi cation of the biotic factors in date palm environment and its application for designing IPM-program of date palm pests in Iraq, University of Aden. *Journal of Natural and Applied Sciences*, 11: 423–457.

Al Kharusi, L., Al Yahyai, R. and Yaish, M.W. (2019a). Antioxidant response to salinity in salt-tolerant and salt-susceptible cultivars of date palm. *Agric. Basel*, 9: 8. Doi: 10.3390/agriculture9010008.

Allaith, A.A.A. (2008). Antioxidant activity of Bahreini date palm (*Phoenix dactylifera*) fruit of various cultivars. *International Journal of Food Science and Technology*, 43(6): 1033–1040.

Allbed, A., Kumar, L. and Shabani, F. (2017). Climate change and agriculture research paper climate change impacts on date palm cultivation in Saudi Arabia. *Journal of Agricultural Science*, 155: 1203–1218, Cambridge University, Doi: 10.1017/S0021859617000260.

Al-Mssallem, M.Q., Al-Qurashi, R.M. and Al-Khayri, J.M. (2019). Bioactive compounds of date palm (*Phoenix dactylifera* L.). pp. 1–15. *In*: Murthy, H.N. and Bapat, V.A. (eds.). *Bioactive Compounds in Underutilized Fruits and Nuts, Reference Series in Phytochemistry*. Springer, Cham. https://doi.org/10.1007/978-3-030-06120-3_6-1.

Amber, P. (2017). Leading Countries Growing Dates (Fresh Date Palm Fruits) April 25, 2017 in Economics. *World Atlas*. https://www.worldatlas.com/articles/world-leading-countries-growing-fresh-dates.html.

Anup, P.U. and Bijalwan, A. (2015). Climate change adaptation: services and role of information communication technology (ICT) in India. *American Journal of Environmental Protection*, 4(1): 70–74. Doi: 10.11648/j.ajep.20150401.20.

Arab, L., Kreuzwieser, J., Kruse, J., Zimmer, I., Ache, P., Alfarraj, S. et al. (2016). Acclimation to heat and drought—lessons to learn from the date palm (*Phoenix dactylifera*). *Environmental and Experimental Botany*, 125: 20–30. Doi: 10.1016/j.envexpbot.2016.01.003.

Al-Yahyai, R. and Khan, M.M. (2015). Date palm status and perspective in Oman. pp. 207–240. *In*: *Date Palm Genetic Resources and Utilization*. Springer: Dordrecht, The Netherlands.

BCSnoticias. (2019). *BCS sufre su peor incendio forestal: fuego esfumó casi 72 hectáreas de palmares en San Ignacio*. Available online: https://www.bcsnoticias.mx/bcs-sufre-su-peor-incendio-forestal-fuego-esfumo-casi-72-hectareas-de-palmares-en-san-ignacio/; accessed on 1 October 2020; (in Spanish).

Blumberg, D. (2008). Review: Date palm arthropod pests and their management in Israel. *Phytoparasitica*, 36(5): 411–448.

Borchert, D.M. (2009). *New Pest Advisory Group Report on Rhynchophorus Ferrugineus (Olivier): Red Palm Weevil*. USDA-APHIS-PPQ-CPSHTPERAL.

Boubaker, D., Salah, M.B. and Frija, A. (2018). Date Palm value chain analysis and marketing opportunities for the Gulf Cooperation Council (GCC) countries. *Agricultural Economy*. Doi: 10.5772/intechopen.82450.

Chao, T.C. and Krueger, R.R. (2007). The date palm (*Phoenix dactylifera* L.): Overview of biology, uses, and cultivation. *Horticulture Science*, 42(5): 1077–1082.

Darfaoui, E. and Assiri, A. (2009). *Response to Climate Change in the Kingdom of Saudi Arabia*. Internal Working Paper FAO RNE, Cairo, Egypt: FAO.

Diaz, J., De las Cagigas, A. and Rodriguez, R. (2003). Micronutrient deficiencies in developing and affluent countries. *Eur. J. Clin. Nutr.*, 57: S70–S72.

Djibril, S., Mohamed, O.K., Diaga, D., Diégane, D., Abaye, F.B., Maurice, S. and Alain, B. (2005). Growth and development of date palm (*Phoenix dactylifera* L.) seedlings under drought and salinity stresses. *Afri. J. Biotechnol.*, 4: 968–972.

Elinformanate, B.C.S. (2019). *Se cumplen 60 años del destructiveciclon de 1959 en baja California sur*. Available online: https://elinformantebcs.mx/se-cumplen-60-anos-del-destructivo-ciclon-de-1959-en-baja-california-sur/; accessed on 1 October 2020 (in Spanish).

El-Shafie, H.A.F. (2012). Review: List of arthropod pests and their natural enemies identified worldwide on date palm, *Phoenix dactylifera* L. *Agriculture and Biology Journal of North America*, 3(12): 516–524.

Elshibli, S., Elshibli, E.M. and Korpelainen, H. (2016). Growth and photosynthetic CO_2 responses of date palm plants to water availability. *Emirates Journal of Food and Agriculture*, 28: 58–65.

Emad, A.A. and Al-Helaly, N.A. (2018). A Count of Palm Trees from Satellite Image. *Contemporary Engineering Sciences*. https://www.researchgate.net/publication/327471227.

Erskine, W., Mostafa, A.T. and Osman, A.E. (2004). *Date Palm in GCC Countries of the Arabian Peninsula*. International Center for Agriculture Research in Dry Area (ICARADA).

htpp://www.icarda.org/APRP/Date palm/Introduction/intro-body.htm; accessed on Oct 10, 2006.

Farooqi, A.B., Khan, A.H. and Mir, H. (2005). Climate change perspective in Pakistan. *Pakistan J., Meteorol.*, p. 2.

Food and Agriculture Organization of the United (FAO). (2012). *Climate Change Adaptation and Mitigation, Challenges and Opportunities in the Food Sector*. Natural Resources Management and Environment Department.

FAO, Food and Agriculture Organization of the United Nations. (2020). Available online: http://www.fao.org/news/story/es/item/1184721/icode/; accessed on 5 October 2020) (in Spanish).

FAOSTAT. (2010). *Crop Production*. Statistics Division, Food and Agriculture Organization of the United Nations, Rome. http://faostat.fao.org.

FAOSTAT. (2012). *Food and Agricultural Commodities Production*. Available at: http://faostat.fao.org/site/567/default.aspx#ancor; accessed on 29 Apr. 2015.

Geoff, O., Rose, P.O.J. and Wisner, B. (2006). Climate change and disaster management. *Disasters*, 30(1): 64–80, Overseas Development Institute, Published by Blackwell Publishing, 9600 Garsington Road, Oxford, OX4 2DQ, UK and 350 Main Street, Malden, MA 02148, USA.

Glasner, B., Botes, A., Zaid, A. and Emmens, J. (2002). Date harvesting, packinghouse management and marketing aspects. pp. 177–208. *In*: Zaid, A. (ed.). *Date Palm Cultivation*. Food and Agriculture Organization, Plant Production and Protection paper No. 156. Food and Agriculture Organization of the United Nations, Rome, Italy.

Gotch, T., Noack, D. and Axford, G. (2006). Feral tree invasions of desert springs. *Abstracts*, Third International Date Palm Conference, Abu Dhabi, United Arab Emirates, 19–21 Feb., 2006, p. 40, United Arab Emirates University, Al-Ain, U.A.E.

Howari, F.M. (2003). The use of remote sensing data to extract information from agricultural land with emphasis on soil salinity. *Australian Journal of Soil Research*, 41: 1243–1253.

Hussain, N., Al-Rasbi, S., Al-Wahaibi, N.S., Al-Ghanum, G. and El-Sharief Abdalla, O.A. (2012). *Salinity Problems and Their Management in Date Palm Production. In*: Manickavasagan, A., Essa, M.M. and Sukumar, E. (eds.). *Dates: Production, Processing, Food and Medicinal Value*. Boca Raton, FL: CRC Press, 442.

Hussein, J.S., Abdi, G. and Fahad, S. (2020). Change in photosynthetic pigments of Date palm offshoots under abiotic stress factors. *Folia Oecologica*, 47(1): 45–51.

Hussein, J.S. and AL-Khayri, J.M. (2021). Salt and drought stress exhibits oxidative stress and modulated protein patterns in roots and leaves of date palm (*Phoenix dactylifera* L.). *Acta Agriculturae Slovenica*, 117(1): 1–10, Ljubljana.

IPCC: Climate Change. (2007). The physical science basis. *In*: Solomon, S.D. et al. (eds.). *Contribution of Working Group I to the Fourth Assessment Report of the Intergovernmental Panel on Climate Change*. Cambridge University Press, Cambridge.

Ismail, B., Haffar, I., Baalbaki, R., Mechref, Y. and Henry, J. (2006). Physico-chemical characteristics and total quality of five date varieties grown in United Arab Emirates. *International Food Science and Technology*, 41: 919–926.

Jain, S. (2011). Prospects of in vitro conservation of date palm genetic diversity for sustainable production. *Emirates Journal of Food and Agriculture*, 23: 110–119.

Johnson, D.V. and Hodel, D.R. (2007). Past and Present Date Varieties in the united States, proceedings of the third international date palm conference, IIIrd IC on Date Palm A. Zaid et al. (eds.). *Acta Hort.*, 736: ISHS 2007, 39–586.

Khai, L.C., Kanniah, K.D., Pohl, C. and Tan, K.P. (2017). A review of remote sensing applications for oil palm studies. *Geo-spatial Information Science*, 20(2): 184–200. https://doi.org/10.1080/10095020.2017.1337317.

Khalifa, M.A., Kwan, P., Andrew, N.R. and Welch, M. (2017). Modeling spatiotemporal patterns of Dubas bug infestations on date palms in northern Oman: A geographical information system case study. *Crop Protection*, 93: 113–121.

Kosek, M., Black, R.E. and Keusch, G. (2005). Nutrition and Micronutrients in Tropical Infectious Diseases. *In*: Guerrant, R.L., Walker, D.H. and Weller, P.F. (eds.). *Tropical Infectious Diseases: Principles, Pathogens, and Practice*. 2nd ed. Elsevier, Amsterdam.

Lamaoui, M., Jemo, M., Datla, R. and Bekkaoui, F. (2018). Heat and drought stresses in crops and approaches for their mitigation. *Frontiers in Chemistry*, 6: 1–14. https://doi.org/10.3389/fchem.2018.00026.

Maged, E.A.M., Alhajhoj, M.R., Ali-Dinar, H.M. and Munir, M. (2020). Impact of a novel water-saving subsurface irrigation system on water productivity, photosynthetic characteristics, yield, and fruit quality of date palm under arid conditions. *Agronomy*, 10: 1265. Doi:10.3390/agronomy10091265 www.mdpi.com/journal/agronomy.

Maillard, P. and Gomes, M.F. (2016). Detection and counting of orchard trees from the images using a geometrical-optical model and marked template matching. *ISPRS Ann. Photogramm. Remote Sens. Spat. Inf. Sci.*, 3: 75.

Mansouri, A., Embarek, G., Kokkalou, E. and Kefalas, P. (2005). Phenolic profile and antioxidant activity of the Algerian ripe date palm fruit (*Phoenix dactylifera*). *Food Chem.*, 89: 411–420.

Mathur, S., Agrawal, D. and Jajoo, A. (2014). Photosynthesis: response to high temperature stress. *Journal of Photochemistry and Photobiology B: Biology*, 137: 116–126. https://doi.org/10.1016/j.jphotobiol.2014.01.010.

Munns, R. and Tester, M. (2008). Mechanisms of salinity tolerance. *Annual Review of Plant Biology*, 59: 651–681. Doi: 10.1146/annurev.arplant.59.032607.092911.

Nazir, H.M.H. (2010). *Date Palm Global Development Plan: Global, Regional and Country Level Strategies for Date Palm Development keeping in View Past and Present Scenarios as well as Expected Climatic Changes*; especially formulated for Khalifa international date palm award, UAE, third session. https://www.researchgate.net/publication/279868772.

Nutan, K.K., Rathore, R.S., Tripathi, A.K., Mishra, M., Pareek, A. and Singla-Pareek, S.L. (2020a). Integrating dynamics of yield traits in rice in response to environmental changes. *Journal of Experimental Botany*, 71: 490–506.

Ortiz-Uribe, N., Salomón-Torres, R. and Krueger, R. (2019). Date palm status and perspective in Mexico. *Agriculture*, 9(46). https://doi.org/10.3390/agriculture9030046.

Poulopoulos, S.G. (2016). Chapter 2: *Atmospheric Environment, Environment and Development, Basic Principles, Human Activities and Environmental Implications*, pp. 45–136. https://doi.org/10.1016/B978-0-444-62733-9.00002-2.

Radiokashana. (2019). *Plaga en palmar de San Ignacio*. Available online: https://www.radiokashana.org/noticias/plaga-en-el-palmar-de-san-ignacio/ accessed on 1 October 2020 (in Spanish).

Rashid, H.A.S., Kumar, L., Al-Khatri, S.A.H., Albahri, M.M. and Alaufi, M.S. (2018). *Relationship of Date Palm Tree Density to Dubas Bug Ommatissuslybicus Infestation in Omani Orchards Agriculture*, 8: 64. Doi: 10.3390/agriculture8050064 www.mdpi.com/journal/agriculture.

Rastogi, S., Shah, S., Kumar, R., Vashisth, D., Akhtar, M.Q., Kumar, A. and Shasany, A.K. (2019). Ocimum metabolomics in response to abiotic stresses: cold, flood, drought and salinity. *PLoS ONE*, 14: 1–26. https://doi.org/10.1371/journal.pone.0210903.

Rivera, D., Johnson, D., Delgadillo, J., Carrillo, M.H., Obón, C., Krueger, R., Alcaraz, F., Ríos, S. and Carreño, E. (2013). Historical evidence of the Spanish introduction of date palm (*Phoenix dactylifera* L., Arecaceae) into the Americas. *Genetic Resources and Crop Evolution*, 60(4): 1433–1452. https://doi.org/10.1007/s10722-012-9932-5.

Roda, A., Kairo, M., Damian, M.T., Franken, F., Heidweiller, K., Johanns, C. and Mankin, R. (2011). Red palm weevil (*Rhynchophorus ferrugineus*), an invasive pest recently found in the Caribbean that threatens the region. *Journal Compilation, Bulletin*, 41: 116–121.

Saafi, E.B., El-Arem, A., Issaoui, M., Hammi, H. and Achour, L. (2009). Phenolic potent and antioxidant activity of four date palm (*Phoenix dactylifera* L.) fruit varieties grown in Tunisia. *International Journal of Food Science and Technology*, 44: 2314–2319.

Saidi, Y., Finka, A. and Goloubinoff, P. (2011). Heat perception and signaling in plants: A tortuous path to thermotolerance. *New Phytol.*, 190(3): 556–65. https://doi.org/10.1111/j.1469-8137.2010.03571.x PMID: 21138439.

Saker, M.M. and Moursy, H.A. (2003). *Transgenic Date Palm: A New Era in Date Palm Biotechnology*, Proceedings of the International Conference on Date Palm, King Saud University, Qaseem, Kingdom of Saudi Arabia, 453–471.

Sakina, E. (2009). *Genetic Diversity and Adaptation of Date Palm* (*Phoenix dactylifera* L.). Faculty of Agriculture and Forestry, University of Helsinki, Finland, Academic Dissertation, Electronic version at http://ethesis.helsinki.fi.

Salomon-Torres, R., Ortiz-Uribe, N., Villa-Angulo, R., Villa-Angulo, C., Norzagaray-Plasencia, S. and Garcia-Verdugo, C. (2017). Effect of pollenizers on production and fruit characteristics of date palm (*Phoenix dactylifera* L.) cultivar Medjool en Mexico. *Turkish Journal of Agriculture and Forestry*, 41: 338–347.

Salomón-Torres, R., Ortiz-Uribe, N., Sol-Uribe, J.A., Villa-Angulo, C., Villa-Angulo, R., Valdez-Salas, B., García-González, C., Iñiguez Monroy, C.G. and Norzagaray-Plasencia, S. (2018). Influence of different sources of pollen on the chemical composition of date (*Phoenix dactylifera* L.) cultivar Medjool in México. *Australian Journal of Crop Science*, 12(6): 1008–1015. https://doi.org/10.21475/ajcs.18.12.06.PNE1213.

Samuel, H. and Smith, J. (2004). Estimating global impacts from climate change. *Global Environmental Change*, 14(3): 201–218. https://doi.org/10.1016/j.gloenvcha.2004.04.010.

Sathaye, J., Shukl, P.R. and Ravindranath, N.H. (2006). Climate Change, Sustainable development and India: Global and national concerns. *Current Science*, 90(3): 314–325.

Shabani, F., Kumar, L. and Taylor, S. (2012). Climate change impacts on the future distribution of date palms: A modeling exercise using CLIMEX. *PLoS ONE*, 7: e48021. Doi:10.1371/journal.pone.0048021.

Shabani, F., Kumar, L. and Esmaeili, A. (2013). Use of CLIMEX, land use and topography to refine areas suitable for date palm cultivation in Spain under climate change scenarios. *Earth Science and Climatic Change*, 4: 4.

Shabani, F., Kumar, L. and Taylor, S. (2014b). Projecting date palm distribution in Iran under climate change using topography, physicochemical soil properties, soil taxonomy, land use, and climate data. *Theoretical and Applied Climatology*, 118: 553–567.

Shabani, F., Kumar, L., Nojoumian, A.H., Esmaeili, A. and Togheyani, M. (2015). Projected future distribution of date palm and its potential use in alleviating micronutrient deficiency. *J. Sci. Food Agric.*, 96: 1132–1140.

Shabani, F., Kumar, L., Nojoumian, A.H., Esmaeili, A. and Toghyani, M. (2016). Projected future distribution of date palm and its potential use in alleviating micronutrient deficiency. *J. Sci. Food Agric.*, 96: 1132–1140. www.soci.org.

Shareef, H.J. (2019). Salicylic acid and potassium promote flowering through modulating the hormonal levels and protein pattern of date palm *Phoenix dactylifera* L. *Sayer* offshoots. *Acta Agriculturae Slovenica*, 114: 231–238. https://doi.org/10.14720/aas.2019.114.2.8.

Shawky, M. and Chockalingam, J. (2020). Diagnostically counting palm date trees in Al-Ahssa Governorate of Saudi Arabia: An integrated GIS and remote sensing processing of IKONOS imagery. *Spat. Inf. Res.*, 28(5): 579–588. https://doi.org/10.1007/s41324-020-00318-w.

SIAP. (2020). *Servicio de InformacionAgroalimentaria y Pesquera*. Available online: https://nube.siap.gob.mx/cierreagricola/; accessed on 7 October 2018 (in Spanish).

Soroker, V., Blumberg, D., Haberman, A., Hamburger-Rishard, M., Reneh, S., Talebaev, S., Anshelevich, L. and Harari, A.R. (2005). Current status of RPW in date palm plantations in Israel. *Phytoparasitica*, 33: 97–106.

Suliman, A.A. (2019). *Action Plan for Promoting Date Palm Cultivation in the Kingdom of Saudi Arabia*. Ministry of Environment water and Agriculture, Kingdom of Saudi Arabia. https://www.maff.go.jp/e/policies/env/attach/pdf/climate_smart_ws_2019-14.pdf.

Tarai, N., Mihi, A. and Belhamra, M. (2014). The effects of climate change on the date palm productivity at Biskra Oasis, South Algeria. *Journal of Earth Science and Climatic Change*, 5(6). http://dx.doi.org/10.4172/2157-7617.S1.016.

Tinajero, R. and Rodríguez-Estrella, R. (2015). *Cotorra argentina (Myiopsittamonachus), especie anidando con éxito en el sur de la Península de Baja California. Acta Zoológica Mexicana*, 31(2): 190–197 (in Spanish). http://www.scielo.org.mx/scielo.php?script=sci_arttext&pid=S0065-17372015000200006&lng=es&nrm=iso&tlng=es.

UNDP, United Nations Development Programme. (2011). *Egypt's National Strategy for Adaptation to Climate Change and Disaster Risk Reduction*.

USGCRP. (2014). Hatfield, J., Takle, G., Grotjahn, R., Holden, P., Izaurralde, R.C., Mader, T., Marshall, E. and Liverman, D. (2014). *Ch.* 6: *Agriculture, Climate Change Impacts in the United States: The Third National Climate Assessment*. pp. 150–174. Melillo, J.M., Terese (T.C.) Richmond and Yohe, G.W. (eds.). U.S. Global Change Research Program.

Vyawahare, N., Pujari, R., Khsirsagar, A., Ingawale, D., Partil, M. and Kagathara, V. (2009). Phoenix dactylifera: An update of its indigenous uses, phytochemistry and pharmacology. *Internet Journal of Pharmacology*, 7: 1.

Waqas, W., Faleiro, J.R. and Miller, T.A. (2015). Sustainable pest management in date palm: current status and emerging challenges. *Sustainability in Plant and Crop Protection*. ISBN 978-3-319-24395-5 ISBN 978-3-319-24397-9 (eBook). Doi: 10.1007/978-3-319-24397-9. www.springer.com.

Wheeler, T. and Braun, J.V. (2013). *Climate Change Impacts on Global Food Security*. Walker Institute for Climate System Research, Department of Agriculture, University of Reading, Reading, UK. t.r.wheeler@reading.ac.uk. 2; 341(6145): 508–13. Doi: 10.1126/science.1239402.

Yaish, M.W. (2015). Proline accumulation is a general response to abiotic stress in the date palm tree (*Phoenix dactylifera* L.). *Genet. Mol. Res.*, 14: 9943–9950. Doi: 10.4238/2015. August, 19.30.

Zaid, A. and Arias Jimenez, E.J. (2002). *Date Palm Cultivation. FAO Plant Production and Protection Paper 156*. FAO Plant Production and Protection Division, Rome.

Zaid, A. (2012). *Date Palm Cultivation*. Available: http://www.fao.org/ DOCREP/006/Y4360E/y4360e07.htm/bm07.2. Accessed 2012 Mar. 15.

Zandalinas, S.I., Rivero, R.M., Martínez, V., Gómez-Cadenas, A. and Arbona, V. (2016). Tolerance of citrus plants to the combination of high temperatures and drought is associated to the increase in transpiration modulated by a reduction in abscisic acid levels. *BMC Plant Biology*, 16: 1–16.

Zatari, T.M. (2011). *Second National Communication: Kingdom of Saudi Arabia*, A Report Prepared, Coordinated by the Presidency of Meteorology and Environment (PME), Riyadh, Saudi Arabia, and submitted to the United Nations Framework Convention on Climate Change (UNFCCC), Bonn, Germany: UNFCCC.

7

Grape Cultivation for Climate Change Resilience

Muhammad Salman Haider,[1,2] *Waqar Shafqat,*[3] *Muhammad Jafar Jaskani,*[3] *Summar Abbas Naqvi*[3,*] and *Iqrar Ahmad Khan*[3]

1. Introduction

Climate of the Earth has been changing throughout history. Modern climate era began for human civilization with the sudden end of the last Ice Age about 11,700 years ago (Feynman, 2007). Climate changes are because of very small variations in Earth's orbit that change the amount of solar energy our planet receives. Climate change is the most drastic factor which involves the climate in many dimensions and is a global problem that will be around for decades and centuries to come (Trumbo and Shanahan, 2000). Carbon dioxide, the heat-storing greenhouse gas that caused global warming, has been in the atmosphere for centuries and takes time for the planet (especially the oceans) to respond to the warming. Despite increasing awareness of climate change, greenhouse gas emissions are rising steadily. In 2013, after five million years, the daily carbon dioxide content in the atmosphere was above 400 ppm for the first time in human history. The Intergovernmental Panel on Climate Change (IPCC) predicts that temperatures will rise by 5–10°C in the next century

[1] Department of Horticulture, Ghazi University, Dera Ghazi Khan, 32200, Punjab-Pakistan.
[2] Key Laboratory of Genetics and Fruit Development, College of Horticulture, Nanjing Agricultural University, 210095, Nanjing China.
[3] Institute of Horticultural Sciences, University of Agriculture, Faisalabad, Pakistan.
Emails: salman.hort1@gmail.com; waqar_shafqat@hotmail.com; jjaskani@uaf.edu.pk; iqrarahmadkhan2008@gmail.com
* Corresponding author: summar.naqvi@uaf.edu.pk

(IPCC, 2017). The effects of climate change on individual regions change over time and different social and environmental systems can weaken or adopt the change. Researchers had previously predicted that the result would be global climate change: loss of sea ice, faster sea level rise and more intense and intense heat waves (Guo et al., 2015). Long-term effects of global climate change increased the frost-free season, changed the average precipitation, heat waves and droughts (phases of irregularly hot weather), shortened the cold wave period, summer temperatures to continue rising and reduced the soil moisture (Adam et al., 1998).

Grapes (*Vitis vinifera*) are very responsive to their surrounding environment with a seasonal variation in yield. Climate is one of the significant factors in grape and wine production, affecting the grape varieties adaptability to a specific region with respect to wine type and quality. Wine composition is mainly reliant on the meso-climate and microclimate of vineyards. This means that climate-soil-variety equilibrium is the most important factor for high quality wine production (Intrigliolo and Castel, 2009). Growing grapes also depends on number of variables which directly or indirectly are affected by climate change impact. Plant growth in general raises the level of CO_2 in our atmosphere at an optimum level beneficial for plant, but drought, heat waves and solar radiation at the same time can have a negative effect (Amthor, 1995; Schultz, 2000). Grape yield is observed to be lower in drier and hot areas, while it is optimum or higher in regions at low risk to climate change. Climatic changes not just disturb yield and quality; they also disturb the vine crop phenology (Garcia-de-Cortazar and Seguin, 2004; Faisal, 2008). Temperature above optimum result in early happening of the critical plant phenological stages: budding, flowering, grape set and fruit maturity of the grapes, which create a problem in orchard management, like irrigation time and duration, time of nutrient application, etc. Climate change can completely change the properties of well-known wines (Moriondo et al., 2013). The wine growth under water-deficit or high temperatures is a big challenge for the future grapes industry. According to current climate models, an increase in temperature of 3–8°C can occur during the ripening of the fruit and this continuous temperature rise can lead to an even earlier harvest. The ability to manage the heat and dryness of plants is heavily influenced by the soil on which they grow. The effects of high temperatures and water scarcity on poor soil can be improved. Environmental factors cause stresses on the plant, affecting the grape fruit, ultimately resulting in changes in the wine that is produced. These include:

a) **Sugar content:** Temperature above the optimum stimulates the expansion of sugar contents in grapes and higher alcohol content.

b) **Acidity:** High temperatures and water stress help in malic acid degradation and improve the potassium uptake of vines, resulting in decease of wine crop acidity. Higher acidity in the wine helps in long-term preservation.

c) **Wine taste:** Climatic changes help in grape aromatic components synthesis (e.g., polyphenols), resulting in different wine flavour.

d) **Colour intensity loss:** Anthocyanin, negatively affected by water and heat stress, gives less red colour development in wine.

2. Crop Regulation and Climatic Requirement

Mediterranean region climate is ideal for grape cultivation. In America, Europe, Russia and Australia, it is grown as a temperate fruit crop, while it can be grown in the sup-tropical climate, as in Iran, Iraq, Afghanistan, Pakistan, China, North India and Israel, whereas Argentina, Chile, Kenya, Venezuela and West India grow grapes as a topical fruit crop (Saxena and Gandhi, 2015). Sandy, loamy and clayey soil with pH 6.5–7.5 and better drainage is best for grapes cultivation. Grapes bear fruit during the dry and hot period but during severe winter, the plants are under the dormant resting stage and tolerate frost. However, plants are frost-sensitive during growing periods (Jones, 2010). Plant physiological processes and shoot growth are ideal at temperature ranging between 15–35°C. Grapes at a temperature lower than 10°C do not grow and fruit well. Annual rainfall below 900 mm is the ultimate for its successful cultivation. Humidity linked with rains at the flowering and fruit ripening stages is not advantageous as it accelerates fungal diseases (Delp, 1954; Belli et al., 2010).

2.1 *The Most Urgent Climate Factors Affecting Grapes*

Water stress: Water is a critical component for plant survival, growth and development. Water stress is a physiological condition that grapes experience when there are insufficient water sources during the critical growth cycle (Hardie and Considine, 1976). The physiological response to grapes includes the development of affected cells, the occlusion of the leaf stoma, the reduction in photosynthesis and, in the worst case, the dehydration of cells, leading towards the death of the grapevines as shown in Fig. 1. In the vineyard, water acts as a universal solvent for the nutrients and minerals necessary to fulfill the important physiological functions that the grape receives from the soil by absorbing water with the nutrients (Girona et al., 2009; Basile et al., 2011). Water stress's indirect effects can reduce berry size which increases the skin to juice ratio, and anthocyanin and phenolic concentration of red grapes (Triolo et al., 2018). Hence, most viticulturists agree that proper water management and availability are beneficial for proper growth and development.

Fig. 1. Morphological changes under climate change, i.e., CK presents control plant and DT showing drought stress (*Photo credit*: M.S. Haider).

High temperature stress: Optimum temperature is the most important factor for plant growth and development of grapevine. Climate change causes high temperature extremes that can affect plant productivity (Hathfield and Prurger, 2015). Temperature above 40°C during the plant and fruit growth and development reduces the fruit set and berry size. Temperatures above optimum improve the thickness of fruit skin. The temperature seems to have the most profound impact on viticulture, as the hibernation temperature affects the beginning of the next growing season. Prolonged high temperatures can adversely affect the quality of grapes and wine, as they affect the development of berry components, such as color, taste, sugar accumulation, loss of acidity through breathing and influence the presence of other substances like aromatic compounds that impart the properties of the grapes (Orduna, 2010). The intermediate temperatures with the smallest daily variation are favorable for the growth and maturation of grapes (Ruml et al., 2012).

3. Climate Change Impacts Grapevine Physiology, Anatomy and Genetics

Todorov et al. (2003) analyzed that proteins which bind photosynthetic pigments, Rubisco and other small and large subunits of Rubisco act actively in light conditions and their efficiency becomes low under heat-stress condition. Plant respiration, membrane stability, photosynthetic activity and water relation are also adversely affected by high temperature. Temperature also affects plant metabolites and hormones. Primary gas exchange activity, such as photosynthetic rate, stomatal conductance and transpiration rate decreases under water stress (Flexas and Medrano,

2002; Centritto et al., 2003), but their response can vary with respect to environmental conditions and species (Warren and Adams, 2006; Loreto et al., 1992). In grapes, an increase in atmospheric CO_2 may result in partial stomatal closure, due to which, leaf temperature increases, leading to a negative response to photosynthetic potential (Schultz, 2000). Similarly, Moutinho-Pereira et al. (2009) proposed that higher atmospheric CO_2 reduced the stomatal density in different grapes genotypes. However, the assimilation rate of CO_2 increased significantly and reduced stomatal conductance with a higher CO_2 rate, thereby improving the intrinsic water use efficiency. Different responses to soil water availability and leaf to air vapor deficit have been observed in grapevine varieties, influencing total plant water consumption and adaptation to drought (Costa et al., 2012). These variations are related to leaf stomatal control and hydraulic properties of the shoot during water shortage (Schultz, 2003; Soar et al., 2006; Chaves et al., 2010). On the other hand, all the grapevine varieties with distinct climatic characteristics for berry quality and yield have led to climate-maturity grouping as described in ground-breaking research based on the growing temperature units of Amerine and Winkler (1944), which assembled the cool region varieties of grapes grown in California into hot region varieties and more recently by Jones et al. (2006). This grouping is associated with phenological characteristics, like the time of ripening, different environmental regimes for wine quality, etc. This has a profound impact on the optimal cultivation location of these varieties and the irrigation strategies that need to be followed. Complex physiological response of grapevine to environmental stress varies with differences in variety and affected by rootstock and scion also. The bushy rootstock has deep root system uptake of more water and nutrients (Paranychianakis et al., 2004; Koundouras et al., 2008; Pavlousek, 2011; Tramontini et al., 2013). The plant water relation is mainly influenced by stem, petiole and leaf hydraulics, shoot characters and leaf stomatal regulation (Schultz, 2003; Souza et al., 2005; Rodrigues et al., 2008; Chaves et al., 2003, 2010; Zufferey et al., 2011; Tramontini et al., 2012, 2013). The differences among varieties and their response to drought also depend on the differences in vessel size (hydraulic traits) (Collins and Loveys, 2010). There will be more bubbles under moderate drought conditions if the size of the vessel is large, and during severe drought conditions, bubbling will be continuouos and more, similar to descriptions of other woody species (Sperry and Saliendra, 1994). However, the relationship between susceptibility to embolism and their drought-tolerance mechanisms at leaf level under typical Mediterranean conditions of eight rain-fed field varieties (White Grenache, Cabernet Sauvignon, Alicante Bouschet, Tempranillo, Chardonnay, Sauvignon Blanc, Black Grenache and Parellade) was not found (Alsina et al., 2007).

4. Climate Change and Stress Management

Plants in their natural or domesticated habitat are constantly threatened by a plethora of biotic and abiotic stress factors. These factors individually or collectively may occur at any phase of plant growth and eventually limit plant development and productivity (Haider et al., 2017). In grapevine, climate change mainly influences vine growth and development and ultimately the yield and productivity. Therefore, grapevine alters their developmental activities, physiology, metabolic processes and gene expression to mitigate the stress adversity. The most common response of grapevine to biotic and abiotic stresses is the accumulation of Reactive Oxygen Species (ROS) scavenging enzymes and regulatory proteins (transcription factors).

4.1 Agronomic Practices

Cover crops are one of the primary tools for soil management and are being adopted to conserve soil water moisture for improving grapevine vigor and wine characteristics (Monteiro and Lopes, 2007; Schultz, 2007). Cover crops have advantages in minimizing the effect of intense rainfall, soil erosion, especially in sloppy lands and reduce the maintenance cost. However, the cover crop may improve soil and water quality. It increases the adaptive ability of grapevine to climate change and may also play a vital role as an ecosystem service provider (Garcia et al., 2018). According to Lopez et al. (2008), an excessive water competition was observed between swards and grapevines in Mediterranean non-irrigated vineyards. Thus, in a few regions with summer rainfall and deep soils, the swards have an integral role in using additional water.

4.2 Varietal Diversification/Rootstock Adaptation

Increasing warmer temperature is a threat to viticulture which provides an opportunity to produce very late ripening genotypes or genotypes that are ble to produce high quality wines under elevated temperatures (Duchene et al., 2010). Dutchene et al. (2010) also explored genetic variability in 120 genotypes of Riesling × Gewurztraminer cross and 14 European varieties based on phenological parameters. As a result, they created a virtual late ripening genotype, derived from a cross between Riesling and Gewurztraminer.

4.3 Breeding Strategies

The genetic improvement of the grapevine is of utmost importance to enhance yield and productivity and to nourish the remunerative wine industry worldwide. Several attempts have been made to develop hybrids and modify their genetics, but these have been largely unsuccessful because of their long lifespan, inbred depression and poor understanding of genetic control of complex enological traits (Strange and Scott, 2005; Gray et al., 2014). Rootstock breeding has been given priority since the outbreak of Phylloxera in 1879. Afterwards, much attention was devoted to develop new hybrid rootstocks that not only possess resistance to pests and diseases but also adapt to varying environmental conditions. However, the conventional breeding method is also not reliable to get promising results of improved hybrids so that they can withstand and adapt to biotic and abiotic environments. A few of the elite grapevine cultivars grown worldwide are maintained predominantly by vegetative propagation. However, often lack of durability and resistance to meet modern intensive viticulture conditions remains (Bowers et al., 1999; Myles et al., 2011).

Most of the cultivated grapevine varieties (about 80%) in commercial vineyards are asexually propagated: where a scion, taken from *V. vinifera*, is grafted on to any of the American *Vitis* rootstocks having desirable traits. The rootstocks used in commercial vineyards are the hybrids of three *Vitis* species, including *V. rupestris*, *V. riparia* and *V. berlandieri* (Whiting, 2005). Apart from Phylloxera, rootstocks can also resist other soil-borne pests (nematodes) and withstand the environmental stresses (i.e., drought, salinity, cold, heat, etc.). In the recent context of climate change, rootstocks are deliberated as the vital element of stress adaptation. Rootstocks can modify phenology, root structure and physiology of the plant (Guan et al., 2019). However, breeding of resistant rootstocks and evaluation is a long-term process. Therefore, modern biotechnology has made it possible to expedite the genetic modifications of already existing cultivars via precision breeding (Fig. 2).

4.3.1 Functional Genomics of Abiotic Stress Tolerance

Being immobile, plants alter their morphological, physiological and molecular response towards varying environmental conditions and this is termed as 'phenotypic plasticity' (Haider et al., 2019). The availability of the grapevine genome combined with next-generation sequence data, such as microarray and RNA-sequencing (RNA-seq), enables the researchers to analyze the differential gene response of different *Vitis* species in response to abiotic stress stimuli. Drought, salts and temperature (high/low) are omnipresent stressors affecting grapevine yield and quality worldwide (Haider et al., 2017; Guan et al., 2018). The transcriptome and proteome analysis revealed that grapevine exhibits diverse gene and protein patterns under an abiotic stress environment. Such information may further be used to interpret the underlying phenotypic and molecular response to improve the performance of grapevine under diverse climatic conditions (Fasoli et al., 2012).

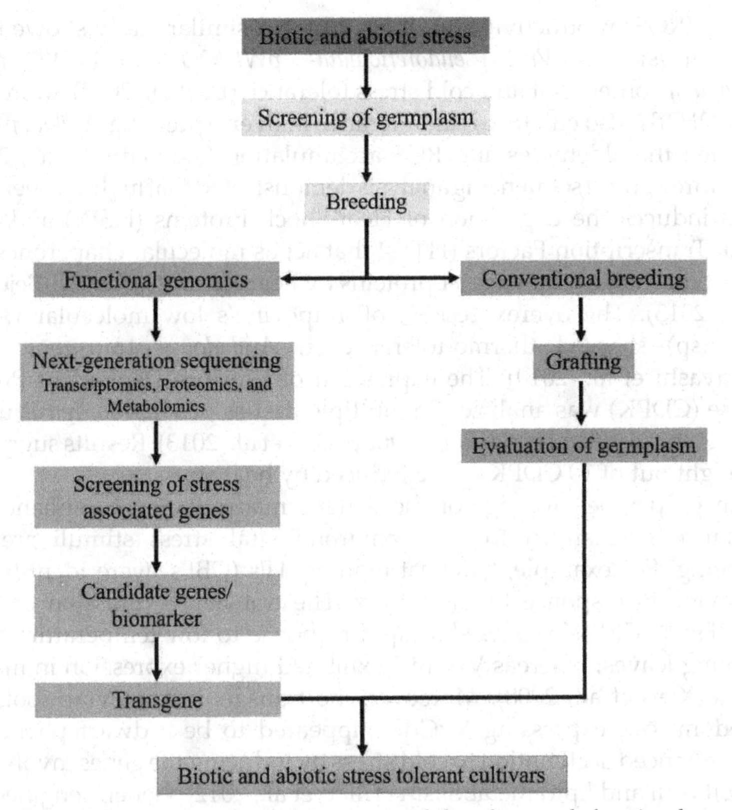

Fig. 2. A step-wise presentation of conventional breeding and functional genomics approaches to develop biotic and abiotic stress-tolerant cultivars (*Credit*: M.S. Haider).

Microarray and RNA-seq analysis of 'Cabernet Sauvignon' and 'Summer Black' grapevine cultivars in response to salts and drought stress revealed transcriptional activation of more than 2,000 genes (Cramer et al., 2007; Haider et al., 2017). Many of these genes act as transcription factors and play a critical role in various signal transductions (hormones) and metabolic (carbohydrate) pathways, which may contribute to abiotic tolerance. Diverse expressional behavior of proteins was observed in cvs. Cabernet Sauvignon and Summer Black when subjected to drought and salt stress. The up-regulation of genes involved in ROS-detoxification was correlated with the accumulation of antioxidant enzyme activities, while proteins regulating the photosynthesis process were down-regulated. This was consistent with the inhibited growth and photosynthetic process (Haider et al., 2017).

Various drought-inducible proteins were isolated from *Vitis* species, such as the overexpression of grapevine *VvNAC08* in *Arabidopsis* showed enhanced drought tolerance and limited reactive oxygen species (ROS) productivity (Ju et al., 2020). A similar study showed that overexpression of *Vitis pseudoreticulata VpWRKY1* and *VpWRKY2* in *A. thaliana* confers salt and cold stress tolerance (Li et al., 2010). Moreover, *VpALDH2B4* also enhances salt tolerance in overexpressing *Arabidopsis* by reducing the aldehydes and ROS accumulation (Skopelitis et al., 2006). High-throughput sequencing analysis demonstrated that high-temperature stress induces the expression of Heat-Shock Proteins (HSPs) and Heat Shock Transcription Factors (HTFs) that act as molecular chaperones and protect mechanical damage of proteins by heat stress (Carbonell-Bejerano et al., 2013). The overexpression of grapevine's low molecular weight (17k hsp) showed thermo-tolerance in *Arabidopsis* transgenic lines (Kobayashi et al., 2010). The expression of Calcium-Dependent Protein Kinase (CDPK) was analyzed in multiple tissues of *Vitis amurensis* under various abiotic stress conditions (Dubrovina et al., 2013). Results suggested that eight out of 10 CDPKs were induced by heat stress.

In grapevine, findings of the genetic manipulation to enhance the resistance mechanism towards environmental stress stimuli are just emerging. For example, C-repeat binding TFs (CBFs) were identified in grapevine in response to cold stress. The transient expression of three VvCBFs (VvCBF1-3) showed a rapid response to low-temperature stress in young leaves, whereas VvCBF4 exhibited higher expression in mature leaves (Xiao et al., 2008). Moreover, the transgenic grapevine rootstock 'Freedom' overexpressing VvCBF4 appeared to be a dwarf phenotype with enhanced acclimation to cold stress by inducing the genes involved in the cell wall and lipid metabolism (Tillett et al., 2012). Genetic engineering of grapevine with precocious fertility and berry characteristics is being assessed (Thomas et al., 2000; Costantini et al., 2007). However, deeper insights into genome mining along with expression profiling will provide a list of candidate genes that can be used to improve grapevine cultivars.

4.3.2 Biotic Stress Resistance-related Genes

Grapevine is susceptible to various fungal pathogens (i.e., downy and powdery mildew) and requires intensive fungicides for sustainable viticulture production worldwide (De-Francesco, 2008). The Next-Generation Sequencing (NGS) response in leaves of resistant *V. aestivalis* and susceptible *V. vinifera* grapevine species demonstrated that endogenous salicylic acid level was higher in resistant species than that of susceptible after 12 h of pathogen inoculation. Moreover, various transcripts, such as WRKY TF, Mitogen-Activated Protein Kinase Kinase (MAPKK), and Pathogen-Related (PR) proteins were significantly induced (Borges et al., 2013). A similar study on metabolomic comparison between *V. vinifera* cultivars 'Regent' (resistant to fungal infection) and 'Trincaderia' (susceptible) suggested amino acids and caffeic acid as important metabolic markers against fungal infection (Figueiredo et al., 2008). Application of *Trichoderma harzianum* also attenuates the grapevine susceptibility to powdery-mildew by triggering the transcripts encoded as Myb TFs, PR proteins along with activation enzymes involved in oxidation and phenylpropanoid pathway (Perazzolli et al., 2012). Overexpression of grapevine WRKY1 (*VvWRKY1*) in rootstock (41B) can enhance the resistance to downy-mildew by triggering the jasmonic acid (JA) signaling (Marchive et al., 2013). Other genes obtained apart from the grapevine, such as lytic peptides, are also found to enhance fungal infection resistance (Agüero et al., 2005; Vidal et al., 2006).

Grapevine productivity is also affected by bacterial diseases, including Pierce's disease (PD; *Xylella fastidiosa*), bois noir (*Candidatus Phytoplasma solani*) and crown gall disease (*Agrobacterium vitis*). However little is known about the underlying molecular events. Constitutive expression of animal-derived *Agrobacterium virE2*, magainin and lytic peptides showed less crown gall symptoms in grapevine (Vidal et al., 2006; Krastanova et al., 2010). Overexpression of hybrid derivatives of lytic peptides, including LIMA1 and LIMA2, in the table and wine grape cultivars showed enhanced resistance to PD in repeated *in-vitro* and *in-vivo* tests (Gray et al., 2010). Another experiment indicated enhanced PD resistance when a non-modified grapevine scion cultivar 'Cabernet Sauvignon' was grafted on animal-derived lytic peptide overexpressing rootstock cultivar 'Thompson Seedless' (Dutt et al., 2007). Many attempts have also been devoted to viral resistance in grapevine. The transcriptome analysis was performed to compare the artificially inoculated leafroll virus (GLRaV-3) with virus-free *V. vinifera* cultivars. The genes related to Systemic Acquired Resistance (SAR), including beta-1, 3-glucanase (GLU), and lipid transfer proteins showed distinct gene expression upon viral infection (Espinoza et al., 2007). The field trial of grafting non-modified scion on rootstock overexpressing GFLP CP showed no visual symptoms even after three years of artificial infection at two different growing sites (Vigne et al., 2004). Singh et al. (2012) conducted microRNA-seq analysis on grapevine infected with the virus and identified six novel virus-associated miRNAs.

4.4 Role of Omics in Biotic and Abiotic Stress Resilience

In plants, the biotic and abiotic stress response is governed by the regulatory mechanism of genes involved in various biological pathways and signaling cascades. Omics approaches (such as genomics, proteomics, metabolomics, metagenomics and transcriptomics) characterize the pool of plant's biomolecules and their function that play a critical role in homeostasis and signaling to alter stress effectivity (Parida et al., 2018). To date, various omics-based approaches, including transcriptomics, proteomics, metabolomics, etc. have been widely used to provide novel insights into the regulatory mechanism involved in plant response to biotic and abiotic stress factors and utilize this information in plant improvements.

4.4.1 Transcriptomics

The use of NGS (microarray and RNA-seq) has become the method of choice to study the grapevine transcriptional response to various biotic and abiotic stress factors. Most of the researchers revealed that various transcription factors act as key regulatory proteins to mitigate the stress effects. The short- and long-term drought and salinity effects were studied in shoots of grapevine cultivar 'Cabernet Sauvignon' under greenhouse conditions. Various transcripts showed higher variability in their expression after short- and long-term stress. However, many were not expressed in long-term stress, signifying that plants themselves can modify their response to withstand the stress adversity (Tattersall et al., 2007; Cramer et al., 2007). Another study demonstrated that five out of 13 ZFP TFs (zinc finger) showed differential response to drought and salinity and also regulated the expression of other stress-associated proteins (Wang et al., 2014a; Yu et al., 2016). Similarly, the *Arabidopsis* homolog of *AtMYB60* (*VvMYB60*) responds to ABA and regulates guard cell activity under drought stress in the grapevine (Galbiati et al., 2011).

Another class of transcriptional regulators (WRKY) is involved in signaling associated with multiple biotic (innate immunity) and abiotic stress factors, such as drought, heat, salinity, cold and ozone (Wang et al., 2014b). Members of this multigene are known to play defense responses as both positive and negative regulators in Effectors-Triggered Immunity (ETI) and Pathogen-Triggered Immunity (PTI) (Rushton et al., 2010). Transcriptomic analysis revealed that WRKY22, 25, 29 and 33 showed increased expression upon fungal infection (*Lasiodiplodia theobromae*) in grapevine shoots (Zhang et al., 2019), while similar genes were also involved against the downy and powdery mildew of grapevine (Merz et al., 2015).

A comprehensive study on identification and characterization of WRKY TFs in grapevine cultivar 'Pinot Noir' revealed that members of subgroup-II, including VvWRKY7-8 and VvWRKY28, were induced by drought stress, while VvWRKY8 was up-regulated in response to salt and cold stresses (Wang et al., 2014b). Not all transcription factors act as positive regulators against stress response, such as WRKY-b, MYB (MybB1-2), SPF1 and NAC1 were repressed by heat stress treatment in 'Cabernet Sauvignon' grapevine cultivar (Liu et al., 2012). However, HSP20 family was found to be up-regulated and has heat stress mitigation effect. The microarray analysis between salt-resistant 'Razegui' and susceptible 'Syrah' demonstrated that various DEGs encoding ZPFs, NACs and

Ethylene Response Factor (ERFs) were significantly up-regulated in salt-resistant variety (Daldoul et al., 2010).

4.4.2 Proteomics

The proteomic approach has been successfully implicated in many crops to investigate and identify biotic and abiotic stress-associated proteins and comprehend the role of specific proteins in stress signaling (Kosová et al., 2018). The proteomic (iTRAQ) analysis of grapevine 'Cabernet Sauvignon' leaves after heat treatment showed that proteins associated with photosynthesis Electron Transport Chain (ETC), HSPs, antioxidant enzymes and glycolysis played a vital role in grapevine adaptation to heat stress (Liu et al., 2014). Likewise, a proteomics-based study was also carried out in grapevine 'Chardonnay' in response to herbicide stress. The protein related to oxygen-evolving enhancers (photosystem II) and chlorophyll-binding protein in light-harvesting component (LHCII) were badly repressed, while PR10, glycine cleavage T-protein (photorespiration) and antioxidant enzymes were activated and encoded as biochemical markers to monitor herbicide effects in grapevine (Castro et al., 2005). Another study demonstrated the activation of ROS scavenging proteins (GRXS17, ASR2 and APX) and photosynthesis-related proteins (RuBisCO and RCA) in proteomic (2D-PAGE) analysis of two grapevine cultivars—'Houamdia' (salt-sensitive) and 'Tebaba' (salt resistant) (Azri et al., 2020).

Most of the proteomic work that has been done focusing on grapevine response to pathogen infection revealed the identification of protein associated with defense response (i.e., PR) and signal transduction (Zhao et al., 2011; Dadakova et al., 2015). The Mn-induced resistance mechanism of grapevine against powdery mildew was analyzed. The proteomic analysis suggested nucleotide-binding site leucine-rich repeat (NBS-LRR) and PR-like proteins in triggering grapevine response to pathogens (Yao et al., 2012). The analysis of differentially-expressed proteins was carried out, using the LC/MS/MS method in *X. fastidiosa* (PD) inoculated grapevine stem in the sibling pair of 9621–94 (resistant) and 9621–67 (susceptible) crosses of *V. rupestris* × *V. arizonica*. Various proteins, including thaumatin-like protein and PR10 in both genotypes, and 40S ribosomal protein S25 was up-regulated in susceptible genotypes. Moreover, formate dehydrogenase and glycoproteins were identified in resistant genotype, signifying their role in PD-resistance (Yang et al., 2011). Spagnolo et al. (2012) investigated grapevine 'cv. Chardonnay' trunk disease using proteomic and transcriptomic analysis and observed a strong correlation between the findings. Identified proteins were functionally involved in defense response, energy metabolism and stress-tolerance. The proteomics of *A. tumefaciens*-mediated transformation showed significant up-regulation of PR and resistant (*R*) proteins (Zhao et al., 2011). Recent advancements in LC/MS and the availability of grape genome data are continually improving the proteomic approach to study the tolerance mechanism of biotic and abiotic stresses.

4.4.3 Metabolomics

Metabolomics is one of the advanced branches of the omics approach executed to detect, identify and quantify different metabolites in different cells, tissues, or organs of plants under varying environmental conditions (Parida et al., 2018). Hochberg et al. (2013) compared metabolic profiles of two grapevine cultivars 'Cabernet Sauvignon' and 'Shiraz' under progressive water stress. The result indicated a significant increase in amino acids (e.g., proline, valine, leucine and threonine), while most of the organic acids were repressed. A similar study showed that secondary volatile compounds (i.e., ketones, alcohol and aldehydes) were detected, of which 3-hexenal and (E)-2-hexenal showed a marked increase in grapevine leaves under water stress (Ju et al., 2018). Moreover, proline, MDA, H_2O_2 and antioxidant enzymes were considered metabolic markers to monitor water stress. The application of ultraviolet (UV) radiations on grape berries led to the synthesis of terpenes and carotenoids, which are the vital components of good-quality wine (Matus, 2016). A recent study performed a transcriptomic and metabolic comparison on cold and anaerobic stresses in grapevine cv. 'Superior Seedless' after storage. Significant accumulation of stilbenes (i.e., piceatannol, *E*-miyabenol and *E*-ε-viniferin) and volatile compounds (i.e., diacetyl and ethyl acetate) was in stored berries (Maoz et al., 2019).

Botrytis cinerea is a causal organism of fungal disease (grey mold) in grape leaves and berries. Multi-omics analysis, including transcriptomics, proteomics and metabolomics were carried out to understand the resistance mechanism in grapevine (Agudelo-Romero et al., 2015; Dadakova et al., 2015). The results showed that carbohydrate and lipid metabolism regulated the synthesis of secondary metabolites, which is involved in plant defense response. The grapevine cultivar 'Chardonnay' inoculated with *P. minimum* was subjected to water stress and xylem sap was used for proteomic analysis. Various metabolites were modulated by a fungal infection, such as leucine, isoleucine, proline, phenylalanine, valine, tyrosine, asparagine, methionine, trigonelline and sarcosine in the xylem sap, possibly involved in esca (trunk) disease tolerance in grapevine (Lima et al., 2017). Based on these findings, the metabolic response of the grapevine to mitigate biotic and abiotic stress factors is represented in Fig. 3. Generally, metabolomics has opened the gateway to identify viable metabolic markers to uplift the breeding programs of biotic and abiotic stress-tolerant grapevine cultivars with respect to climate change.

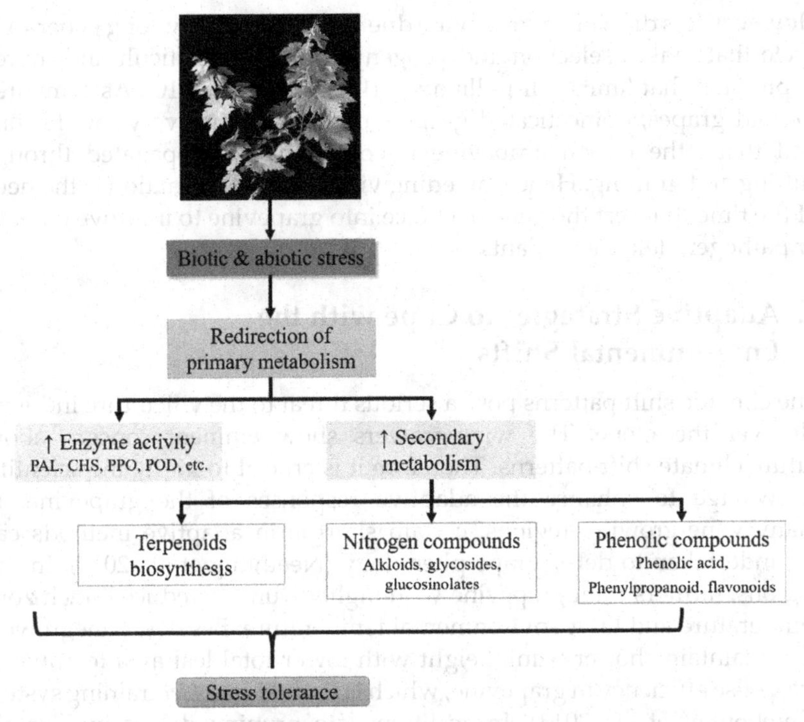

Fig. 3. A schematic complex regulatory physiological mechanism of grapevine in response to biotic and abiotic stress factors. The physiological response impacts primary metabolism that induces biosynthesis of secondary metabolism, with concomitant effect on biosynthesis of bioactive compounds (*Credit*: M.S. Haider). Abbreviations: PAL: phenylalanine ammonia-lyase, CHS: chalcone synthase, PPO: Polyphenolic oxidase, POD: peroxidase dismutase.

5. Conventional Breeding

Hybridization of resistant cultivars is one of the convenient conventional breeding techniques to develop resistant grapevine cultivars. Within genus *Vitis*, *Vitis vinifera* is abundantly used, especially in the wine industry. *V. labrusca* is used to develop hybrids for table and wine purposes while other species are used as rootstocks, according to their resistance against biotic and abiotic stress factors. The selection of rootstock with enhanced capability to cope with unfavorable environmental conditions has enhanced in the last two decades. The inter-specific grapevine cultivars (e.g., *V. vinifera*, *V. labrusca* and *V. riparia*) were bred for cold hardiness and disease resistance in cold regions of the USA since the 80s (Hazelrigg et al., 2018, 2021). Recently, breeding of scion cultivars has been observed to obtain different goals, such as quality improvement, abiotic stress resistance and resistance against pathogens (Saporta et al., 2016). It is possible to improve grape cultivar by conventional breeding techniques.

However, it is difficult in grapevine due to heterozygosity, long generation cycle that makes selection and progeny evaluation difficult and inbred depression that limits self-pollination (Louime et al., 2010). As compared to wild grapes, domesticated grape varieties conduce very low fertility and that's the reason grapevine is commercially propagated through cutting and grafting. Hence, breeding via gene transformation is the need of the time to insert the gene of choice into grapevine to improve climatic or pathogen defense in plants.

6. Adaptive Strategies to Cope with the Environmental Shifts

The climatic shift patterns pose a serious threat to the viticulture industry all over the globe. The wine-growers show eminent concern about future climate shift patterns. Therefore it is crucial to imply the scientific knowledge to enhance the adaptive responses of the grapevine. In many wine-growing regions, certain short-term adaptive methods can be undertaken to delay grape phenology (Neethling et al., 2017). In dry regions, farmers train grapevine with higher trunk to reduce bunch zone temperature and limit environmental temperature. Besides, winegrowers also maintain shorter trunk height with lower total leaf area to enhance water use efficiency in grapevine, which is termed as *goblet* training system (Lereboullet et al., 2013). In addition, late pruning delays bud break, which in turn delays the overall phenology of the grapevine (Parker et al., 2015). The vine vigor can also be improved by growing cover crops during the wet growing season and can further be used as mulching material to conserve moisture during dry season (Teixeira et al., 2013). An effective irrigation strategy that modifies vine water uptake conditions is another method to avoid climate change effects. In dry regions, deficit irrigation has shown to be effective and produced high-quality grapes (van Leeuwen and Seguin, 2006). Moreover, it has also been observed that mild water stress may increase berry quality potential; however, severe water stress affects yield but might improve the fine red color of wine (van Leeuwen and Seguin, 2006). The use of shading panels in the vineyard has been depicted to reduce canopy temperature and delay ripening (Greer and Weedon, 2013).

Balanced plant nutrition plays a critical role in improving yield and stress tolerance of crops. Generally, the nutrition requirements of grapevine are moderate; though, unfavorable environmental conditions can compromise the nutritional balance. Excessive fertilization has been reported to decrease acidity levels in wine grapes, while low nutrition increases plant oxidative state (Waraich et al., 2012). Thus balanced nutrition is a key to plant structural integrity to drive various physiological processes. Experimental evidence showed that grapevine plants grown under high-light and high nitrogen (N) showed more photo-oxidative tolerance as compared to plants grown under high-light and low N, signifying that sufficient N level can improve the plant defense mechanism (Kato et al., 2003). Moreover, exogenous application of hormones (ABA, JA, SA and BR), osmoprotectants (glycine betaine and proline) and organic compounds (melatonin) have been shown to decrease the summer stress damage in plants. Many authors demonstrated that exogenous application of ABA, SA and BRs enhanced graft union formation and improved stress tolerance (Zhang et al., 2011; Degaris et al., 2017). Another study suggested that treatment of ethylene-releasing compounds increased endogenous auxin biosynthesis, which delayed the berry ripening process and may assist to avoid summer stress (Böttcher et al., 2013). Moreover, kaolin application is a sustainable approach in a challenging climate that optimizes leaf and berry surface temperature and improves antioxidant defense system (Dinis et al., 2015). Overall, this study elucidates stress-based plant responses and underlying mechanisms that will help to fully comprehend these processes for the sustainable viticulture industry.

7. Conclusion

Climate change impacts the viticulture industry, thus alleviating the multidisciplinary research to cope with the environmental shifts. In the near future, climate change patterns will make it difficult to maintain high-quality fruit with traditional varieties. Many researchers suggested that climate change affects both the biochemistry and physiology of the grapevine along with the methodology used for the wine-making process. Grapevine is cultivated throughout the globe for raw or wine purposes; warmer temperature and water scarcity are the main issues for future industry. Therefore, devised adaptation strategies should be focused to maintain berry yield and quality under challenging environments. The precise evaluation of the adaptation artifices for changing climate patterns assists to perceive site-specific adaptation adjustments. The fanatical strategy for wine-growers is to use new varieties specially designed for the particular region or to find suitable places for already existing varieties to maintain the berry and wine quality. Previously work had been done on several aspects, i.e., agronomic practices, modeling-based varietal selection, functional genomics to omics but there is a need for trait-specific characterization of different varieties at a molecular level to understand the varietal adaptation capability against changing climate threats.

Despite the research work focusing on grapevine response to environmental stresses in controlled conditions, future research must concentrate on the data regarding how grapevine, pathogens and microorganisms will counter the high/low temperature, water deficit/logged and high CO_2 concentration under field conditions. The environmental stresses, such as high temperature and UV-radiations, primarily affect the grapevine physiology, berry biochemical composition and ultimately the wine processing. However, there are secondary effects associated with high temperature and UV radiations; for instance, the incidence of forest or bushfire can cause harm to green areas and consequently to the viticulture sector. The development of mixed knowledge (local and scientific) on the spatial and temporal response of grapevine to climate change is vital to determine the local adaptation of viticulture to climate shifts.

References

Adam, R.M., Hurd, B.H., Lenhart, S. and Leary, N. (1998). Effects of global climate change on agriculture: An interpretative review. *Climate Research*, 11: 19–30.

Agudelo-Romero, P., Erban, A., Rego, C., Carbonell-Bejerano, P., Nascimento, T., Sousa, L. and Fortes, A.M. (2015). Transcriptome and metabolome reprogramming in *Vitis vinifera* cv. Trincadeira berries upon infection with *Botrytis cinerea*. *Journal of Experimental Botany*, 66: 1769–1785.

Agüero, C.B., Uratsu, S.L., Greve, C., Powell, A.L.T., Labavitch, J.M., Meredith, C.P. and Dandekar, A.M. (2005). Evaluation of tolerance to Pierce's disease and Botrytis in transgenic plants of *Vitis vinifera* L. expressing the pear PGIP gene. *Molecular Plant Pathology*, 6: 43–51.

Alsina, M.M., De-Herralde, F., Aranda, X., Save, R. and Biel, C.C. (2007). Water relations and vulnerability to embolism are not related: experiments with eight grapevine cultivars. *Vitis-Geilweilerhof*, 46: 11–19.

Amerine, M. and Winkler, A. (1944). Composition and quality of musts and wines of California grapes. *Hilgardia*, 15: 493–675.

Amthor, J.S. (1995). Terrestrial higher-plant response to increasing atmospheric (CO_2) in relation to the global carbon cycle. *Global Change Biology*, 1: 243–274.

Azri, W., Cosette, P., Guillou, C., Rabhi, M., Nasr, Z. and Mliki, A. (2020). Physiological and proteomic responses to drought stress in leaves of two wild grapevines (*Vitis sylvestris*): A comparative study. *Plant Growth Regulation*, 91: 37–52.

Basile, B., Marsal, J., Mata, M., Vallverdú, X., Bellvert, J. and Girona, J. (2011). Phenological sensitivity of Cabernet Sauvignon to water stress: vine physiology and berry composition. *American Journal of Enology and Viticulture*, 62: 452–461.

Bellí, N., Marín, S., Coronas, I., Sanchis, V. and Ramos, A.J. (2007). Skin damage, high temperature and relative humidity as detrimental factors for *Aspergillus carbonarius* infection and ochratoxin A production in grapes. *Food Control*, 18: 1343–1349.

Böttcher, C., Burbidge, C.A., Boss, P.K. and Davies, C. (2013). Interactions between ethylene and auxin are crucial to the control of grape (*Vitis vinifera* L.) berry ripening. *BMC Plant Biology*, 13(1): 1–14.

Borges, A.F., Ferreira, R.B. and Monteiro, S. (2013). Transcriptomic changes following the compatible interaction *Vitis vinifera* necator: Paving the way towards an enantioselective role in plant defence modulation. *Plant Physiology and Biochemistry*, 68: 71–80.

Bowers, J., Boursiquot, J.M., This, P., Chu, K., Johansson, H. and Meredith, C. (1999). Historical genetics: The parentage of chardonnay, gamay, and other wine grapes of northeastern France. *Science*, 285: 1562–1565.

Castro, A.J., Carapito, C., Zorn, N., Magné, C., Leize, E., Dorsselaer, A.V. and Clément, C. (2005). Proteomic analysis of grapevine (*Vitis vinifera* L.) tissues subjected to herbicide stress. *Journal of Experimental Botany*, 56: 2783–2795.

Carbonell-Bejerano, P., María, E.S., Torres-Pérez, R., Royo, C., Lijavetzky, D., Bravo, G. and Martínez-Zapater, J.M. (2013). Thermotolerance responses in ripening berries of *Vitis vinifera* L. cv Muscat Hamburg. *Plant and Cell Physiology*, 54: 1200–1216.

Centritto, M., Loreto, F. and Chartzoulakis, K. (2003). The use of low (CO_2) to estimate diffusional and non-diffusional limitations of photosynthetic capacity of salt-stressed olive saplings. *Plant, Cell & Environment*, 26: 585–594.

Chaves, M.M., Maroco, J.P. and Pereira, J.S. (2003). Understanding plant responses to drought—from genes to the whole plant. Functional Plant Biology, 30(3): 239–264.

Chaves, M.M., Zarrouk, O., Francisco, R., Costa, J.M., Santos, T., Regalado, A.P., Rodrigues, M.L. and Lopes, C.M. (2010). Grapevine under deficit irrigation: Hints from physiological and molecular data. *Annals of Botany*, 105: 661–676.

Collins, M. and Loveys, B. (2010). Optimizing irrigation for different cultivars. GWRDC Project No. CSP 05/02, CSIRO Plant Industry, 118 pp.

Costa, R.O., Ferreiro, E., Oliveira, C.R. and Pereira, C.M. (2012). Inhibition of mitochondrial cytochrome oxidase potentiates Aβ-induced ER stress and cell death in cortical neurons. *Molecular and Cellular Neuroscience*, 52: 1–8.

Costantini, E., Landi, L., Silvestroni, O., Pandolfini, T., Spena, A. and Mezzetti, B. (2007). Auxin synthesis-encoding transgene enhances grape fecundity. *Plant Physiology*, 143: 1689–1694.

Cramer, G.R., Ergül, A., Grimplet, J., Tillett, R.L., Tattersall, E.A.R., Bohlman, M.C. and Osborne, C. (2007). Water and salinity stress in grapevines: Early and late changes in transcript and metabolite profiles. *Functional and Integrative Genomics*, 7: 111–134.

Dadakova, K., Havelkova, M., Kurkova, B., Tlolkova, I., Kasparovsky, T., Zdrahal, Z. and Lochman, J. (2015). Proteome and transcript analysis of *Vitis vinifera* cell cultures subjected to *Botrytis cinerea* infection. *Journal of Proteomics*, 119: 143–153.

Daldoul, S., Guillaumie, S., Reustle, G.M., Krczal, G., Ghorbel, A., Delrot, S. and Höfer, M.U. (2010). Isolation and expression analysis of salt induced genes from contrasting grapevine (*Vitis vinifera* L.) cultivars. *Plant Science*, 179: 489–498.

De-Francesco, L. (2008). Vintage genetic engineering. *Nature Biotechnology*, 26: 261–263.

Degaris, K.A., Walker, R.R., Loveys, B.R. and Tyerman, S.D. (2017). Exogenous application of abscisic acid to root systems of grapevines with or without salinity influences water relations and ion allocation. *Australian Journal of Grape and Wine Research*, 23(1): 66–76.

Delp, C.J. (1954). Effect of temperature and humidity on the grape powdery mildew fungus. *Phytopathology*, 44: 11–20.

Dinis, L.T., Ferreira, H., Pinto, G., Bernardo, S., Correia, C.M. and Moutinho-Pereira, J. (2015). Kaolin-based, foliar reflective film protects photosystem II structure and function in grapevine leaves exposed to heat and high solar radiation. *Photosynthetica*, 54(1): 47–55.

Dubrovina, A.S., Kiselev, K.V. and Khristenko, V.S. (2013). Expression of calcium-dependent protein kinase (CDPK) genes under abiotic stress conditions in wild-growing grapevine *Vitis amurensis*. *Journal of Plant Physiology*, 170: 1491–1500.

Duchêne, E., Huard, F., Dumas, V., Schneider, C. and Merdinoglu, D. (2010). The challenge of adapting grapevine varieties to climate change. *Climate Research*, 41: 193–204.

Dutt, M., Li, Z.T., Kelley, K.T., Dhekney, S.A., Aman, M.V., Tattersall, J. and Gray, D.J. (2007). Transgenic rootstock protein transmission in grapevines. *International Symposium on Biotechnology of Temperate Fruit Crops and Tropical Species*, 38: 749–753.

Espinoza, C., Vega, A., Medina, C., Schlauch, K., Cramer, G. and Arce-Johnson, P. (2007). Gene expression associated with compatible viral diseases in grapevine cultivars. *Functional and Integrative Genomics*, 7: 95–110.

Faisal, A.M. (2008). Climate change and phenology. *New Age*, 3: 1–6.

Fasoli, M., Santo, S.D., Zenoni, S., Tornielli, G.B., Farina, L., Zamboni, A. and Murino, V. (2012). The grapevine expression atlas reveals a deep transcriptome shift driving the entire plant into a maturation program. *The Plant Cell*, 24: 3489–3505.

Feynman, J. (2007). Has solar variability caused climate change that affected human culture? *Advances in Space Research*, 40: 1173–1180.

Figueiredo, A., Fortes, A.M., Ferreira, S., Sebastiana, M., Choi, Y.H., Sousa, L. and Pais, M.S. (2008). Transcriptional and metabolic profiling of grape (*Vitis vinifera* L.) leaves unravel possible innate resistance against pathogenic fungi. *Journal of Experimental Botany*, 59: 3371–3381.

Flexas, J. and Medrano, H. (2002). Drought-inhibition of photosynthesis in C3 plants: stomatal and non-stomatal limitations revisited. *Annals of Botany*, 89: 183–189.

Galbiati, M., Matus, J.T., Francia, P., Rusconi, F., Cañón, P., Medina, C. and Arce-Johnson, P. (2011). The grapevine guard cell-related VvMYB60 transcription factor is involved in the regulation of stomatal activity and is differentially expressed in response to ABA and osmotic stress. *BMC Plant Biology*, 11: 142–142.

Garcia, L., Celette, F., Gary, C., Ripoche, A., Valdes-Gomez, H. and Metay, A. (2018). Management of service crops for the provision of ecosystem services in vineyards: A review. *Agriculture, Ecosystems & Environment*, 251: 158–170.

Garcia-de-Cortazar, I. and Seguin, B. (2004). Climate warming: consequences for viticulture and the notion of terroirs in Europe. *In: VII International Symposium on Grapevine Physiology and Biotechnology*, 689: 61–70.

Girona, J., Marsal, J., Mata, M., Del-Campo, J. and Basile, B. (2009). Phenological sensitivity of berry growth and composition of Tempranillo grapevines (*Vitis vinifera* L.) to water stress. *Australian Journal of Grape and Wine Research*, 15: 268–277.

Gray, D., Zhijian, L., Dhekney, S. and Zimmerman, T. (2010). Identification of genetically engineered grapevines in the field for pierce's disease and fungal disease resistance. *In Vitro Cellular and Developmental Biology – Animal*, Springer, USA, 233 pp.

Gray, D.J., Li, Z.T. and Dhekney, S.A. (2014). Precision breeding of grapevine (*Vitis vinifera* L.) for improved traits. *Plant Science*, 228: 3–10.

Greer, D.H. and Weedon, M.M. (2013). The impact of high temperatures on *Vitis vinifera* cv. Semillon grapevine performance and berry ripening. *Frontiers in Plant Science*, 4: 491.

Guan, L., Haider, M.S., Khan, N., Nasim, M., Jiu, S., Fiaz, M. and Fang, J. (2018). Transcriptome sequence analysis elaborates a complex defensive mechanism of grapevine (*Vitis vinifera* L.) in response to salt stress. *International Journal of Molecular Sciences*, 19: 20–39.

Guan, L., Fan, P., Li, S.H., Liang, Z. and Wu, B.H. (2019). Inheritance patterns of anthocyanins in berry skin and flesh of the interspecific population derived from teinturier grape. *Euphytica*, 215(4): 1–14.

Guo, H.D., Zhang, L. and Zhu, L.W. (2015). Earth observation big data for climate change research. *Advances in Climate Change Research*, 6: 108–117.

Haider, M.S., Zhang, C., Kurjogi, M.M., Pervaiz, T., Zheng, T., Zhang, C. and Fang, J. (2017). Insights into grapevine defense response against drought as revealed by biochemical, physiological and RNA-Seq analysis. *Scientific Reports*, 7: 131–134.

Haider, M.S., Jogaiah, S., Pervaiz, T., Yanxue, Z., Khan, N. and Fang, J. (2019). Physiological and transcriptional variations inducing complex adaptive mechanisms in grapevine by salt stress. *Environmental and Experimental Botany*, 162: 455–467.

Hardie, W.J. and Considine, J.A. (1976). Response of grapes to water-deficit stress in particular stages of development. *American Journal of Enology and Viticulture*, 27: 55–61.

Hathfield, J.L. and Prueger, J.H. (2015). Temperature extremes: Effect on plant growth and development. *Weather and Climate Extremes*, 10: 4–10.

Hazelrigg, A., Bradshaw, T.L., Berkett, L.P., Maia, G. and Kingsley-Richards, S.L. (2018). Disease susceptibility of cold-climate grapes in Vermont, U.S.A. *Acta Hort.*, 1205: 477–482.

Hazelrigg, A.L., Bradshaw, T.L. and Maia, G.S. (2021). Disease susceptibility of interspecific cold-hardy grape cultivars in Northeastern U.S.A. *Horticulturae*, 7: 216. https://doi.org/10.3390/horticulturae7080216.

Hochberg, U., Degu, A., Toubiana, D., Gendler, T., Nikoloski, Z., Rachmilevitch, S. and Fait, A. (2013). Metabolite profiling and network analysis reveal coordinated changes in grapevine water stress response. *BMC Plant Biology*, 13: 184–187.

Intrigliolo, D.S. and Castel, J.R. (2009). Response of *Vitis vinifera* cv. 'Tempranillo' to partial rootzone drying in the field: Water relations, growth, yield and fruit and wine quality. *Agricultural Water Management*, 96: 282–292.

IPCC, 2013. Climate Change. (2013): *The Physical Science Basis. Contribution of Working Group I to the Fifth Assessment Report of the Intergovernmental Panel on Climate Change.* Stocker, T.F., Qin, D., Plattner, G.-K., Tignor, M., Allen, S.K., Boschung, J., Nauels, A., Xia, Y., Bex, V. and Midgley, P.M. (eds.). Cambridge University Press, Cambridge, UK and New York, USA, 1535 pp.

Jones, G.V. (2007). Climate change: Observations, projections, and general implications for viticulture and wine production. *Economics Department Working Paper*, 7: 14–15.

Jones, G.V. (2010). Climate, grapes, and wine: Structure and suitability in a changing climate. pp. 19–28. *In*: *XXVIII International Horticultural Congress on Science and Horticulture for People (IHC2010)*.

Ju, Y., Yue, X., Zhao, X., Zhao, H. and Fang, Y. (2018). Physiological, micro-morphological and metabolomic analysis of grapevine (*Vitis vinifera* L.) leaf of plants under water stress. *Plant Physiology and Biochemistry*, 130: 501–510.

Ju, Y., Min, Z., Yue, X., Zhang, Y., Zhang, J., Zhang, Z. and Fang, Y. (2020). Overexpression of grapevine VvNAC08 enhances drought tolerance in transgenic Arabidopsis. *Plant Physiology and Biochemistry*, 151: 214–222.

Kato, M.C., Hikosaka, K., Hirotsu, N., Makino, A. and Hirose, T. (2003). The excess light energy that is neither utilized in photosynthesis nor dissipated by photoprotective mechanisms determines the rate of photoinactivation in photosystem II. *Plant and Cell Physiology*, 44(3): 318–325.

Kobayashi, M., Katoh, H., Takayanagi, T. and Suzuki, S. (2010). Characterization of thermotolerance-related genes in grapevine (*Vitis vinifera*). *Journal of Plant Physiology*, 167: 812–819.

Kosová, K., Vítámvás, P., Urban, M.O., Prášil, I.T. and Renaut, J. (2018). Plant abiotic stress proteomics: The major factors determining alterations in cellular proteome. *Frontiers in Plant Science*, 9: 113–122.

Koundouras, S., Tsialtas, I.T., Zioziou, E. and Nikolaou, N. (2008). Rootstock effects on the adaptive strategies of grapevine (*Vitis vinifera* L. cv. Cabernet-Sauvignon) under contrasting water status: leaf physiological and structural responses. *Agriculture, Ecosystems & Environment*, 128: 86–96.

Krastanova, S.V., Balaji, V., Holden, M.R., Sekiya, M., Xue, B., Momol, E.A. and Burr, T.J. (2010). Resistance to crown gall disease in transgenic grapevine rootstocks containing truncated virE2 of agrobacterium. *Transgenic Research*, 19: 949–958.

Lereboullet, A.L., Bardsley, D. and Beltrando, G. (2013). Assessing vulnerability and framing adaptive options of two Mediterranean wine growing regions facing climate change: Roussillon (France) and McLaren Vale (Australia). *EchoGéo*, (23).

Li, H., Xu, Y., Xiao, Y., Zhu, Z., Xie, X., Zhao, H. and Wang, Y. (2010). Expression and functional analysis of two genes encoding transcription factors, VpWRKY1 and VpWRKY2, isolated from Chinese wild *Vitis pseudoreticulata*. *Planta*, 232: 1325–1337.

Lima, M.R.M., Machado, A.F. and Gubler, W.D. (2017). Metabolomic study of chardonnay grapevines double stressed with esca-associated fungi and drought. *Phytopathology*, 107: 669–680.

Liu, G.T., Wang, J.F., Cramer, G., Dai, Z.W., Duan, W., Xu, H.G., Wu, B.H., Fan, P.G., Wang, L.J. and Li, S.H. (2012). Transcriptomic analysis of grape (*Vitis vinifera* L.) leaves during and after recovery from heat stress. *BMC Plant Biology*, 12: 174.

Liu, G.T., Ma, L., Duan, W., Wang, B.C., Li, J.H., Xu, H.G. and Wang, L.J. (2014). Differential proteomic analysis of grapevine leaves by iTRAQ reveals responses to heat stress and subsequent recovery. *BMC Plant Biology*, 14: 110–112.

Lopez, C.M., Monteiro, A., Machado, J.P., Fernandes, N. and Araújo, A. (2008). Cover cropping in a slopping non-irrigated vineyard: Effects on vegetative growth, yield, berry and wine quality of 'Cabernet Sauvignon' grapevines. *Ciência e Técnica Vitivinícola*, 23: 37–43.

Loreto, F., Harley, P.C., Di-Marco, G. and Sharkey, T.D. (1992). Estimation of mesophyll conductance to CO_2 flux by three different methods. *Plant Physiology*, 98: 1437–1443.

Louime, C., Vasanthaiah, H.K., Basha, S.M. and Lu, J. (2010). Perspective of biotic and abiotic stress research in grapevines (*Vitis* sp.). *International Journal of Fruit Science*, 10(1): 79–86.

Maoz, I., Rosso, M.D., Kaplunov, T., Vedova, A.D., Sela, N., Flamini, R. and Lichter, A. (2019). Metabolomic and transcriptomic changes underlying cold and anaerobic stresses after storage of table grapes. *Scientific Reports*, 9: 17–29.

Marchive, C., Leon, C., Kappel, C., Coutos-Thevenot, P., Corio-Costet, M.F., Delrot, S. and Lauvergeat, V. (2013). Over-expression of VvWRKY1 in grapevines induces expression of jasmonic acid pathway-related genes and confers higher tolerance to the downy mildew. *PLoS One*, 8: e54185.

Matus, J.T. (2016). Transcriptomic and metabolomic networks in the grape berry illustrate that it takes more than flavonoids to fight against ultraviolet radiation. *Frontiers in Plant Science*, 7: 13–37.

Merz, P.R., Moser, T., Höll, J., Kortekamp, A., Buchholz, G., Zyprian, E. and Bogs, J. (2015). The transcription factor VvWRKY33 is involved in the regulation of grapevine (*Vitis vinifera*) defense against the oomycete pathogen *Plasmopara viticola*. *Physiologia Plantarum*, 153: 365–380.

Monteiro, A. and Lopes, C.M. (2007). Influence of cover crop on water use and performance of f Monteiro vineyard in Mediterranean Portugal. *Agriculture, Ecosystems & Environment*, 121: 336–342.

Moriondo, M., Jones, G.V., Bois, B., Dibari, C., Ferrise, R., Trombi, G. and Bindi, M. (2013). Projected shifts of wine regions in response to climate change. *Climatic Change*, 119: 825–839.

Moutinho-Pereira, J., Gonçalves, B., Bacelar, E., Cunha, J.B., Coutinho, J. and Correia, C.M. (2009). Effects of elevated CO_2 on grapevine (*Vitis vinifera* L.): Physiological and yield attributes. *Vitis*, 48(4): 159–165.

Myles, S., Boyko, A.R., Owens, C.L., Brown, P.J., Grassi, F., Aradhya, M.K. and Ware, D. (2011). Genetic structure and domestication history of the grape. *Proceedings of the National Academy of Sciences of the United States of America*, 108: 3530–3535.

Neethling, E., Barbeau, G., Séverine, J., Le Roux, R. and Quénol, H. (2017). Local-based approach for assessing climate change adaptation in viticulture. *OENO One*.

Orduña, R.M. (2010). Climate change associated effects on grape and wine quality and production. *Food Research International*, 43: 1844–1855.

Paranychianakis, N.V., Aggelides, S. and Angelakis, A.N. (2004). Influence of rootstock, irrigation level and recycled water on growth and yield of Soultanina grapevines. *Agricultural Water Management*, 69: 13–27.

Parida, A.K., Panda, A. and Rangani, J. (2018). Metabolomics-guided elucidation of abiotic stress tolerance mechanisms in plants. pp. 89–131. *In*: *Plant Metabolites and Regulation under Environmental Stress*. Academic, San Diego, CA.

Parker, A.K., Hofmann, R.W., van Leeuwen, C., McLachlan, A.R. and Trought, M.C. (2015). Manipulating the leaf area to fruit mass ratio alters the synchrony of total soluble solids accumulation and titratable acidity of grape berries. *Australian Journal of Grape and Wine Research*, 21(2): 266–276.

Pavlousek, P. (2011). Evaluation of drought tolerance of new grapevine rootstock hybrids. *Journal of Environmental Biology*, 32: 543–547.

Perazzolli, M., Moretto, M., Fontana, P., Ferrarini, A., Velasco, R., Moser, C. and Pertot, I. (2012). Downy mildew resistance induced by *Trichoderma harzianum* T39 in susceptible grapevines partially mimics transcriptional changes of resistant genotypes. *BMC Genomics*, 13: 660–662.

Rodrigues, M.L., Nimrichter, L., Oliveira, D.L., Nosanchuk, J.D. and Casadevall, A. (2008). Vesicular trans-cell wall transport in fungi: a mechanism for the delivery of virulence-associated macromolecules? *Lipid Insights*, 2: 327–332.

Rushton, P.J., Somssich, I.E., Ringler, P. and Shen, Q.J. (2010). WRKY transcription factors. *Trends in Plant Science*, 15: 247–258.

Ruml, M., Vuković, A., Vujadinović, M., Djurdjević, V., Ranković-Vasić, Z., Atanacković, Z. and Petrović, N. (2012). On the use of regional climate models: Implications of climate change for viticulture in Serbia. *Agricultural and Forest Meteorology*, 158: 53–62.

Saporta, R., San Pedro, T. and Gisbert, C. (2016). Attempts at grapevine (*Vitis vinifera* L.) breeding through genetic transformation: The main limiting factors. *Vitis*, 55(4): 173–186.

Saxena, M. and Gandhi, C.P. 2015. National Horticulture Board. Indian Horticulture Database, pp. 286.

Schultz, H. (2000). Climate change and viticulture: A European perspective on climatology, carbon dioxide and UV-B effects. *Australian Journal of Grape and Wine Research*, 6: 2–12.

Schultz, P.W. (2000). New environmental theories: Empathizing with nature: The effects of perspective taking on concern for environmental issues. *Journal of Social Issues*, 56: 391–406.

Schultz, G.S., Sibbald, R.G., Falanga, V., Ayello, E.A., Dowsett, C., Harding, K., Romanelli, M., Stacey, M.C., Teot, L. and Vanscheidt, W. (2003). Wound bed preparation: A systematic approach to wound management. *Wound Repair and Regeneration*, 11: 11–28.

Schultz, H.R. (2007). Climate change: Implications and potential adaptation of vine growth and wine composition. *In*: *Proceedings of Congress on Climate and Viticulture*. Centro Transferencia Agroalimentaria, Saragoza, 12: 87–92.

Singh, K., Talla, A. and Qiu, W. (2012). Small RNA profiling of virus-infected grapevines: Evidences for virus infection-associated and variety-specific miRNAs. *Functional and Integrative Genomics*, 12: 659–669.

Skopelitis, D.S., Paranychianakis, N.V., Paschalidis, K.A., Pliakonis, E.D., Delis, I.D., Yakoumakis, D.I., Kouvarakis, A., Papadakis, A.K., Stephanou, E.G. and Roubelakis-Angelakis, K.A. (2006). Abiotic stress generates ROS that signal expression of anionic glutamate dehydrogenases to form glutamate for proline synthesis in tobacco and grapevine. *Plant Cell*, 18: 2767–2781.

Soar, C.J., Speirs, J., Maffei, S.M., Penrose, A.B., McCarthy, M.G. and Loveys, B.R. (2006). Grape vine varieties shiraz and grenache differ in their stomatal response to VPD: Apparent links with ABA physiology and gene expression in leaf tissue. *Australian Journal of Grape and Wine Research*, 12: 2–12.

Souza, C.R., Maroco, J.P., dos Santos, T.P., Rodrigues, M.L., Lopes, C., Pereira, J.S. and Chaves, M.M. (2005). Control of stomatal aperture and carbon uptake by deficit irrigation in two grapevine cultivars. *Agriculture, Ecosystems & Environment*, 106: 261–274.

Spagnolo, A., Magnin-Robert, M., Alayi, T.D., Cilindre, C., Mercier, L., Schaeffer-Reiss, C. and Fontaine, F. (2012). Physiological changes in green stems of *Vitis vinifera* L. cv. Chardonnay in response to esca proper and apoplexy revealed by proteomic and transcriptomic analyses. *Journal of Proteome Research*, 11: 461–475.

Sperry, J.S. and Saliendra, N.Z. (1994). Intra- and inter-plant variation in xylem cavitation in Betula occidentalis. *Plant, Cell & Environment*, 17: 1233–1241.

Strange, R.N. and Scott, P.R. (2005). Plant disease: A threat to global food security. *Annual Review of Phytopathology*, 43: 83–116.

Tattersall, E.A.R., Grimplet, J., De-Luc, L., Wheatley, M.D., Vincent, D., Osborne, C. and Schlauch, K.A. (2007). Transcript abundance profiles reveal larger and more complex responses of grapevine to chilling compared to osmotic and salinity stress. *Functional and Integrative Genomics*, 7: 317–333.

Teixeira, A., Eiras-Dias, J., Castellarin, S.D. and Gerós, H. (2013). Berry phenolics of grapevine under challenging environments. *International Journal of Molecular Sciences*, 14(9): 18711–18739.

Thomas, M.R., Iocco, P. and Franks, T. (2000). Transgenic grapevines: Status and future. pp. 279–287. *Proceedings of the Seventh International Symposium on Grapevine Genetics and Breeding*. Montpellier, France, 6–10 July 1998.

Tillett, R.L., Wheatley, M.D., Tattersall, E.A.R., Schlauch, K.A., Cramer, G.R. and Cushman, J.C. (2012). The *Vitis vinifera* C-repeat binding protein 4 (VvCBF4) transcriptional factor enhances freezing tolerance in wine grape. *Plant Biotechnology Journal*, 10: 105–124.

Todorov, A. and Uleman, J.S. (2003). The efficiency of binding spontaneous trait inferences to actors faces. *Journal of Experimental Social Psychology*, 39: 549–562.

Tramontini, C.C. and Graziano, K.U. (2012). Factors related to body heat loss during the intraoperatory period: Analysis of two nursing interventions. *Ciênc Cuid Saúde*, 11: 220–225.

Tramontini, S., Vitali, M., Centioni, L., Schubert, A. and Lovisolo, C. (2013). Rootstock control of scion response to water stress in grapevine. *Environmental and Experimental Botany*, 93: 20–26.

Triolo, R., Roby, J.P., Plaia, A., Hilbert, G., Buscemi, S., Di-Lorenzo, R. and Leeuwen, C. (2018). Hierarchy of factors impacting grape berry mass: separation of direct and indirect effects on major berry metabolites. *American Journal of Enology and Viticulture*, 69: 103–112.

Trumbo, C.W. and Shanahan, J. (2000). Social research on climate change: Where we have been, where we are, and where we might go. *Public Inderstanding of Science*, 9: 199–204.

Van Leeuwen, C. and Seguin, G. (2006). The concept of terroir in viticulture. *Journal of Wine Research*, 17(1): 1–10.

Vidal, J.R., Kikkert, J.R., Malnoy, M.A., Wallace, P.G., Barnard, J. and Reisch, B.I. (2006). Evaluation of transgenic 'Chardonnay' (*Vitis vinifera*) containing magainin genes for resistance to crown gall and powdery mildew. *Transgenic Research*, 15: 69–82.

Vigne, E., Komar, V. and Fuchs, M. (2004). Field safety assessment of recombination in transgenic grapevines expressing the coat protein gene of grapevine fanleaf virus. *Transgenic Research*, 13: 165–179.

Waraich, E.A., Ahmad, R., Halim, A. and Aziz, T. (2012). Alleviation of temperature stress by nutrient management in crop plants: A review. *Journal of Soil Science and Plant Nutrition*, 12(2): 221–244.

Wang, H., Yin, X., Li, X., Wang, L., Zheng, Y., Xu, X. and Wang, X. (2014a). Genome-wide identification, evolution and expression analysis of the grape (*Vitis vinifera* L.) zinc finger-homeodomain gene family. *International Journal of Molecular Sciences*, 15: 5730–5748.

Wang, M., Vannozzi, A., Wang, G., Liang, Y.H., Tornielli, G.B., Zenoni, S. and Cheng, Z.M. (2014b). Genome and transcriptome analysis of the grapevine (*Vitis vinifera* L.) WRKY gene family. *Horticulture Research*, 1: 14016–14019.

Warren, C.R. and Adams, M.A. (2006). Internal conductance does not scale with photosynthetic capacity: Implications for carbon isotope discrimination and the economics of water and nitrogen use in photosynthesis. *Plant, Cell & Environment*, 29: 192–201.

Whiting, J.R. (2005). *Grapevine Rootstocks in Viticulture*. pp. 167–188. *In*: Dry, P.R. and Coombe, B.G. (eds.). Winetitles Pty Ltd, Ashford, Australia.

Xiao, H., Tattersall, E.A.R., Siddiqua, M.K., Cramer, G.R. and Nassuth, A. (2008). CBF4 is a unique member of the CBF transcription factor family of *Vitis vinifera* and *Vitis riparia*. *Plant Cell and Environment*, 31: 1–10.

Yao, Y.A., Wang, J., Ma, X., Lutts, S., Sun, C., Ma, J. and Xu, G. (2012). Proteomic analysis of Mn-induced resistance to powdery mildew in grapevine. *Journal of Experimental Botany*, 63: 5155–5170.

Yang, L., Lin, H., Takahashi, Y., Chen, F., Walker, M.A. and Civerolo, E.L. (2011). Proteomic analysis of grapevine stem in response to *Xylella fastidiosa* inoculation. *Physiological and Molecular Plant Pathology*, 75: 90–99.

Yu, Y.H., Li, X.Z., Wu, Z.J., Chen, D.X., Li, G.R., Li, X.Q. and Zhang, G.H. (2016). VvZFP11, a Cys2His2-type zinc finger transcription factor, is involved in defense responses in *Vitis vinifera*. *Biologia Plantarum*, 60: 292–298.

Zhao, F., Chen, L., Perl, A., Chen, S. and Ma, H. (2011). Proteomic changes in grape embryogenic callus in response to *Agrobacterium tumefaciens*-mediated transformation. *Plant Science*, 181: 485–495.

Zhang, Y., Mechlin, T. and Dami, I. (2011). Foliar application of abscisic acid induces dormancy responses in greenhouse-grown grapevines. *HortScience*, 46(9): 1271–1277.

Zhang, W., Yan, J., Li, X., Xing, Q., Chethana, K.W.T. and Zhao, W. (2019). Transcriptional response of grapevine to infection with the fungal pathogen *Lasiodiplodia theobromae*. *Scientific Reports*, 9: 53–87.

Zufferey, V., Cochard, H., Ameglio, T., Spring, J.L. and Viret, O. (2011). Diurnal cycles of embolism formation and repair in petioles of grapevine (*Vitis vinifera* cv. Chasselas). *Journal of Experimental Botany*, 62: 3885–3894.

8

Climate Change and Its Implications for the Cultivation of Olive

Fabíola Villa,[1,*] *Daniel Fernandes da Silva*[1] and *Glacy Jaqueline da Silva*[2]

1. Introduction

With climate change, all species are forced to adapt to ensure their existence. One of the ways to ensure this adaptation is to seek new areas in which edaphoclimatic conditions allow good development as also ceasing to exist in areas where current conditions are different from previous ones. Thus, this new geospatial design of the regions favorable to the development of each species configures the agricultural region for the exploitation of species of economic interest.

The olive tree is a fruit species which favors a temperate climate for its full development. Its origin and cultivation for agricultural production are heavily concentrated in the Mediterranean region (Fig. 1), precisely where, according to projections by scientists, it will suffer greater damage from climate change, with an increase in temperature and a reduction in precipitation. This species is widely affected by climatic conditions

[1] Centro de Ciências Agrárias, Western Parana State University (Unioeste), Campus de Marechal Cândido Rondon. Rua Pernambuco, 1777, Centro, Marechal Cândido Rondon/PR, Brazil.

[2] Molecular Biology Department, University of Paraná (Unipar), Umuarama Campus I - Sede. Praça Mascarenhas de Moraes, 4282, Zona III, Umuarama/PR, Brazil.

Emails: daniel_eafi@yahoo.com.br; glacyjaqueline@prof.unipar.br

* Corresponding author: fvilla2003@hotmail.com

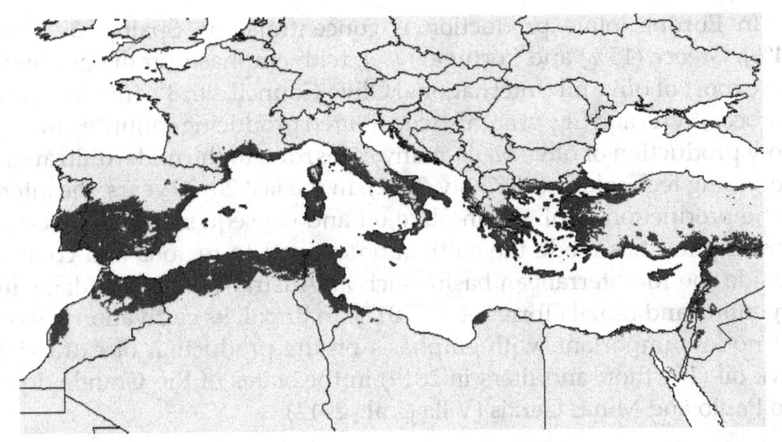

Fig. 1. Potential distribution of olive tree over the Mediterranean Basin (Oteros et al., 2014).

in its various phases, especially during the reproductive period when flowers and fruits are formed—a phase that requires ideal conditions of temperature and humidity.

Considering the importance of olive cultivation in the world economy and from a nutritional point of view, due to its various components essential to human health, changes in the olive production scenario arouse the interest of scientists who use computational tools and mathematical models to predict future scenarios for the production of that fruit. The following content deals with the edaphoclimatic needs of the olive tree, correlating them with the possibilities of cultivation in different regions of the world and how climate change will affect the distribution of this fruit around the globe. This review brings the predictions and behavioral studies of the species, based on the work of different specialists in the area, in order to help in the understanding and future planning of olive tree cultivation.

1.1 Origin, Dispersion and Cultivation of the Olive Tree

The olive tree (*Olea europaea* L.) is one of the oldest species cultivated in the Mediterranean region. Being a perennial and evergreen crop, with a strong socioeconomic impact in many countries of this region, it covers 80% of the world olive area. The Mediterranean is responsible for 95% of world oil production. The geographical origin of the cultivated olive tree can be traced to areas along the Mediterranean basin, covering countries such as Turkey, Syria, Lebanon, Palestine and Israel. Some records indicate that olive trees have been grown in these areas since at least 3000 B.C. (Fraga et al., 2019). The cultivation of olive trees has spread widely across southern Europe, northern Africa and the Iberian Peninsula. In the 16th century, Spanish colonization took olive trees to the Americas.

In Europe, olive production is concentrated in Spain (53%), Italy (24%), Greece (15%) and Portugal (7%), with emphasis on the production and export of olive oil (International Olive Council, 2018). Tunisia, Turkey, Morocco, Syria and Egypt are also considered producing countries, but with a low production of olive oil, in many cases, due to the underutilization of processing technology (El-Kholy, 2012). In the last 20–30 years, the interest in the production of olives, in olive oil and consequently the increase in consumption has made the cultivation expand to regions and countries outside the Mediterranean basin, such as Australia, China, India, Chile, Argentina and Brazil (Torres et al., 2017). In Brazil, its cultivation is recent, but no less important, with emphasis on the production of extra virgin olive oil (160 thousand liters in 2019) in the states of Rio Grande do Sul, São Paulo and Minas Gerais (Villa et al., 2012).

1.2 Agricultural Zoning and Edaphoclimatic Requirements of Cultivation

Agricultural zoning is considered a very important resource for the activity in Brazil, as it reduces losses due to climatic adversities, identifies and delimits the best areas for planting and defines the best planting time with the least climate risk. The knowledge of the climatic factors that influence the development, production, limits and interrelationships with the phenology of the culture, is of great importance to define the best areas and define the most suitable planting time, avoiding questions about where to produce with safety. Therefore, precipitation, temperature, relative humidity, altitude and soil types must be analyzed in time and space (Filipini Alba et al., 2014; Oliveira et al., 2010).

In the Mediterranean climate, during the winter, accumulation of chilling units occurs, which is considered indispensable for the olive tree for achieve uniform flowering (Coutinho et al., 2009). According to planting experiences in Mediterranean countries, the appropriate temperature for normal effective fruiting to occur should be 25–35°C. The plants, however, are able to withstand high temperatures in the summer – close to 40°C, without the branches and leaves being burned. However, photosynthetic activity is inhibited when temperature exceeds 35°C. The olive tree is more sensitive to the cold than other fruit species. However, there is a gradual increase in tolerance, caused by low autumn temperatures responsible for stimulating dormancy. Thus, the olive tree can resist temperatures just below 0°C, with the appearance of small lesions on new shoots and branches if the temperature is between 0°C and –5°C. If the temperature drops further, down to –10°C, permanent damage to the shoots and branches can occur. Below –10°C, the plant can suffer irreversible damage and die (Bienes et al., 2018).

In general, the olive tree is a plant with a temperate climate and needs low temperatures in the period before flowering for satisfactory production to occur. The crop requires average winter temperatures between 10–12°C, not exceeding 21°C, variable altitudes (800–1.900 m) and rainfall above 800 mm per year. Preference is given to flat or gently undulating soils to facilitate crop treatment and harvesting. However, planting can be done in places with greater slope and greater exposure to sunlight, preferably the north face of the land (Torres et al., 2017).

1.3 Climate Change in Europe

Global climate change, from the emission of greenhouse gases, may have implications for plant life in the future (IPCC, 2013). The subject received much attention from scientists around the world, according to some reviews and scientific papers on the subject (Kallarackal and Renuka, 2014). Studies on changes in the duration of phenophases in the cultivation of fruit species are of great importance and may have direct impact on production, fruit quality and reproductivity (Ramírez and Kallarackal, 2015), mainly for olives and olive oil (Ponti et al., 2014; Fraga et al., 2019), in addition to the spread of some diseases, such as those caused by bacteria, as *Xylella fastidiosa* (Schneider et al., 2020).

The Mediterranean region and South America are considered by some researchers to be a climate change hotspot (Torres et al., 2014; Ponti et al., 2014), that is, future projections point to considerable warming trends (IPCC, 2012) and increase in aridity in these places. In this context, climate change can become particularly challenging for olive growers (Moriondo et al., 2015). This increase in temperature can have a strong impact on the crop, advancing the phenophases, particularly related to flowering (Orlandi et al., 2013). Fraga et al. (2019) point to a sudden change in the thermal conditions of olive trees in Europe until the end of this century.

1.4 Climate Change in Brazil

In Brazil, olive cultivation is expanding to promising areas, where it is possible to adapt, especially in regions with mild climate, different from those in the Center of Origin (Martins et al., 2014). Adverse climatic conditions to cultivation, such as water stress and irregular temperatures, can cause physiological disturbances (Taiz and Zeiger, 2017) in plants, directly interfering with their development, growth and yield (Tanasijevic et al., 2014; Moriondo et al., 2015).

In Brazil and more specifically in the state of Minas Gerais, some climate studies point to changes in seasonal averages and variability in various meteorological elements, mainly in precipitation and air temperature (Torres and Marengo, 2014). Santos et al. (2017) found that the probable increase in temperature and change in the rainfall regime may modify the water balance and harm the olive harvest at Minas Gerais state. The same authors also stated that a large part of the area suitable for the cultivation of olives will be reduced, with marginal displacement and inadequate areas for the south of the state. They also concluded that towards the end of the 21st century (2071–2100), in the period of intense climate change scenario, olive-growing in Minas Gerais will only be viable in the extreme south of the state. And to conclude this study, they observed that the increase in temperature and the change in the rainfall regime in the state interfered the flowering and fruiting stages.

For the north of the state of Minas Gerais, some models project a warmer climate, up to 5°C, towards the end of the century, with heterogeneous precipitation and reduction trends, and a slight increase in precipitation in the center-south regions (IPCC, 2013). Changes in climatic patterns can impact and possibly make olive cultivation unfeasible, mainly due to the physiological and phenological changes during the development stages (García-Mozo et al., 2015).

1.5 Cultivation Models

In the agrarian sector, cultivation models gradually become reliable tools to support decision making (Challinor et al., 2014). Cultivation models can be statistical, empirical or dynamic, with processes based on their nature (Challinor et al., 2016).

Although cultivation models are applied to a wide variety of crops worldwide (wheat, corn and rice), they are not yet widely used for olive trees. Some statistical models studied and published relate the seasons and temperature to the plant phenological phases and its production (Oteros et al., 2014; Aguilera et al., 2015a; Aguilera et al., 2015b).

Regarding dynamic models, some are dedicated to the phenological stages, relating them to the growth and development of the olive tree (Florêncio et al., 2019; Moriondo et al., 2019), while others aim to predict only the growth of the plant's biomass (Maselli et al., 2012). Given their great complexity, dynamic models generally tend to be more used in statistical approaches, as they simulate the physiology of plants and their relationship with the environment, in addition to being updated with some frequency. They also lead to reliable and robust future projections of productivity, periods of plant growth and stress indicators in a given region, correlating the data obtained with those of edaphoclimatic models (Aybar et al., 2015).

1.6 Climatic Changes Related to Olive Tree Physiology

Temperature changes, as indicated in climate change projections, negatively influence the cultivation of the olive tree, especially by altering plant evapotranspiration (Ahmadi and Baaghideh, 2020). The increase in temperature tends to enhance evapotranspiration (Knežević et al., 2017), causing water loss in plants to the atmosphere at high rates. With the reduction of precipitation, the water supply to the plants is less than the evaporation rate, resulting in a situation of water deficit, reduction of leaf area and stomatal closure, which are the main defense strategies of the plant (Taiz and Zeiger, 2017). As a consequence, there is a reduction in gas exchange (transpiration and CO_2 assimilation for photosynthesis).

Evapotranspiration (ET) is the amount of water transpired by plants during their growth period, together with the evaporated moisture from the soil surface and vegetation (Mundo-Molina, 2015). The rate of ET depends on the energy, the water available at the evaporation site and the vapor pressure gradient. These three conditions are highly influenced by the temperature change and, as a result, provide changes in the ET rate (Sarkar and Sarkar, 2018).

In addition, high temperatures, above 35°C (Wrege et al., 2015), at the beginning of flowering (Oliveira et al., 2012), are harmful to olive growing, as they can cause floral abortion (García-Mozo et al., 2010). García-Mozo et al. (2015) evaluated phenological trends in southern Spain in response to climate change and observed that the increase in the average annual temperature also impaired pollination, decreased the duration of flowering and fruiting stages and increased respiration of cellular tissue. This happens particularly in plants with C3 metabolism, as is the case with the olive tree, leading to a lower yield of olives.

High air temperatures (above the ideal temperature) (Martins et al., 2014) also reduce the photosynthetic rate, mainly due to the reduction in the efficiency of photosystem II, increasing maintenance respiration and reducing the leaf area. In addition, changes in the developmental stages of the olive tree will impact the management of the crop, which may accelerate flowering, maturation and harvest, with a probable occurrence of adverse weather events during these stages (Sghaier et al., 2019; Taiz and Zeiger, 2017).

Regarding precipitation, a more pronounced reduction in production can occur when the months of flowering start until total flowering of the olive tree (Oliveira et al., 2012), which can cause a reduction in the number of inflorescences, increased production of imperfect flowers and floral abortion (González et al., 2018). During this phenophase, it is not ideal for precipitation to be in excess, as it can cause a reduction in the oil content contained in the olives (Mousavi et al., 2019).

1.7 Plant Response to Increased Carbon Dioxide (CO_2)

Among the main abiotic factors (light, water, nutrients and CO_2) that interfere with plant growth, carbon dioxide is one that is directly linked to global warming (Richardson et al., 2013). Any change in the availability of the elements mentioned above impacts not only the plants, but the entire living system. With climate change, an increase of 0.5% of CO_2 is expected by the end of this century, reaching concentrations above 600 ppm, compared to the current 380 ppm (Houghton et al., 2012; Houghton et al., 2013). This increase in CO_2 levels will certainly interfere globally in the plant photosynthesis process and has become the object of studies in the last half century (Ramirez and Kallarackal, 2015).

It is important to remember that currently, several research groups use the experimental artificial enrichment of CO_2 as a study methodology, thus improving the understanding of the response of plants to the elevation of this gas. All the methods used previously were denominated as chamber-and-enrichment methods with carbon dioxide free air (FACE or Free-Air-Carbon dioxide-Enrichment), and have positive and negative attributes and, therefore, the data obtained through some of these must be treated with caution (Walter et al., 2015). Free-Air-Carbon dioxide-Enrichment (FACE) is a method used by ecologists and plant biologists to increase the concentration of CO_2 in a specified area, thus allowing the plant's growth response to be measured. In addition, there is a lot of interaction of CO_2 with other biotic and abiotic factors, which are ignored in many studies (Carvalho et al., 2014).

The main effects of increasing CO_2 in plants have already been properly studied and discussed and include a reduction in stomatal conductance and transpiration, better efficiency in the use of water, high rates of photosynthesis and greater efficiency in the use of light (Sghaier et al., 2019). According to the world literature, studies with FACE focus mainly on tree plants (Haworth et al., 2016) and there are few studies with fruit species.

Although photosynthesis is stimulated to approximately 37%, in short-term experiments, with high CO_2 (Dusenge et al., 2019), when this gas is increased by 350–550 ppm in environments at 25°C, photosynthetic rates tend to decrease in some species, compared to plants grown in environments with lower CO_2 levels. This phenomenon is known as 'photosynthetic acclimatization'. Although not very common, it is reported in several species. This acclimatization to high levels of CO_2 can be attributed to at least five potential mechanisms at the cellular level, namely: accumulation of sugar and genetic repression, insufficient nitrogen absorption by the plant, binding of inorganic phosphate with accumulation of carbohydrates and subsequent limitation in RuBP regeneration, accumulation of starch in the chloroplast and ability to use phosphate triose (Bagley et al., 2015).

An important point to be discussed regarding the impact of elevated carbon dioxide (eCO_2) on fruit species is the stimulation of productivity, as observed in other cultures (Ghahramani et al., 2019). In general, studies on FACE reported 47% stimulation in photosynthesis in tree species, when compared to annual crops, such as wheat and rice (7–8% stimulation in yield). Most of the increase in productivity reported for studies in FACE showed an increase in vegetative biomass, including the leaf area, and it can be concluded that the increase in this biomass of tree species is directly proportional to the increase in CO_2 in the atmosphere (Montanaro et al., 2018).

When compared to natural vegetation, studies on the impact of eCO_2 on fruit trees are very limited, where some studies have been carried out with temperate and subtropical species, such as grapevine, acid orange and olive (Kimball, 2016; De Ollas et al., 2019).

In olive trees, Chamizo et al. (2017) and Montanaro et al. (2018) observed a significant long-term increase in the biomass of fruits, stems and branches in response to an increase in the concentration of atmospheric CO_2. It was possible to recover a soluble fraction of CO_2 sensitive proteins, known as vegetative storage proteins (VSPs). The amount of these proteins is directly proportional to eCO_2 and their existence may be the key that allows tree plants to temporarily store a nitrogen stock necessary for the formation of new branches and the production of fruit biomass.

One positive aspect of climate change should also be mentioned, i.e., the possible beneficial effect of higher CO_2 atmospheric concentrations in the future. It is known that the increase in CO_2 levels under future climates may have a positive influence on plants, mostly by increasing biomass under CO_2-enriched environments. This effect may partially counteract climate change's detrimental impacts resulting from enhanced heat and water stresses (Ainsworth et al., 2020). The widespread distribution of olive groves in the Mediterranean Basin can be also exploited for their important carbon sequestration capacity to mitigate the impact of climate change (Fraga et al., 2019).

The provision of high levels of nitrogen fertilizer to the soil has the capacity to fully compensate the reduced concentrations of nitrogen (N) present in leaves caused by high levels of atmospheric CO_2. The reduction in the concentration of N in the leaves or in the aerial part of plants in response to high CO_2 is highly dependent on the supply of this element and practically disappears when N is available in the soil for the plant's root system. Probably, with a climate change situation, this means that it is necessary to supplement the soil with more N in order to maintain productivity (Cha et al., 2017).

Abobatta (2019) found in a study with citrus plants grown in greenhouses under eCO_2 that, in the absence of other environmental stresses, photosynthesis occurred satisfactorily with the increase in atmospheric CO_2. These results demonstrate that photosynthetic acclimatization may have occurred for new and older leaves in relation to eCO_2. This photosynthetic acclimation was accompanied by a negative regulation of Rubisco protein concentration and activity, possibly related to the high accumulation of starch and sucrose (Fischer et al., 2012; Tognetti, 2012). The new leaves acclimatized very well to eCO_2, compared to the older leaves, in terms of gas exchange, photosynthetic capacity and sucrose synthesis. Furthermore, the starch accumulation in young leaves during the day was higher than in old leaves under eCO_2.

The detection of CO_2-induced changes in plant quality is a challenge to be faced. Plant quality involves several nutritional elements, such as macronutrients (carbohydrates, proteins and fats) and micronutrients (minerals, vitamins and phytonutrients). Assessing relative changes within and between elements requires significantly more effort than just measuring fruit yield and production (Loladze, 2014).

1.8 Nutritional Value of Olive Fruits in Response to eCO_2

The effects of eCO_2 on plant quality and its possible cascading effects on human nutrition have been largely ignored in estimating the impact on humans. Notably, the IPCC (2013) includes direct effects of CO_2 in its assessments of climate change, but does not mention any effect on crop quality (Franks et al., 2013).

There are some studies related to the impact of eCO_2 on the change of nutritional constituents of plants continuously exposed to this elements. Some experiments with olives produced in areas with possible climate change were conducted, relating the physiological and biochemical characteristics to the quality of extra virgin olive oil and its storage (Rouina, Bem et al., 2020; Nissim et al., 2020; Algataa, 2020).

1.9 Effect of Increased Temperature on Plant Phenology

Phenology is the study of periodic biological events that occur in the cycle of cultivated plants, such as overcoming dormancy (temperate species), flowering, fruit development, closely regulated by climatic and seasonal changes, thus interfering in the development of fruit species in given cultivation regions (Belaj et al., 2020). Higher temperatures generated as a result of global warming are responsible for the reduction or increase in the phenological cycles of fruit species, mainly in olive trees (García-Mozo et al., 2015).

In addition to the olive tree, citations in the literature emphasize that phenological studies of this species (Rai et al., 2015) have been impacted by global warming (García-Mozo et al., 2015; Belaj, 2020; Di Paola et al., 2021).

1.10 Cold Need

Climate change interferes with the amount of cold and the accumulation of heat. These are vital to the flowering and production of temperate fruit species (Rodríguez et al., 2019; Fraga et al., 2019). The reduced amount of cold in winter, due to these changes, will possibly impact with more serious consequences on fruit production, especially on yield (Koubouris et al., 2019). Several modeling approaches are currently used to quantify chilling requirements (Martins et al., 2020; Santos et al., 2017). For olive trees, the Tb values of 7.0, 9.5, and 13°C are commonly adopted for cultivars with the highest, medium and lowest chilling requirements, respectively (Garcia et al., 2018).

A comprehensive climatological analysis over the Mediterranean Basin indicated that olive cultivation areas are nowadays constrained by temperatures of the coldest (mean monthly temperature of January) and warmest months (mean monthly temperature of July), where the optimum monthly mean temperatures for its cultivation are centered on ~ 7°C in January and ~ 25°C in July (Moriondo et al., 2013).

1.11 Precipitation

A few studies demonstrate the possible role of climate change in precipitation and soil moisture, associated with the conduction of phenophases of olive fruits (Conde-Innamorato et al., 2019; Baldi et al., 2019). Garrido et al. (2021) and Conde-Innamorato et al. (2019) observed that rainfall and temperature operate synergistically, influencing the average flowering dates on olive varieties in Spain region and Uruguay, respectively. Finally, long-term temperature changes in the phenological stages of these fruit species can be attributed to the combined impacts of progressive regional warming and reduced rainfall. Studies on the different phonological phases, reproductive blossom, morphology, physiology and flowering stage, influenced by meteorological variables (temperature, radiation and precipitation) that affect induction, differentiation and flowering, have demonstrated that these aspects directly determine the number of fruits that can be harvested (Rojo and Pérez-Badia, 2014).

Koubouris et al. (2019) show that the results of the study may contribute to understanding olive flowering biology and selecting appropriate cultivars for new plantations according to historical meteorological data and predicted climate change scenarios.

1.12 Biodiversity and the Spread of Disease

Due to global warming, experts and scientists project a reduction in the diversity of fruit species in agroforestry vegetation in the long term, where these temperate climate species are likely to be most affected by climate change. However, cultivars in this group of plants may be less or more adapted to these changes in different climatic conditions. The biodiversity of genetic resources of the fruits is in serious threat of extinction due to climate change, large-scale urbanization and urban development projects (Malik et al., 2010).

Pests and diseases are also a major concern in conditions of global warming (Jactel et al., 2019; Deustch et al., 2018; Elad and Perlot, 2014). As a consequence of current and projected climate changes in Europe's temperate regions, agricultural pests and diseases are expected to occur more frequently and possibly extend to previously unaffected regions (Hirschi et al., 2012), such as bacterial diseases caused by *Xylella fastidiosa* (Schneider et al., 2020).

The severity of diseases induced by *Xylella fastidiosa* has recently increased, possibly due to global warming (Godefroid et al., 2019). Low winter temperatures, especially in Europe, can interfere with the survival of the bacteria in the xylem vessels, allowing fruit species to partially recover from the induced diseases (Pereira, 2015). The same was observed in the studies conducted by Gutierrez et al. (2009), where it was possible to predict, through simulations, a change in the geographic distribution and abundance of fruit flies (*Bactrocera oleae*), due to climate change, in olive groves grown in ecological areas of'Arizona (California) and in Italy (Caselli and Petacchi, 2021; Frem et al., 2021) (Fig. 2).

2. Future Directions in Phenological Research

Overall, considerable uncertainty remains about the relative roles of seasonal changes in temperature, precipitation and photoperiod in driving phenological dynamics, thus hampering the ability to predict how phenological phases may or may not change with climate change. More detailed analysis of geographic variation in plant response can help, with strong regional differences (Menzel et al., 2020). This variability opens the possibilities to compare responses of different species in different locations, with their own characteristics, such as the study conducted by Panchen et al. (2015), where the authors monitored the leaf phenology of more than 1,300 deciduous woody species in six botanical gardens of Asia, North America and Europe.

Gill et al. (2015) carried out a meta-analysis of studies on the time of leaf senescence of autumn in the Northern Hemisphere, showing that global warming or temperature increase could contribute to a delay of two days/year in this senescence. The same authors also reported that the occurrence of leaf senescence in places of high latitude is more sensitive to the photoperiod and in low latitude, it is more sensitive to temperature. These patterns contrasted sharply with leaf emergence times, suggesting that leaf senescence is governed by a larger set of local environmental

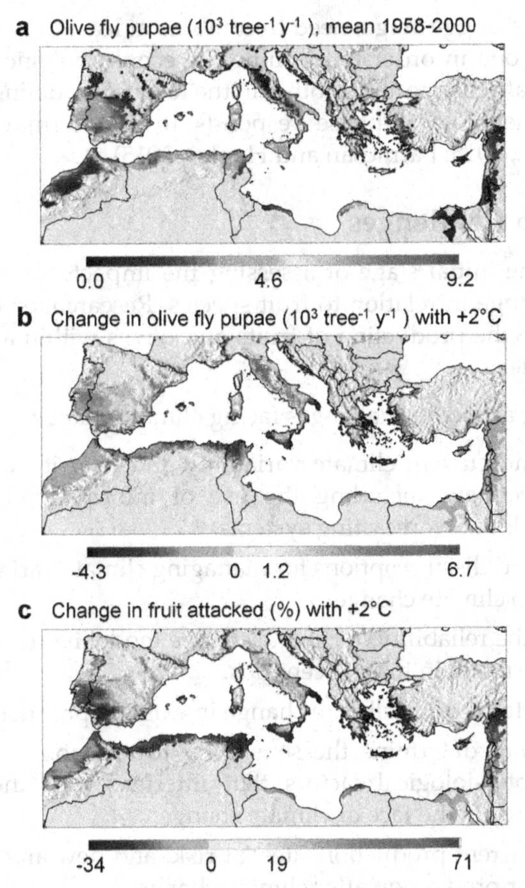

Fig. 2. Simulated dynamics of olive fly (*Bactrocera oleae*) as influenced by bottom-up effects of olive and driven by present climate (years 1958–2000) or in a +2°C climate warming scenario: Cumulative number of pupae (10^3 tree^{-1} year^{-1}) under present climate (a), and change of this number with +2°C climate warming (b); effects of climate warming on per cent olive fruit attacked by the fly (c) (*Source*: Gutierrez and Ponti, 2014).

factors than emergence (Mariën et al., 2019). Although remote sensing techniques have been effective in discerning and analyzing differences between years and regions, such as the 'green-up' (Fitchett et al., 2015), it is still not possible to effectively distinguish the responsible components (Wang et al., 2018).

Leaf phenology is a widely used tool in the assessment of olive plants and a potential indicator of the effects of climate change, especially in the face of the trend towards global warming (Sghaier et al., 2019). It can also be used in the vegetative assessment of fruit species, such as those with temperate climate and the olive tree itself. Climatic fluctuations within and between years should be studied in the phenological phases of fruit species, suggesting a need to combine at least two approaches: an experimental one in order to examine the ecophysiological response of plants to climate change and another in the form of modeling to determine how each phenological phase responds to long-term climate trends (Primack et al., 2015; Parmesan and Hanley, 2015).

3. Research Challenges

Brazil is in the initial stage of assessing the impacts and consequences of climate change in relation to fruit species. Researchers and producers try to manage the production of fruits and olives within a very variable current climate.

The following are some challenges facing climate change:

1) Understand current climate variability and how it can be managed more effectively, including the use of information packages and seasonal climate forecasting systems.
2) Identify and discuss options for managing climate variability in order to adapt to climate change.
3) Improve the reliability of climate change modeling results in order to reduce variation in future scenarios.
4) Continue to monitor climate change in existing production areas.
5) Identify and determine the sensitivity to agronomic, physiological and eco-physiological factors that interfere with the olive tree's performance in the face of climate change.
6) Identify current production sites 'at risk' and new areas that may be suitable for production after climate change.

4. Conclusion

There is little research on the impact of eCO_2 on olive cultivation. However, there is evidence that there is an increase in vegetative and reproductive biomass, due to higher CO_2. Some constituents of the fruit showed an increase, due to eCO_2. Genetic manipulation to improve Rubisco's specificity for CO_2 compared to O_2 and increase Rubisco's catalytic rate in the potential yield of olives, thus increasing efficiency in the use of inputs, is a promising technique. Efficiency in the use of nitrogen is another important aspect to be investigated in the future in a future climate change scenario. Few phenological observations of budding, flowering and fruiting in the olive tree are available in relation to the increase or decrease in production. Phenological models should be developed in future in order to estimate the impact of climate change on the development of plants in different global regions. Due to climate change, there has been a considerable reduction in the number of cold hours, which can be observed in the long term. Changes in precipitation events can also lead to a reduction in the fruit yield. The inadequate function of the pollinator, reduction of the natural population of pollinating agents, inadequate winter cooling, occurrence of frosts, hail and gales, nutritional deficiencies, droughts, etc., can lead to malformation and fixation of the fruit. Requirements of different strategies to adapt to climate change and moving to other cultivars more suited to climate change would be effective options; so would be the use of breeding programs in order to create new varieties better suited to a warmer future. The future of olive cultivation in the different global climate change scenarios will depend on the types of adaptations to be followed.

References

Abobatta, W.F. (2019). Potential impacts of global climate change on citrus cultivation. *MOJ Ecology & Environmental Sciences*, 4: 308–312.

Aguilera, F., Dhiab, A.B., Msallem, M., Orlandi, F., Bonofiglio, T., Ruiz-Valenzuela, L., Galán, C., Díaz-de la Guardia, C., Giannelli, A., del Mar Trigo, M., García-Mozo, H., Pérez-Badia, R. and Fornaciari, M. (2015a). Airborne-pollen maps for olive-growing areas throughout the Mediterranean region: Spatio-temporal interpretation. *Aerobiologia*, 31: 421–434.

Aguilera, F., Fornaciari, M., Ruiz-Valenzuela, L., Galán, C., Msallem, M., Dhiab, A.B., de la Guardia, C.D., Trigo, M.M., Bonofiglio, T. and Orlandi, F. (2015b). Phenological models to predict the main flowering phases of olive (*Olea europaea* L.) along a latitudinal and longitudinal gradient across the Mediterranean region. *International Journal of Biometeorology*, 59: 629–41.

Ahmadi, H. and Baaghideh, M. (2020). Assessment of anomalies and effects of climate change on reference evapotranspiration and water requirement in pistachio cultivation areas in Iran. *Arabian Journal of Geosciences*, 13: 332–337.

Ainsworth, E.A. and Long, S.P. (2020). 30 years of free-air carbon dioxide enrichment (FACE): What have we learned about future crop productivity and its potential for adaptation? *Global Changing Biology*, 27: 27–49.

Algataa, B. (2020). Analysis of the impact of climate change and storage methods on the quality of olive oil in Libya. *International Journal of Scientific and Research Publications*, 10: 968–972.

Aybar, V.E., Abreu, J.P.M., Searles, P.S., Matias, A.C., Del Río, C, Caballero, J.M. and Rousseaux, M.C. (2015). Evaluation of olive flowering at low latitude sites in Argentina using a chilling requirement model. *Spanish Journal of Agricultural Research*, 13: e09–001.

Bagley, J., Rosenthal, D.M., Ruiz-Vera, U.M., Siebers, M.H., Kumar, P., Ort, D.R. and Bernacchi, C.J. (2015). The influence of photosynthetic acclimation to rising CO_2 and warmer temperatures on leaf and canopy photosynthesis models. *Global Biogeochemical Cycles*, 29: 194–206.

Baldi, A., Brandani, G., Petralli, M., Messeri, A., Cecchi, S., Vivoli, R. and Mancini, M. (2019). Termo-pluviometric variability of Val d'Orcia olive orchards area (Italy). *Italian Journal of Agrometeorology*, 2: 11–20.

Belaj, A., de la Rosa, R., León, L., Gabaldón-Leal, C., Santos, C., Porras, R., de la Cruz-Blanco, M. and Loritte, I.J. (2020). Phenological diversity in a Word Olive Germplasm Bank: Potential use for breeding programs and climate change studies. *Spanish Journal of Agricultural Research*, 18: e0701.

Bienes, R., Rodríguez Rastrero, M., Gumuzzio Fernández, J., García-Díaz, A., Sastre, B.E. and Gumuzzio Such, A. (2018). Methodology for edaphoclimatic assessment of olive cultivation: application to the area of the quality mark 'Olive Oil Madrid' (Spain). *Spanish Journal of Soil Science*, 8: 74–101.

Carvalho, C.A.G., Silva, E.O. and Bezerra, M.A. (2014). Impact of climate change on plants, fruits and grains. *Revista Caatinga*, 27: 205–212.

Caselli, A. and Petacchi, R. (2021). Climate change and major pests of Mediterranean olive orchards: Are we ready to face the global heating? *Insects*, 12: 1–13.

Cha, S., Chae, H.M., Lee, S.H. and Shim, J.K. (2017). *Effect of Elevated Atmospheric* CO_2 *Concentration on Growth and Leaf Litter Decomposition of Quercus acutissima and Fraxinus rhynchophylla*, 12: e0171197.

Challinor, A.J., Watson, J., Lobell, D.B., Howden, S.M., Smith, D.R. and Chhetri, N. (2014). A meta-analysis of crop yield under climate change and adaptation. *Nature: Climate Change*, 4: 287–291.

Challinor, A.J., Koehler, A.K., Ramirez-Villegas, J., Whitfield, S. and Das, B. (2016). Current warming will reduce yields unless maize breeding and seed systems adapt immediately. *Nature: Climate Change*, 6: 954–958.

Chamizo, S., Serrano-Ortiz, P., López-Ballesteros, A., Sánchez-Cañete, E.P., Vicente-Vicente, J.L. and Kowalski, A.S. (2017). Net ecosystem CO_2 exchange in an irrigated olive orchard of SE Spain: Influence of weed cover. *Agriculture, Ecosystems & Environment*, 239: 51–64.

Conde-Innamorato, P., Arias-Sibillotte, M, Villamil, J.J., Bruzzone, J., Bernaschina, Y., Ferrari, V., Zoppolo, R., Villamil, J. and Leoni, C. (2019). It is feasible to produce olive oil in temperate humid climate regions. *Frontiers Plant Science*, 27: 01544.

Coutinho, E.F., Wrege, M.S., Reisser Júnior, C., Almeida, I.R. and Steinmetz, S. (2009). *Cultivo de oliveira (Olea europaea L.): Clima, Pelotas: Embrapa Clima Temperado*, 2009 (*Sistema de produção*).

De Ollas, C., Morillón, R., Fotopoulos, V., Puértolas, J., Ollitrault, P., Gómez-Cadenas, A. and Arbona, V. (2019). Facing climate change: biotechnology of iconic Mediterranean woody crops. *Frontiers in Plant Science*, 10: 427.

Deustch, A., Tewksbury, J.J., Tigchelaar, M., Battisti, D.S., Merrill, C., Huey, R.B. and Naylor, R.L. (2018). Increase in crop losses to insect pests in a warming climate. *Science*, 361: 916–919.

Di Paola, A., Chiriacò, M.V., Di Paola, F. and Nieddu, G. (2021). A phenological model for olive (*Olea europaea* L. var *europaea*) growing in Italy. *Plants*, 10: e1115.

Dusenge, M.E., Duarte, A.G. and Way, D.A. (2019). Plant carbon metabolism and climate change: Elevated CO_2 and temperature impacts on photosynthesis, photorespiration and respiration. *New Phytologist*, 221: 32–49.

Elad, Y. and Perlot, I. (2014). Climate change impact on plant pathogens and plant diseases. *Journal of Crop Improvement*, 28.

El-Kholy, M. (2012). *Following Olive Footprints (Olea europaea L.): Cultivation and Culture, Folklore and History, Traditions and Uses*. Córdoba: International Society for Horticultural Science, Series Scripta Horticulturae, Imprenta Luque.

Filipini, Alba, J.M., Flores, C.A., Wrege, M.S., Coutinho, E.F. and Jorge, R.O. (eds.). (2014). *Zoneamento edafoclimático da olivicultura para o Rio Grande do Sul*. Brasília: Embrapa, 80p.

Fischer, G., Almanza-Merchán, P.J. and Ramírez, F. (2012). Source-sink relationships in fruit species: A review. *Revista Colombiana de Ciencias Hortícolas*, 6: 238–253.

Fitchett, J.M., Grab, S.W. and Thompson, D.I. (2015). Plant phenology and climate change: Progress in methodological approaches and application. *Progress in Physical Geography*, 39: 460–482.

Florêncio, G.W.L., Martins, F.B., Ferreira, M.C. and Pereira, R.A.A. (2019). Impacts of climatic changes on the vegetative development of olive cultivars. *Revista Brasileira de Engenharia Agrícola e Ambiental*, 23: 641–647.

Fraga, H., Pinto, J.G, Viola, F. and Santos, J.A. (2019). Climate change projections for olive yields in the Mediterranean Basin. *International Journal of Climatology*, [s.n.]: 1–13.

Franks, P.J., Adams, M.A., Amthor, J.S., Barbour, M.M., Berry, J.A., Ellsworth, D.S., Farquhar, G.D., Ghannoum, O., Lloyd, J., McDowell, N., Norby, R.J., Tissue, D.T. and von Caemmerer, S. (2013). Sensitivity of plants to changing atmospheric CO_2 concentration: from the geological past to the next century. *New Phytologist*, 197: 1077–1094.

Frem, M., Santeramo, F.G., Lamonaca, E., El Moujabber, M., Choueiri, E., La Notte, P., Nigro, F., Bozzo, F. and Fucilli, V. (2021). Landscape restoration due to *Xylella fastidiosa* invasion in Italy: Assessing the hypothetical public's preferences. *NewBiota*, 66: 31–54.

Garcia, S.R., Santos, D.F., Martins, F.B. and Torres, R.R. (2018). *Aspectos climatológicos associados ao cultivo da oliveira (Olea europaea L.) em Minas Gerais. Revista Brasileira de Climatologia*, 22: 188–209.

García-Mozo, H., Oteros, J. and Galán, C. (2015). Phenological changes in olive (*Olea europaea* L.) reproductive cycle in southern Spain due to climate change. *Annals of Agricultural and Environmental Medicine*, 22: 421–428.

Garrido, A., Fernández-González, M., Vázquez-Ruiz, R.A., Rodríguez-Rajo, F.J. and Aira, M.J. (2021). Reproductive biology of olive trees (Arbequina cultivar) at the northern limit of their distribution areas. *Forests*, 12: 1–16.

Ghahramani, A., Howden, S.M., del Prado, A., Thomas, D.T., Moore, A.D., Ji, B. and Ates, S. (2019). Climate change impact, adaptation, and mitigation in temperate grazing systems: A review. *Sustainability*, 11: 1–30.

Gill, A.L., Gallinat, A.S., Sanders-DeMott, R., Rigden, A.J., Gianotti, D.J.S., Mantooth, J.A. and Templer, P.H. (2015). Changes in autumn senescence in northern hemisphere deciduous trees: A meta-analysis of autumn phenology studies. *Annals of Botany*, 116: 875–888.

Godefroid, M., Cruaud, A., Streito, J.C., Rasplus, J.Y. and Rossi, J.P. (2019). *Xylella fastidiosa*: Climate suitability of European continent. *Scientific Reports*, 9: 8844.

González, M.B., Lucas, R.S., Benlloch, M. and Escobar, R.F. (2018). An approach to global warming effects on flowering and fruit set of olive trees growing under field conditions. *Scientia Horticulturae*, 240: 405–410.

Gutierrez, A.P., Ponti, L. and Cossu, Q.A. (2009). Effects of climate warming on olive and olive fly (*Bactrocera oleae* (Gmelin)) in California and Italy. *Climatic Change*, 95: 195–217.

Gutierrez, A.P. and Ponti, L. (2014). Analysis of invasive insects: Links to climate change. pp. 41–61. *In*: Ziska, L.H. and Dukes, J.S. (eds.). *Invasive Species and Global Climate Change*. CABI Publishing, Wallingford, UK.

Haworth, M., Hoshika, Y. and Killi, D. (2016). Has the impact of rising CO_2 on plants been exaggerated by meta-analysis of free air CO_2 enrichment studies? *Frontiers in Plant Science*, 7: e66844.

Hirschi, M., Stoeckli, S., Dubrovsky, M., Spirig, C., Calanca, P., Rotach, M.W., Fischer, A.M., Duffy, B. and Samietz, J. (2012). Downscaling climate change scenarios for apple pest and disease modeling in Switzerland. *Earth System Dynamics*, 3: 33–47.

Houghton, R.A., House, J., Pongratz, J., Van der Werf, G., DeFries, R., Hansen, M., Quéré, C.L. and Ramankutty, N. (2012). Carbon emissions from land use and land-cover change. *Biogeosciences*, 9: 5125–5142.

Houghton, R.A. (2013). Keeping management effects separate from environmental effects in terrestrial carbon accounting. *Global Change Biology*, 19: 2609–2612.

IOC, International Olive Council. (2018). *Statistical Series*. Madrid, Spain: International Olive Council. Disponible in: <http://www.internationaloliveoil.org>. Acess on: 20 May, 2020.

IPCC, Intergovernamental Panel on Climate Change. (2012). Managing the risks of extreme events and disasters to advance climate change adaptation. *In*: Field, C.B., Barros, V., Stocker, T.F., Qin, D., Dokken, D.J., Ebi, K.L., Mastrandrea, M.D., Mach, K.J., Plattner, G.K., Allen, S.K., Tignor, M. and Midgley, P.M. (eds.). *A Special Report of Working Groups*

I and II of the Intergovernmental Panel on Climate Change. Cambridge, UK and New York, NY, USA: Cambridge University Press, p. 582.

IPCC, Intergovernamental Panel on Climate Change. (2013). Summary for policymaker. *In*: Stocker, T.F., Qin, D., Plattner, G.K., Tignor, M., Allen, S.K., Boschung, J., Nauels, A., Xia, Y., Bex, V. and Midgley, P.M. (eds.). *Climate Change 2013: The Physical Science Basis*. Cambridge University Press, 2013, 1535p. Contribution of Working Group II to the Fourth Assessment Report of the Intergovernmental Panel on Climate Change.

Jactel, H., Koricheva, J. and Castagneyrol, B. (2019). Responses of forest insect pests to climate change: Not so simple. *Insect Science*, 35: 103–108.

Kallarackal, J. and Renuka, R. (2014). Phenological implications for the conservation of forest trees. pp. 90–109. *In*: Kapoor, R., Kaur, I. and Koul, M. (eds.). *Plant Reproductive Biology and Conservation*. I.K. International, Delhi.

Kimball, B.A. (2016). Crop responses to elevated CO_2 and interactions with H_2O, N, and temperature. *Current Opinion in Plant Biology*, 31: 36–43.

Knežević, M., Zivotic, L., Perović, V. and Topalovic, A. (2017). Impact of climate change on olive growth suitability, water requirements and yield in Montenegro. *Italian Journal of Agrometeorology*, 2: 39–52.

Koubouris, G., Limperaki, I., Darioti, M. and Sergentani, C. (2019). Effects of various winter chilling regimes on flowering quality indicators of Greek olive cultivars. *Biologia Plantarum*, 63: 504–510.

Loladze, I. (2014). Hidden shift of the ionome of plants exposed to elevated CO_2 depletes minerals at the base of human nutrition. *eLife*, 3: e02245.

Malik, S.K., Chaudhury, R., Dhariwal, O.P. and Bhandari, D.C. (2010). Genetic resources of tropical under-utilized fruits in India. *NBPGR*, New Delhi.

Mariën, B., Balzarolo, M., Dox, I., Leys, S., Lorene, M.J., Geron, C., Portillo-Estrada, M., AbdElgawad, H., Asard, H. and Campiol, M. (2019). Detecting the onset of autumn leaf senescence in deciduous forest trees of the temperate zone. *New Phytologist*, 224: 166–176.

Martins, F.B., Pereira, R.A.A., Pinheiro, M.V.M. and Abreu, M.C. (2014). *Desenvolvimento foliar em duas cultivares de oliveira estimado por duas categorias de modelos. Revista Brasileira de Meteorologia*, 29: 505–514.

Martins, F.B., Pereira, R.A.A., Torres, R.R. and Santos, D.F. (2020). Climate projections of chill hours and implications for olive cultivation in Minas Gerais, Brazil. *Pesquisa Agropecuária*, 55: e01852.

Maselli, F., Chiesi, M., Brilli, L. and Moriondo, M. (2012). Simulation of olive fruit yield in Tuscany through the integration of remote sensing and ground data. *Ecological Modelling*, 244: 1–12.

Menzel, A., Yuan, Y., Matiu, M., Sparks, T., Scheifinger, H., Gehrig, R. and Estrella, N. (2020). Climate change fingerprints in recent European plant phenology running. *Climate Change Attribution in Phenology*, 26: 2599–2612.

Montanaro, G., Nuzzo, G., Xiloyannis, C. and Dichio, B. (2018). Climate change mitigation and adaptation in agriculture: the case of olive. *Journal of Water and Climate Change*, 9: 633–642.

Moriondo, M., Trombi, G., Ferrise, R., Brandani, G., Dibari, C., Ammann, C.M., Lippi, M.M. and Bindi, M. (2013). Olive trees as bio-indicators of climate evolution in the Mediterranean Basin. *Global Ecology and Biogeography*, 22: 818–833.

Moriondo, M., Ferrise, R., Trombi, G., Brilli, L. and Dibari, C. (2015). Modelling olive trees and grape vines in a changing climate. *Environmental Modelling & Software*, 63: 1–15.

Moriondo, M., Leolini, L., Brilli, L., Dibari, C., Tognetti, R., Giovannelli, A., Rapi, B., Battista, P., Caruso, G., Gucci, R., Argenti, G., Raschi, A., Centritto, M., Cantini, C. and Bindi, M. (2019). A simple model simulating development and growth of an olive grove. *European Journal of Agronomy*, 105: 129–145.

Mousavi, S., de la Rosa, R., Moukhli, A., El Riachy, M., Mariotti, R., Torres, M., Pierantozzi, P., Stanzione, V., Mastio, V., Zaher, H., El Antari, A., Ayoub, S., Dandachi, F., Youssef, H., Aggelou, N., Contreras, C., Maestri, D., Belaj, A., Bufacchi, M., Baldoni, L. and Leon, L. (2019). Plasticity of fruit and oil traits in olive among different environments. *Scientific Reports*, 9: e16968.

Mundo-Molina, M. (2015). Climate change effects on evapotranspiration in Mexico. *American Journal of Climate Change*, 4: 163–172.

Nissim, Y., Shloberg, M., Biton, I., Many, Y., Doron-Faigenboim, A., Zemach, H., Hovav, R., Kerem, Z., Avidan, B. and Ben-Ari, G. (2020). High temperature environment reduces olive oil yield and quality. *PLoS ONE* (s.n.): 1–24.

Oliveira, A.F., Villa, F., Gonçalves, E.D., Vieira Neto, J., Silva, L.F.O., Cruz, M.C.M. and Mesquita, H.A. (2010). *Estudos preliminares para o zoneamento agroclimático da cultura da oliveira no estado de Minas Gerais. Circular Técnica*, 8: 1–4.

Oliveira, M.C., Ramos, J.D., Pio, R. and Cardoso, M.G. (2012). *Características fenológicas e físicas e perfil de ácidos graxos em oliveiras no sul de Minas Gerais. Pesquisa Agropecuária Brasileira*, 47: 30–35.

Orlandi, F., Garcia-Mozo, H., Dhiab, A.B., Galán, C., Msallem, M., Romano, B., Abichou, M., Dominguez-Vilches, E. and Fornaciari, M. (2013). Climatic indices in the interpretation of the phenological phases of the olive in Mediterranean areas during its biological cycle. *Climatic Change*, 116: 263–284.

Oteros, J., Orlandi, F., García-Mozo, H., Aguilera, F., Dhiab, A.B., Bonofiglio, T., Abichou, M., Ruiz-Valenzuela, L., del Trigo, M.M., Díaz de la Guardia, C., Domínguez-Vilches, E., Msallem, M., Fornaciari, M. and Galán, C. (2014). Better prediction of Mediterranean olive production using pollen-based models. *Agronomy for Sustainable Development*, 34: 685–694.

Panchen, Z.A., Primack, R.B., Gallinat, A.S., Nordt, B., Stevens, A.D., Du, Y. and Fahey, R. (2015). Substantial variation in leaf senescence times among 1360 temperate woody plant species: Implications for phenology and ecosystem processes. *Annals of Botany*, 116: 865–873.

Parmesan, C. and Hanley, M.E. (2015). Plants and climate change: Complexities and surprises. *Annals of Botany*, 116: 849–864.

Pereira, P.S. (2015). *Xylella fastidiosa*: A new menace for Portuguese agriculture and forestry. *Revista de Ciências Agrárias*, 38: 149–154.

Ponti, L., Gutierrez, A.P., Ruti, P.M. and Dell'Aquila, A. (2014). Fine-scale ecological and economic assessment of climate change on olive in the Mediterranean Basin reveals winners and losers. *Proceedings of the National Academy of Sciences*, 111: 5598–5603.

Primack, R.B., Laube, J., Gallinat, A.S. and Menzel, A. (2015). From observations to experiments in phenology research: investigating climate change impacts on trees and shrubs using dormant twigs. *Annals of Botany*, 116: 889–897.

Rai, R., Joshi, S., Roy, S., Singh, O., Samir, M. and Chandra, A. (2015). Implications of changing climate on productivity of temperate fruit crops with special reference to apple. *Journal of Horticulture*, 2: e1000135.

Ramírez, F. and Kallarackal, J. (2015). Responses of fruit trees to global climate change. *Briefs in Plant Science*. Springer, 42p.

Richardson, A.D., Keenana, T.F., Migliavacca, M., Ryua, Y., Sonnentaga, O. and Toomeya, M. (2013). Climate change, phenology, and phenological control of vegetation feedbacks to the climate system. *Agricultural and Forest Meteorology*, 169: 156–173.

Rodríguez, A., Pérez-López, D., Sánchez, E., Centeno, A., Gómara, I., Dosio, A. and Ramos, M.R. (2019). Chilling accumulation in fruit trees in Spain under climate change. *Natural Hazards and Earth System Sciences*, 19: 1087–1103.

Rojo, J. and Pérez-Badia, R. (2014). Effects of topography and crown-exposure on olive tree phenology. *Trees*, 28: 449–459.

Rouina, Y.B., Zouari, M., Zouari, N., Rouina, B.B. and Bouaziz, M. (2020). Olive tree (*Olea europaea* L. cv. Zelmati) grown in hot desert climate: Physiobiochemical responses and olive oil quality. *Scientia Horticulturae*, 261: e108915.

Santos, D.F., Martins, F.B. and Torres, R.R. (2017). Impacts of climate projections on water balance and implications on olive crop in Minas Gerais. *Revista Brasileira de Engenharia Agrícola e Ambiental*, 21: 77–82.

Sarkar, S. and Sarkar, S. (2018). A review on impact of climate change on evapotranspiration. *The Pharma Innovation Journal*, 7: 387–390.

Schneider, K., Werf, van der W., Cendoya, M., Mourits, M., Navas-Cortes, J.A., Vicent, A. and Lansinka, A.O. (2020). Impact of *Xylella fastidiosa* subspecies *Pauca* in European olives. *PNAS Latest Articles*, 117: 9250–9259.

Sghaier, A., Perttunen, J., Sievaènen, R., Boujnah, D., Ouessar, M., Ayed, R.B. and Naggaz, K. (2019). Photosynthetical activity modelisation of olive trees growing under drought conditions. *Scientific Reports*, 9: e15536.

Taiz, L. and Zeiger, E. (2017). *Fisiologia Vegetal*. 4th ed., *Porto Alegre: Artmed*, 848p.

Tanasijevic, L., Todorovic, M., Pereira, L.S., Pizzigalli, C. and Lionello, P. (2014). Impacts of climate change on olive crop evapotranspiration and irrigation requirements in the Mediterranean region. *Agricultural Water Management*, 144: 54–68.

Tognetti, R. (2012). Adaptation to climate change of dioecious plants: Does gender balance matter? *Tree Physiology*, 32: 1321–1324.

Torres, M., Pierantozzi, P., Searles, P., Rousseaux, M.C., García-Inza, G., Miserere, A., Bodoira, R., Contreras, C. and Maestri, D. (2017). Olive cultivation in the southern hemisphere: Flowering, water requirements and oil quality responses to new crop environments. *Frontiers in Plant Science*, 8: 1–12.

Torres, R.R. and Marengo, J.A. (2014). Climate change hotspots over South America: From CMIP3 to CMIP5 multi-model datasets. *Theoretical and Applied Climatology*, 117: 579–587.

Villa, F. and Oliveira, A.F. (2012). *Origem e expansão da oliveira na América Latina*. pp. 21–38. *In*: Oliveira, A.F. (ed.). *Oliveira no Brasil Tecnologias de Produção*. EPAMIG.

Walter, L.C., Rosa, H.T. and Streck, N.A. (2015). Mecanismos de aclimatação das plantas à elevada concentração de CO_2. *Ciência Rural*, 45: 1564–1571.

Wang, L., Tian, F., Wang, Y., Wu, Z., Schurgers, G. and Fensholt, R. (2018). Acceleration of global vegetation greenup from combined effects of climate change and human land management. *Global Change Biology*, 24: 5484–5499.

Wrege, M.S., Coutinho, E.F., Jorge, R.O., Fritzsons, E. and Pantano, A.P. (2015). Regiões de clima homogêneo no Brasil para produção comercial de oliveiras. *Revista Brasileira de Climatologia*, 16: 142–158.

9

Prospects of Climate Resilience in Pomegranate

Muhammad Nafees,[1] *Sajjad Hussain,*[2] *Muhammad Usman*[3,*] and *Muhammad J Jaskani*[3]

1. Introduction

1.1 Biology and Distribution

Pomegranate (*Punica granatum* L.) is a deciduous to semi-deciduous shrub or medium tree belonging to family Punicaceae (Melgarejo et al., 2009); however, Graham et al. (2005) included it in family Lythraceae, based on its molecular characterization. There are three species in *Punica* (genus): *P. granatum* (cultivated pomegranate), *P. protopunica* (ancestor of the genus, *Punica*) and *P. nana* (ornamental pomegranate) (Melgarejo et al., 2009). The fruit is included in the list of the five oldest cultivated fruits, including date palm, olive, grapes and fig (Aslanova and Magerramov, 2012). Moreover, references to pomegranate are also present in the Quran and the Bible.

Pomegranate has five to six leaves arranged as opposite, sub-opposite or in whorls and are narrow-oblong-lanceolate in shape. It has white or bright red colored flowers arranged as terminally single or in groups of

[1] Department of Horticultural Sciences, Faculty of Agriculture & Environment, The Islamia University of Bahawalpur, Bahawalpur, Pakistan.
[2] Department of Horticulture, Faculty of Agricultural Sciences & Technology, Bahauddin Zakariya University, Multan.
[3] Institute of Horticultural Sciences, University of Agriculture, Jail Road, Faisalabad 38040, Pakistan.
Emails: muhammad.nafees@iub.edu.pk; sajjad.hussain@bzu.edu.pk
* Corresponding author: m.usman@uaf.edu.pk

two to three, with the flexible sum of flowery whorls in cultivated or wild genotypes. Newly emerged leaves in pomegranate are reddish and turn to receding green at maturity. Moreover, polygonal young branches (thorny) turn round in shape at old age (Holland et al., 2009). Wild pomegranate shrubs develop various trunks with bushy look; however, cultivated plants are trained as small trees with a single stem and a sidling bush in a harsh environment (Levin, 2006). Pomegranate bark color fades pink to grey with old age branches and further changes to dark gray in subsequent years.

Flowers appear after 20–30 days of bud break on current season growth with pink to orange-red petals. There are two types of flowers—bell shape (male flower) and vase shape (perfect flowers); however, the percentage of vase shape flowers may be variable across genotypes and could be enhanced, using good cultural practices (Wetzstein, 2011). Self-pollination is predominant with 13% open pollination and sepals develop into fruit crown during fruit development. The fruit is a fleshy berry, aggregate with a great number of seeds, round shaped with open or close globular calyx, depending upon the cultivar. Fruit color in most of the pomegranate genotypes is pink to deep red, black or yellow to green, depending on genotype and concentration of the anthocyanins. Aril size and seed hardiness vary among varieties and is dependent upon environmental conditions (Holland et al., 2009).

The pomegranate plant is highly adaptive to variable agro-climatic conditions and is genetically diverse. Hence, it can be successfully grown in a wide variety of variable environments with a wider circulation in Himalayas and Eurasia (Holland et al., 2009). History of pomegranate domestication and civilization of the Mediterranean region are interconnected for considerable vital nutritional and medicinal use of its fruit in human life. However, despite its nutraceutical properties and economic benefits, still its commercial cultivation is limited and it is being dealt as a minor fruit. Since 2010, due to rapid climatic changes, commercial production of pomegranate is challenging. Despite these issues, global pomegranate production has increased from 3 million tons (2014) to 3.8 million tons in 2017 (Kahramanoglu, 2019). The largest pomegranate exporters are Persia and Spain. California produced > 90% of the pomegranates in the US with 400 tons of total production and 40% exports to Mexico, Australia and Japan in 2017. The demand of pomegranate fruit is rising globally due to its greater nutraceutical and economic significance (Sturgeon and Ronnenberg, 2010) and its cultivation is increasing; however, its production is declining due to many stress factors, including climate change.

The pomegranate originated in Central Asia and spread to southwest Asia from Iran. Moreover, wild pomegranates were native to Iran and spread to different parts of the world (Mars, 2000). Commercial pomegranate cultivars in Israel were developed in 2000 B.C. (Stover and Mercure, 2007) and domesticated (Still, 2006). Today, it is believed that Iran, Afghanistan, China and India are the main centers of origin and Pakistan is declared as the secondary center of origin of wild pomegranate with great genetic diversity (Nafees et al., 2015). Though pomegranate is a minor fruit crop, it is successfully grown in warm tropics to subtropics and in arid to desert zones.

1.2 Climatic Requirements for Cultivation

Pomegranate is a warm temperate crop, cultivated in the Mediterranean region, requiring high temperatures for proper fruit growth and development. So, commercial production is narrow in the coastal to mild summer regions of the world (Gozlekci et al., 2011b). The species are tolerant to wider soil conditions, grow well under mild winters with temperatures not lower than −12°C, and hot dry summer for fruit ripening (Levin, 2006). A temperature range of 32–38°C and dry climate during fruit maturity and ripening result in the best quality fruits. Areas with high relative humidity (RH) are unsuitable for pomegranate production as fruits produced in these conditions are less sweet and face bacterial blight and cracking problems (Ozguven and Yilmaz, 2000). Also, pomegranate is a fairly drought-tolerant crop; however, it require usual amount of water to produce good quality fruit. Over-irrigation results in soft fruits with poorly developed color. Pomegranate trees are amenable to irrigation with saline water having salinity range between 1600–2500 ppm (Maas, 1990).

The effect of regional environment on plant development accounts for most of the morphological and physiological differences in adults, within a specie. Plant responses to the changing environment not only support the plant to survive but also determine the flowering and fruit production spell (Raven and Johnson, 1986). Most woody deciduous fruit trees in the subtropical environments, including pomegranate, require specific chilling units to break endo-dormancy before onset of active vegetative growth in the spring. Therefore, planting cultivars with different chilling requirements, in distinct climatic zones, can affect the growing and marketing period (Gratacos and Cortés, 2007). Initiation of flower bud and its development is less clearly understood; however, floral initiation is associated with early spring growth (Porter and Wetzstein, 2014). Flower initiation takes place at different times under tropical conditions (Babu, 2010). In regions with subtropical conditions, an inadequate number of chilling units causes problems, such as erratic and uneven bud-break and a decrease in the quantity and quality of fruit produced (Allan and Burnett, 1995). Variation in the chilling requirements of 20 pomegranate cultivars had also been reported previously (Soloklui et al., 2017). The study of dormancy had a considerable economic influence on control and production of plants. Moreover, the bud dormancy of deciduous fruit trees including pomegranate in the temperate climates also allow them to survive the harsh environmental conditions during winter (Faust et al., 1991). Chilling requirement is one of the main factors controlling the adaption of plants to the appropriate climate (Aslamarz et al., 2009) and has to be fully satisfied to obtain the desired shoot growth and maximize fruit production (Samish and Lavee, 1982).

Iran, India, China and Turkey have been leading global producers of pomegranate. Turkey has more diverse agro-climatic conditions, ranging from temperate to subtropical areas. The Mediterranean, Aegean and southern Anatolia regions have more suitable climate for pomegranate cultivation. Average temperature during vegetation period is always higher in southeastern Anatolia as compared to Mediterranean and Aegean regions (Ercisli et al., 2007; Ak et al., 2009). Once the chilling requirement is achieved, bud break will start and full bloom stage will be attained with proper heat units. The deficiency in meeting the chilling in deciduous plants leads to uneven leaf emergence and blooming and causes variation in fruit size and maturity which reduce quantity and productivity (Luedeling et al., 2009; Guo et al., 2014). Pomegranate plants need 100–300 h of chilling (Westwood, 1999); however, heat unit requirement of the crop is still not clear. According to Jackson (1999), a temperature of about 10°C, required for bud growth, is considered as base temperature for calculation of Growing Degree Days (GDD) of pomegranate.

1.3 *Impact of Climate Change on Crop Productivity and Nutritional Values*

Knowledge gap regarding the impact of potential climate change is negatively affecting plant productivity and causing nutrient depletion. Adverse environmental conditions negatively affect woody plants' productivity and performance, like citrus, grapes and pomegranate. Pomegranate is usually grown under semi-arid or subtropical climate where abiotic stresses cause serious problems of poor fruit quality and variation in transcriptional activity (De Ollas et al., 2019).

Pomegranate plant is hardy in nature for vegetative growth; however, reproductive stages are very sensitive to water scarcity, uneven heavy rains, heat shock and salinity. Pomegranate production is becoming difficult since 2010, in Egypt, due to heat stress and in Pakistan and India, due to onset of extreme weather conditions including rains; the fruit quality is getting poor (Kahramanoglu, 2019). Overall, pomegranate yield is decreasing in almost all fruit producing countries due to higher disease incidence and unfavorable weather conditions, although production area and market demand is increasing rapidly (Venkatesha and Yogish, 2016). Pomegranate production in California has also reduced due to five years of drought spell in the recent past and growers have uprooted pomegranate plants growing in the almond and pistachio orchards for efficient utilization of water (Packer, 2014). Chinese pomegranate growers are also facing problems and the fruit prices are rising due to decrease in the supply (Kahramanoglu, 2019).

In addition to enhancing productivity, scientists also need to focus on the impact of climate change on the nutritional values of the crops (Leisner, 2020). Drastic changes in the climatic factors, i.e., temperature, humidity and rainfall showed negative impact on secondary metabolites (phenolics, fatty acids, terpenoids and alkaloids) in pomegranate and strawberry fruit crops (Ahmed and Stepp, 2016).

1.4 Germplasm Resources, Characterization and Natural Adaptation

There are about 760 genotypes of pomegranate, conserved in Persia and have been raised to 1,157 accessions (Levin, 1996). Turkmenistan, Russia and Iran have conserved 1,100, 800 and 770 accessions of pomegranate, respectively (Levin, 2006; Rahimi et al., 2012; Holland and Baryakov, 2018). Ukraine has 370 accessions conserved at Yalta (Yezhov et al., 2005). China conserved 238 pomegranate cultivars in a variety improvement program for sustainable pomegranate supply in the market (Fang et al., 2006) and has established germplasm repository having 289 accessions (Table 1) in Shandong (Yuan and Zhao, 2019). There are 280 genotypes conserved in National Clonal Germplasm Repository, Davis, California (Preece, 2017). Uzbekistan, Tajikistan and Azerbaijan have conserved about 200–300 pomegranate accessions (Levin, 2006). Horticultural Research Institute in Alata, had 180 pomegranate genotypes composed from Turkey and Aegean regions (Holland et al., 2009). Moreover, 158 pomegranate accessions were reported by Frison and Servinski (1995) in Izmir. There are 87 and 63 accessions conserved in Spain and Tunisia, respectively. Pakistan has collected, characterized and conserved 40 different pomegranate genotypes (Fig. 1) at University of Agriculture, Faisalabad (Nafees et al., 2018).

Pomegranate descriptors were successfully used in morphological (Mars and Marrakchi, 1999) and biochemical (Hasnaoui et al., 2011) characterization which provided the base to compare genotypes for selection and breeding programs (Dafny-Yalin et al., 2010). A high-level diversity was recorded in pomegranate accessions for numerous fruit characters (Ercisli et al., 2007). Fruits are generally grouped as soft, semi-soft, hard or semi-hard, based on seed hardiness (Zamani, 2007). Hasnaoui et al. (2011) measured fruit color of Tunisian pomegranate genotypes scoring yellow green (01), red purple (16) with juice colors as white, yellow to dark red purple. There was 103–505 g and 23–274 g

Table 1. An overview of major germplasm resources of pomegranate.

Country	Number of accessions	Organization/Institute	References
Turkmenistan	1100	Turkmenistan Experimental Station of Plant Genetic Resources, Garrygala	Levin (2006), Holland and Baryakov (2018)
Russia	800	Vavilov Research Institute of Plant Industry, St. Petersburg	Frison and Servinsky (1995)
Iran	770	Iranian National Pomegranate Collection, Yazd Horticulture Research Center, University of Tehran, Tehran	Rahimi et al. (2012)
Ukraine	370	Yalta	Yezhov et al. (2005)
China	289	National Pomegranate Germplasm Repository Yichang, Shandong	Yuan and Zhao (2019)
Azerbijan	200–300	-	Levin (2006)
Tajikistan	200–300	-	Levin (2006)
USA	280	National Clonal Germplasm Repository, Davis, California	Preece (2017)
India	190	Central Inst. of Arid Horticulture, Rajasthan Indian Council of Agricultural Research Solapur, Maharashtra	Chandra et al. (2010)
Turkey	180	Alata Horticultural Research Institute, Mersin Aegean Agricultural Research Institute, Izmir	Onur and Kaska (1985), Frisen and Servinsky (1995)
Spain	87	Spanish Germplasm Collection, *Instituto Valenciano de Investigacions Agrarias* (IVIA), Moncada	Badanes et al. (2017)
Tunisia	63	Tunisian National Pomegranate Germplasm Collection, Zerkin	Hasnaoui et al. (2012)
Thailand	29	Bangkok and Chiang Mai	Thongtham (1986)
Pakistan	40	Institute of Horticultural Sciences, University of Agriculture, Faisalabad	Nafees et al. (2018)
Afghanistan	-	Horticulture Germplasm Collections, Kabul	www.ku.edu.af/en

fruit weight in wild and cultivated Irani and Pakistani pomegranates (Akbarpour et al., 2009; Nafees et al., 2015), respectively.

In breeding programs, molecular markers are used for gene sequencing and to understand the basis of genetic mechanisms. DNA markers consist of secondary metabolites, deoxyribonucleic acids, protein contents and

Fig. 1. Pomegranate germplasm of Pakistan collected from Punjab, Khyber Pakhtunkhwa and Baluchistan with fruit and aril color, shape and size variation, and conservation activities at Institute of Horticultural Sciences, University of Agriculture, Faisalabad-Pakistan (Nafees, 2015, Ph.D dissertation).

macromolecules. Molecular studies are not depending on morphological characters (Kumar, 1999). However, phenotype and expression of genes for aril and rind color were different in various pomegranate genotypes (Holland et al., 2009). Moreover, morphological variations of some characters are not reproduced in polymorphism of molecular markers. This argument was resolved by separating more pomegranate markers and using them in evolutionary studies. Mehranna et al. (2006) reported that a lot of work is required to build pomegranate genomic libraries with 26 SSR primers which could be successfully used in screening of pomegranate genotypes. It was proved that SSR markers could successfully be used in genetic diversity assessment (Hasnaoui et al., 2010b) for its reliability, reproducibility and specificity.

Specific sequence markers could accurately be used in genetic diversity and population mapping processes in pomegranates (Xue et al., 2006). DNA-based fingerprinting techniques, like AFLP was used by Fang et al. (2006) and Yuan et al. (2007) to investigate 85 Chinese pomegranate genotypes for genetic diversity and population mapping. Use of molecular markers could be useful for characterization, genetic mapping and sequencing-based studies could help to underpin genes of interest that may regulate the abiotic stress tolerance and enable mitigation towards climatic resilience. Moreover, there is need to identify stress-tolerant genotypes for future crop improvement programs.

2. Impact of Extreme Temperature and Elevated Atmospheric Carbon Dioxide (CO_2) on Plant Productivity and Management

Climate change has emerged as a serious global challenge faced by humanity (Pachauri et al., 2014). It is predicted that due to rapid climatic changes, world agrosystem will be seriously affected, causing a reduction in the plant health and ultimately declining productivity. Significant increase in greenhouse emission gases, such as CO_2 is affecting the global climate, particularly distribution of the rainfall and its frequency due to the elevated temperature of Earth's surface and oceans (Cicerone and Nurse, 2014). By the end of 21st century, a rise of 2.5–7.8°C temperature on the global surface is expected and the increase in CO_2 level up to 1000 ppm (Pachauri et al., 2014). All environmental factors are involved in decline of the plant health, which ultimately results in a decrease in crop production.

Moreover, the decline in yield is not solely affected by the climatic factors; several other factors including plant species, topography, agricultural practices and management also influence in that scenario (Olesen and Bindi, 2002). Pomegranate fruit cracking is associated with high temperatures and significant alterations in the irrigation requirements. It is well cited that increase in temperatures reduces net carbon gain by increasing the respiration in plant more than the photosynthetic rate. The increase in heat above the maximum limit could significantly reduce plant photosynthesis due to the heating ability of Rubisco (an enzyme involved in the first major step of carbon fixation), whereas it also caused limitation of electron transport in the chloroplast (Sage et al., 2008). Also, by increasing the moisture content in the atmosphere, the higher temperature often results in closure of the stomata, which further decreases photosynthesis due to less CO_2 flux into the leaves. The rise in overall global temperature will enhance crop growth and development, while negatively affecting the yield as compared to the total biomass. These adverse effects are also correlated with other environmental conditions, like severe water deficiency, more intense rainfall, salinity and enhanced CO_2 (Sage et al., 2008).

Flowering and fruiting are the most important events in fruit crops which are regulated by the climatic conditions. The alterations in climate change attributes disturb the pattern of flowering, pollination and fruit setting in many fruit crops, mainly due to reduction in the pollinator activities and pollen viability. Rain during flowering period washes out pollens from the stigma, resulting in poor or no fruit set. Crop duration or maturity indices are important physiological processes in fruit crops as they directly affect the fruit quality and yield. Various parameters of climate change also alter the fruit maturation. Production timing of the crops will change because of rise in temperature, leading to rapid crop development and early maturity (Hatfield and Prueger, 2015). Pomegranates are mainly grown in the Mediterranean region with quite warm summers. These high temperatures cause sunburn to the skin of the fruit. Melgarejo et al. (2004) have shown that the use of kaolin dye treatments can ultimately decrease sunburn effect. The antioxidants and fruit quality are also affected by high temperatures and sunburn. Heat stress in plants causes more production of Reactive Oxygen Species (ROS), leading to occurrence of oxidative stress (Ma et al., 2008). Plants protect themselves from the negative effects of the active oxygen species by increasing the accumulation of antioxidant enzymes or metabolites. These enzymes and metabolites may cause reduction in ROS (Ma et al., 2008). The pomegranate fruit has polyphenols which help to cope with ROS (Mousavinejad et al., 2009).

3. Sensitivity to Air Pollutants and Nutritional Contaminants

Different air pollutants from anthropogenic sources, such as oxides of sulfur and nitrogen, particulate emissions and other contaminants may impair the plant health, including trees and crops. Air-borne pollution causes either direct or indirect effects through leaf and soil acidification (Jones et al., 2012; Rai et al., 2013). When exposed to air pollution, most plants experience physiological changes before exhibiting visible damage to leaves (Seyyednejad et al., 2011). It has been reported that vegetation along the roadside could reduce the concentration of various air pollutants. This might be caused by the absorption of pollutants by the vegetation canopy and pollutant dispersion into the planting area. Also, plant morphological structure might change due to the acclimatization of plants to air pollutants; for example, thicker epidermal cells and -longer trichomes (Liu and Ding, 2008). On leaves, the effect of airborne pollution is most apparent since these are the primary receptors of air pollutants. Such responses to air pollution have made urban vegetation increasingly important in recent years due to their effect on local and regional air quality.

The increase in industrial activities and urbanization are the major factors contributing to air pollution. Among the air pollutants, SO_2 is the main pollutant which is directly produced from the pollution source as a product of processing of raw material that contains sulfur. The primary sources of SO_2 are the burning of coal and oil with high sulfur content. The direct toxic effect of pollutants and SO_2 on plants has been well documented (Winner et al., 1985). The mode and extent of damage caused by SO_2 to pomegranate has not been extensively studied. The various studied traits showed leaf injury, which includes both leaf number and its area enhanced significantly in pomegranate with increasing SO_2 and time of fumigation. Moreover, the damage was high when leaves were exposed at the highest concentration of SO_2 as compared to control (Swain and Padhi, 2015). This kind of injury mainly occurs due to the faster accumulation of SO_2 than its oxidation and assimilation in the plant tissues exceeding threshold accumulation in the intercellular spaces of the leaf and causing cell injury. A decrease in fresh and dry weight of the tissues increased by increasing levels of SO_2. The increase in SO_2 significantly affects the chloroplast machinery of cell. The chlorophyll content decreased with increasing SO_2 concentration and the effect is being more accentuated when the simulation period increases (Swain and Padhi, 2015). The decrease in chlorophyll content is caused by the degradation of chloroplast membrane due to the phytotoxic nature of SO_2, resulting in leaching of pigments (Rath et al., 1994). Such interference of SO_2 is believed to promote secondary processes that break down chlorophyll and kill the cells. The sulphur content in pomegranate plant tissues increased with an increase in levels of SO_2 fumigation. This increase in sulphur content of tissues with rising levels of SO_2 is probably due to inability of plants to metabolize and assimilate the excess SO_2 at cellular level, thus resulting in its manifold accumulation (Swain and Padhi, 2015).

4. Abiotic Stress and Tolerance Mechanism

Reactive Oxygen Species (ROS) are expected chemicals released in the living organisms during various stresses and produced during aerobic metabolism (Halliwell, 2006). Abiotic stress causes the release of ROS, resulting in a wide range of physiological changes (Nafees et al., 2018). Under abiotic stress, ROS production could be a crucial indication of phytotoxicity (Choudhury et al., 2013). Aerobic metabolism (respiration and photosynthesis) generates ROS, like hydrogen peroxide, superoxide and hydroxyl radicals. These may cause oxidative damage to native enzymes, proteins and nucleic acids.

Under the current climate change scenario, water stress restricts the distribution of plant species and adversely affects plant growth and performance (Liu et al., 2011). There are various morpho-chemical and molecular changes in plants in response to environmental stresses (Reddy et al., 2004). The plants adopt a drought escape or drought-resistance mechanism to address water deficiency during any stage of its life span. Resistance mechanism can be further classified into drought avoidance or drought tolerance which could be achieved through dehydration avoidance or dehydration tolerance through the addition of protective proteins and solutes, metabolic processes, detoxification of ROS species and change in gene expression (Verslues et al., 2006; Wang et al., 2012). Reactive oxygen species are released in response to various environmental stresses, like high temperature, chilling, salinity and drought (Sung et al., 2003; Jung, 2004; Khan and Panda, 2008; Bian and Jiang, 2009). Excessive production of ROS damages the cellular structures through protein oxidation, peroxidation of lipids of cell membrane, degradation of various pigments, damage to nucleic acids and enzyme inactivation which result in destruction of normal plant metabolism (Gholami et al., 2012; Silva et al., 2012). Plants have developed defense system against ROS to normalize its level in various compartments of cells and with oxidative damage the balance of ROS is disturbed in cells (Menezes-Benavente et al., 2004). A brief overview of pomegranate varietal responses to drought stress is presented (Table 2).

Table 2. An overview of water-stress tolerance in pomegranate.

S. No.	Cultivars/Accessions	Key Traits and Responses	References
1.	Gho-jagh and M-Saveh	Recovery of Chlorophyll a + b.	Liu et al. (2011)
2.	Gho-jagh and M-Saveh	Increase in leaf carotenoid content.	Pourghayoumi et al. (2017)
3.	Rabab, Gho-jagh, Shishecap, M-Saveh and M-Yazdi	Higher enzyme activity (SOD, G-POD & CAT) and more proline production.	
4.	Rabab, Gho-jagh, Shishecap, M-Saveh and M-Yazdi	More proline production.	Ahmed et al. (2009); Liu et al. (2011), Pourghayoumi et al. (2017)
5.	Rabab and M-Saveh	Increase in malondialdehyde contents.	Pourghayoumi et al. (2017)
6.	Rabab, Gho-jagh, Shishecap, M-Saveh M-Yazdi (sensitive to drought stress)	Increase in GPX and cytosolic GR.	

5. Mitigation and Yield Improvement Approaches

Naturally growing pomegranate accessions are being successfully used in breeding and variety improvement programs. Objectives of pomegranate breeding in the world remained for yield, seed softness, fruit quality, disease resistance, aril color and juice production (Samadia and Pareek, 2006; Jalikop et al., 2006). There are cultivars available with soft seeds, good-quality fruit and early fruit ripening (Wang et al., 2006; Zhao et al., 2006; Zhao et al., 2007) developed through characterization and selection process. Matuskovic and Micudova (2006) used colchicine to produce tetraploid pomegranates in a variety improvement program. *Agrobacterium* was used for the establishment of genetic transformation and development of transgenics (Terakami et al., 2007). Phenotypic and genotypic differences shall be well understood to engineer the future crops for more resilience to the changing climatic scenarios. Gene-mapping techniques, like Quantitative Trait Loci (QTL) and Genome-Wide Association Studies (GWAS) are effective tools to link a phenotypic character with its genetics and functional validation work (Leisner, 2020). Application of Next-Generation Sequencing (NGS) and GWAS have allowed for more precise localization of associated genetic signals (Varshney et al., 2014).

Knowledge about pomegranate breeding approaches for better climate resilience is limited though this is an emerging issue in the world, especially in the developing countries. Turkish Government has launched a joint effort for the adoption of pomegranate production and its relationship to climate-related agricultural concerns.

Pomegranate orchards are facing huge water stress which is one of the causes of the reduction in yield, fruit quality and cracking. It is necessary to screen out most water-deficit tolerant pomegranate varieties through characterization and breeding programs. Studies on the response of pomegranate to water stress conditions to investigate various physiological aspects, like photosynthesis, stomatal conductance, transpiration and leaf osmotic potential (Parvizi et al., 2016) are of great concern under rapidly changing climatic conditions to screen out drought resistance in pomegranate germplasm. Pomegranate plants with an increased level of super dismutase, peroxidase and catalase under drought conditions showed that these enzymes are controlling the cellular level of ROS (Pourghayoumi et al., 2017). They marked pomegranate cultivar 'Rabab' as 'excellent' under drought conditions in the expression level of cytosolic glutathione reductase and glutathione peroxidase genes in comparison to other cultivars in this study. Pourghayoumi et al. (2017) concluded that 'Ghojagh' and 'Rabab' had higher tolerance against water stress and advised growing these in regions with water scarcity. This information could be successfully used to develop new drought- and heat-tolerant pomegranate cultivars.

High density plantation system is an emerging trend through advance canopy management practices to develop microclimatic conditions for plants with efficient utilization of farm inputs. Haneef et al. (2014) planted cv. Bhagwa at a dimension of 2 m × 2 m which increased the yield. High density plantation is possible in cv. Mridula with three sprays of Auxins (NAA 10 ppm + GA_3 50 ppm) to support productivity and fruit quality (Shanmugasundaram and Balakrishnamurthy, 2017). Marathe et al. (2017) worked on the bedding system in pomegranate trees to ensure best aeration in root zone and increased nutrient uptake in arid and semi-arid regions. Pruning of pomegranate branches to about 15–30 cm in length proved effective against yield (Hiremath et al., 2018; Kabuli et al., 2018) with strong source-sink relationship and effective utilization of nutrients. Use of amino and humic acids is important to address the climatic stress and increase fruit size with yield of cv. Manfalouty (Khattab et al., 2012). Water requirements of the pomegranate plant could be predicted by recording the trunk diameter with the midday stem water potential and leaf stomatal conductance (Galindo et al., 2013). Fruit thinning improved the efficacy of insecticide to reduce pest population in the orchard (Kahramanoglu and Usanmaz, 2013). Pomegranate fruits with black heart disease were observed after 2010, in Italy (Faedda et al., 2015), Spain (Berbegal et al., 2014) and other European countries (Kahramanoglu et al., 2014) because of climate change in the region. Azole group fungicides were reported to be effective to inhibit the pathogen growth (Kumar et al., 2017). Physiological disorders (sunburn and sunscald) are because of fruit exposure to direct sunlight which could be controlled through proper canopy management and using some covering materials (Gundeşli et al., 2019). Covering fruits with 18/22 holed covering materials led to 50% reduction in sunburn (Ghorbani et al., 2015); however, Meena et al. (2016) used red nets of 35% shade levels to prevent sunburn on fruit skin. Fruit cracking in pomegranate orchards is also directly linked with climate change (drought and humidity fluctuation) and was successfully managed with application of paclobutrazol (Khalil et al., 2013). Foliar calcium fertilization also reduced fruit cracking (Davarpanah et al., 2013). Suitable storage temperature for pomegranate fruits is 5–7°C with > 90% relative humidity for preservation of the quality of fresh pomegranate fruits (Okatan et al., 2018).

6. Conclusion and Way Forward

Pomegranate is a minor fruit that originated in Asia and is a predominantly deciduous large shrub. Its widespread cultivation in the world shows its higher adaptation in temperate to tropical and subtropical to arid climates of the world. Moreover, in changing climate scenario, measures should be taken towards germplasm resources conservation, abiotic stresses management and fruit quality and yield improvement. Tree modeling under the current climate change scenario is the dire need to continue the sustainable supply of pomegranate fruit in the international market and to address malnutrition and food security issues of the world.

References

Ahmed, S. and Stepp, J.R. (2016). Beyond yields: Climate change effects on specialty crop quality and agro-ecological management. *Elementa Science of the Anthropocene*, 4: 92–92.

Ak, B.E., Ikinci, A., Parlakci, H., Özgüven, A.I. and Yilmazand, C. (2009). Some pomological traits of different pomegranate varieties grown in Sanliurfa-Turkey. *Acta Horticulturae*, 818: 115–119.

Akbarpour, V., Hemmati, K. and Sharifani, M. (2009). Physical and chemical properties of pomegranate (*Punica granatum* L.) fruit in maturation stage. *American-Eurasian Journal of Agricultural and Environmental Sciences*, 6(4): 411–416.

Allan, P. and Burnett, M.J.I. (1995). Peach production in an area with low winter chilling. *Journal of Southern African Society for Horticultural Sciences*, 5: 15–18.

Aslamarz, A.A., Vahdati, K., Rahemi, M. and Hassani, D. (2009). Estimation of chilling and heat requirements of some Persian walnut cultivars and genotypes. *HortScience*, 44(3): 697–701.

Aslanova, M.S. and Magerramov, M.A. (2012). Physicochemical parameters and amino acid composition of new pomological sorts of pomegranate fruits. *Chemistry of Plant Raw Material*, 1: 165–169.

Babu, D.K. (2010). Floral biology of pomegranate (*Punica granatum* L.). *Pomegranate*, 4(2): 45–50.

Badanes, M.L., Zuriaga, E., Bartual, J. and Pintova, J. (2017). Genetic diversity among pomegranate germplasm assessed by microsatellite markers. pp. 19. *In*: Baratual, J. and Badenes, M.L. (eds.). *The Proceedings of the IV International Symposium on Pomegranate and Minor Mediterranean Fruits*. 18–22 September 2017, Elche, Spain.

Berbegal, M., Lopez Cortes, I., Salazar, D., Gramaje, D., Perez-Sierra, A., Garcia-Jimenez, J. and Armengol, J. (2014). First report of Alternaria Black Spot of pomegranate caused by *Alternaria alternata* in Spain. *Plant Disease*, 98(5): 689–689.

Bian, S. and Jiang, Y. (2009). Reactive oxygen species, antioxidant enzyme activities and gene expression patterns in leaves and roots of Kentucky bluegrass in response to drought stress and recovery. *Scientia Horticulturae*, 120: 264–270.

Chandra, R., Jadhav, V.T. and Sharma, J. (2010). Global scenario of pomegranate (*Punica granatum* L.) culture with special reference to India. *Fruit Vegetable and Cereal Science and Biotechnology*, 4(2): 7–18.

Choudhury, S., Panda, P., Sahoo, L. and Panda, S.K. (2013). Reactive oxygen species signaling in plants under abiotic stress. *Plant Signaling Behavior*, 8(4): e23681.

Cicerone, R.J. and Nurse, P. (2014). *Climate Change Evidence & Causes*. National Academy of Sciences Press, Washington, DC.

Dafny-Yalin, M., Glazer, I., Bar-Ilan, I., Kerem, Z., Holland, D. and Amir, R. (2010). Color, sugars and organic acids composition in aril juices and peel homogenates prepared from different pomegranate accessions. *Journal of Agricultural and Food Chemistry*, 58: 4342–4352.

Davarpanah, S., Tehranifar, A., Abadía, J., Val, J., Davarynejad, G., Aran, M. and Khorassani, R. (2018). Foliar calcium fertilization reduces fruit cracking in pomegranate (*Punica granatum* cv. Ardestani). *Scientia Horticulturae*, 230: 86–91.

De Ollas Valverde, C.J., Morillon, R., Fotopoulos, V., Puértolas, J., Ollitrault, P., Gómez Cadenas, A. and Arbona, V. (2019). *Facing Climate Change: Biotechnology of Iconic Mediterranean Woody Crops.*

Ercisli, S., Agar, G., Orhan, E., Yildirim, N. and Hizarci, Y. (2007). Inter specific variability of RAPD and fatty acid composition of some pomegranate cultivars (*Punica granatum* L.) growing in Southern Anatolia Region in Turkey. *Biochemical Systematics and Ecology*, 35: 764–769.

Faedda, R., Granata, G., Pane, A., Evoli, M., Giudice, V.L., Lio, G.M.S. and Cacciola, S.O. (2015). Heart rot and soft rot of pomegranate fruit in southern Italy. *Acta Horticulturae*, 1144: 198–198.

Fang, J., Tao, J. and Chao, C.T. (2006). Genetic diversity in fruiting-mei, apricot, plum and peach revealed by AFLP analysis. *Journal of Horticultural Science and Biotechnology*, 81(5): 898–902.

Faust, M., Liu, D., Millard, M.M. and Stutte, G.W. (1991). Bound versus free water in dormant apple buds—A theory for endodormancy. *HortScience*, 26(7): 887–890.

Frisen, E.A. and Servinsky, J. (1995). *Directory of European Institutions Holding Crop Genetic Resources Collections.* International Plant Genetic Resources Institute, Rome, Italy, 87 pp.

Galindo, A., Rodríguez, P., Mellisho, C.D., Torrecillas, E., Moriana, A., Cruz, Z.N., Conejero, W., Moreno, F. and Torrecillas, A. (2013). Assessment of discretely measured indicators and maximum daily trunk shrinkage for detecting water stress in pomegranate trees. *Agricultural and Forest Meteorology*, 180: 58–65.

Gholami, M., Rahemi, M., Kholdebarin, B. and Rastegar, S. (2012). Biochemical responses in leaves of four fig cultivars subjected to water stress and recovery. *Scientia Horticulturae*, 148: 109–117.

Ghorbani, M., Dabbagh, G.R., Yousefi, D., Khademi, S. and Taki, M. (2015). The effect of application of different kinds of covers on the sunburn and internal qualities of pomegranate in Iran. *Biological Forum—An International Journal*, 7(1): 64–68.

Gozlekci, S., Saracoglu, O., Onursal, E. and Ozgen, M. (2011b). Total phenolic distribution of juice, peel, and seed extracts of four pomegranate cultivars. *Pharmacognosy Magazine*, 7(26): 161–164.

Graham, S.A., Hall, J., Sytsma, K. and Shi, S. (2005). Phylogenetic analysis of the Lythraceae based on four gene regions and morphology. *International Journal of Plant Sciences*, 166(6): 995–1017.

Gratacós, E. and Cortés, A. (2007). Chilling requirements of cherry cultivars. *Compact Fruit Tree*, 40: 7–9.

Gundeşli, M.A., Kafkas, S., Zarifikhosroshahi, M. and Kafkas, N.E. (2019). Role of endogenous polyamines in the alternate bearing phenomenon in pistachio. *Turkish Journal of Agriculture and Forestry*, 43(3): 265–274.

Guo, L., Dai, J.H., Ranjitkar, S., Yu, H.Y., Xu, J.C. and Luedeling, F. (2014). Chilling and heat requirements for flowering in temperate fruit trees. *International Journal of Biometeorology*, 58(6): 1195–1206.

Halliwell, B. (2006). Reactive species and antioxidants. Redox biology is a fundamental theme of aerobic life. *Plant Physiology*, 141: 312–322.

Haneef, M., Kaushik, R.A., Sarolia, D.K., Mordia, A. and Dhakar, M. (2014). Irrigation scheduling and fertigation in pomegranate cv. Bhagwa under high density planting system. *Indian Journal of Horticulture*, 71(1): 45–48.

Hasnaoui, N., Buonamici, A., Sebastian, F., Mars, M., Trifi, M. and Vendramin, G.G. (2010a). Development and characterization of SSR markers for pomegranate (*Punica granatum* L.) using an enriched library. *Conservation Genetic Resources*, 2: 283–285.

Hasnaoui, N., Mars, M., Chibani, J. and Trifi, M. (2010b). Molecular polymorphisms in Tunisian pomegranate (*Punica granatum* L.) as revealed by RAPD fingerprints. *Diversity*, 2: 107–114.

Hasnaoui, N., Jbir, R., Mars, M., Trifi, M., Kamal-Eldin, A., Melgarejo, P. and Hernandez, F. (2011). Organic acids, sugars and anthocyanins contents in juices of Tunisian pomegranate fruit. *International Journal of Food Properties*, 14: 741–757.

Hasnaoui, N., Buonamici, A., Sebastiani, F., Mars, M., Zhang, D. and Vendramin, G.G. (2012). Molecular genetic diversity of *Punica granatum* L. (pomegranate) as revealed by microsatellite DNA markers (SSR). *Gene*, 493(1): 105–112.

Hatfield, J.L. and Prueger, J.H. (2015). Temperature extremes: Effect on plant growth and development. *Weather and Climate Extremes*, 10: 4–10.

Hiremath, A., Patil, S.N., Hipparagi, K., Gandolkar, K. and Gollagi, S.G. (2018). Influence of pruning intensity on growth and yield of pomegranate (*Punica granatum* L.) CV. Super Bhagwa under organic conditions. *Journal of Pharmacognosy and Phytochemistry*, 7(2): 1027–1031.

Holland, D., Hatib, K. and Bar-Ya'akov, I. (2009). Pomegranate: botany, horticulture, breeding. pp. 127–191. *In*: Janick, J. (ed.). *Horticultural Reviews*. vol. 35, John Wiley & Sons, Ramat Yishay, Israel.

Holland, D. and Bar-Ya'akov, I. (2018). The pomegranate: New interest in an ancient fruit. *Chronica Horticulturae*, 48(3): 12–15.

Jackson, D. (1999). Climate and Fruit Plants, *Temperate and Subtropical Fruit Production*. CAB International Publications, Cambridge, UK, 321 pp.

Jalikop, S.H., Kumar, P.S., Rawal, R.D. and Kumar, R. (2006). Breeding pomegranate for fruit attributes and resistance to bacterial blight. *Indian Journal of Horticulture*, 63(4): 352–358.

Jones, D.L., Rousk, J., Edwards-Jones, G., DeLuca, T.H. and Murphy, D.V. (2012). Biochar-mediated changes in soil quality and plant growth in a three-year field trial. *Soil Biology and Biochemistry*, 45: 113–124.

Jung, S. (2004). Variation in antioxidant metabolism of young and mature leaves of Arabidopsis thaliana subjected to drought. *Plant Science*, 166: 459–466.

Kabuli, K., Sharma, D.P. and Singh, N. (2018). Effect of rejuvenation pruning on the growth, productivity and disease incidence in declining trees of pomegranate (*Punica granatum* L.) cv. Kandhari Kabuli. *Journal of Applied and Natural Science*, 10(1): 358–362.

Kahramanoglu, İ. and Usanmaz, S. (2013). Management strategies of fruit damaging pests of pomegranates: *Planococcus citri, Ceratitis capitata* and *Deudorix* (*Virachola*) *livia*. *African Journal of Agricultural Research*, 8(49): 6563–6568.

Kahramanoglu, I., Usanmaz, S. and Nizam, I. (2014). Incidence of heart rot at pomegranate fruits caused by *Alternaria* spp. in Cyprus. *African Journal of Agricultural Research*, 9(10): 905–907.

Kahramanoglu, I. (2019). Trends in pomegranate sector: production, postharvest handling and marketing. *International Journal of Agriculture, Forestry and Life Sciences*, 3(2): 239–246.

Khalil, H.A. and Aly, H.S.H. (2013). Cracking and fruit quality of pomegranate (*Punica granatum* L.) as affected by pre-harvest sprays of some growth regulators and mineral nutrients. *Journal of Horticultural Science and Ornamental Plants*, 5(2): 71–76.

Khan, M. and Panda, S. (2008). Alterations in root lipid peroxidation and antioxidative responses in two rice cultivars under NaCl-salinity stress. *Acta Physiologia Plantarum*, 30: 81–89.

Khattab, M.M., Shaban, A.E., El-Shrief, A.H. and Mohamed, A.S.E. (2012). Effect of humic acid and amino acids on pomegranate trees under deficit irrigation. I: Growth, Flowering and Fruiting. *Journal of Horticultural Science and Ornamental Plants*, 4(3): 253–259.

Kumar, A., Chahal, T.S., Hunjan, M.S., Kaur, H. and Rawal, R. (2017). Studies of Alternaria black spot disease of pomegranate caused by *Alternaria alternata* in Punjab. *Journal of Applied and Natural Science*, 9(1): 156–161.

Kumar, L.S. (1999). DNA markers in plant improvement: An overview. *Biotechnology Advances*, 17(2): 143–182.

Leisner, C.P. (2020). Review: Climate change impacts on food security-focus on perennial cropping systems and nutritional value. *Plant Science*, 293: 110412.

Levin, G.M. (1996). Pomegranate (*Punica granatum* L.) collection research in Turkmenistan. *Plant Genetic Resources Newsletter*, 106: 47–49.

Levin, G.M. (2006). *Pomegranate Roads: A Soviet Botanist's Exile from Eden*. Floreat Press, Forestville, CA, USA, 183 pp.

Liu, C., Liu, Y., Guo, K., Fan, D., Li, G., Zheng, Y., Yu, L. and Yang, R. (2011). Effect of drought on pigments, osmotic adjustment and antioxidant enzymes in six woody plant species in karst habitats of southwestern China. *Environmental and Experimental Botany*, 71: 174–183.

Liu, Y.J. and Ding, H.U.I. (2008). Variation in air pollution tolerance index of plants near a steel factory: Implication for landscape-plant species selection for industrial areas. *Transactions on Environment and Development*, 4(1): 24–32.

Luedeling, E., Zhang, M. and Girvetz, E.H. (2009). Climatic changes lead to declining winter chill for fruit and nut trees in California during 1950–2099. *PLoS ONE*, 4: e6166.

Ma, Y.H., Ma, F.W., Zhang, J.K., Li, M.J., Wang, Y.H. and Liang D. (2008). Effects of high temperature on activities and gene expression of enzymes involved in ascorbate-glutathione cycle in apple leaves. *Plant Science*, 175: 761–766.

Maas, E.V. (1990). Crop salt tolerance. pp. 262–304. *In*: Tanji, K.K. (ed.). *Agricultural Salinity Assessment and Management*. ASCE Manual Reports on Engineering Practices, New York, USA.

Marathe, R.A., Sharma, J., Babu, K.D. and Murkute, A.A. (2017). Bedding system: A unique plantation method of pomegranate in arid and semi-arid region. *National Academy Science Letters*, 40(4): 249–251.

Mars, M. and Marrakchi, M. (1999). Diversity of pomegranate (*Punica granatum* L.) germplasm in Tunisia. *Genetic Resources and Crop Evolution*, 46(5): 461–467.

Mars, M. (2000). Pomegranate plant material: Genetic resources and breeding, A review. *Serie A Seminaires Mediterraneans*, 42: 55–62.

Matuskovic, J. and Micudova, O. (2006). Practices with chemical mutagen (*Natrium azid*) on growth habit (*Punica granatum* L.). pp. 41. *The Proceedings of the International Society of Horticultural Sciences*. 1st International Symposium, Pomegranate and Minor Mediterranean Fruits, 16–19 October 2006, Adana, Turkey.

Meena, V.S., Kashyap, Pö., Nangare, D.D. and Singh, J. (2016). Effect of colored shade nets on yield and quality of pomegranate (*Punica granatum*) cv. Mridula in semi-arid region of Punjab. *Indian Journal of Agricultural Sciences*, 86(4): 500–505.

Mehranna, K.D., Badraldin, E.S.T. and Masoud, B. (2006). Characterization of microsatellites and development of SSR markers in pomegranate. pp. 66. *The Proceedings of the International Society of Horticultural Sciences*. 1st International Symposium, Pomegranate and Minor Mediterranean Fruits, 16–19 October 2006, Adana, Turkey.

Melgarejo, P., Martı´nez, J.J., Herna´ndez, F., Martı´nez-Font, R., Barrows, P. and Erez, A. (2004). Kaolin treatment to reduce pomegranate sunburn. *Scientia Horticulturae*, 100: 349–353.

Melgarejo, P., Martinez, J.J., Hernandez, F., Martinez, R., Legua, P., Oncina, R. and Martinez-Murcia, A. (2009). Cultivar identification using 18S–28S rDNA intergenic spacer-RFLP in pomegranate (*Punica granatum* L.). *Scientia Horticulturae*, 120: 500–503.

Menezes-Benavente, L., Teixeira, F.K., Kamei, C.L.A. and Margis-Pinheiro, M. (2004). Salt stress induces altered expression of genes encoding antioxidant enzymes in seedlings of a Brazilian indica rice (*Oryza sativa* L.). *Plant Science*, 166: 323–331.

Mousavinejad, G., Emam-Djomeh, Z., Rezaei, K. and Khodaparast, M.H.H. (2009). Identification and quantification of phenolic compounds and their effects on antioxidant activity in pomegranate juices of eight Iranian cultivars. *Food Chemistry*, 115: 1274–1278.

Nafees, M. (2015). *Estimation of Morphological, Biochemical and Genetic Diversity in Pomegranate (Punica granatum L.) Germplasm.* Ph.D. Thesis, University of Agriculture Faisalabad, Pakistan.

Nafees, M., Jaskani, M.J., Ahmed, S. and Awan, F.S. (2015). Morpho-molecular characterization and phylogenetic relationship in pomegranate germplasm of Pakistan. *Pakistan Journal of Agricultural Sciences*, 52(1): 97–106.

Nafees, M., Jaskani, M.J., Naqvi, S.A., Haider, M.S. and Khan, I.A. (2018). Evaluation of elite pomegranate genotypes of balochistan based on morphological, biochemical and molecular traits. *International Journal of Agriculture and Biology*, 20: 1405–1412.

Okatan, V., Çolak, A.M., Güçlü, S.F. and Gündoğdu, M. (2018). The comparison of antioxidant compounds and mineral content in some pomegranate (*Punica granatum* L.) genotypes grown in the east of Turkey. *Acta Scientiarum Polonorum, Hortorum Cultus*, 17(4): 201–204.

Olesen, J.E. and Bindi, M. (2002). Consequences of climate change for European agricultural productivity, land use and policy. *European Journal of Agronomy*, 16(4): 239–262.

Onur, C. and Kaska, N. (1985). Selection of pomegranate (*Punica granatum* L.) from Mediterranean region of Turkey. *Doga Bilim Dergisi D2 Tarm ve Ormancilik*, 9: 25–33.

Ozguven, A.I. and Yilmaz, C. (2000). Pomegranate growing in Turkey. *Serie A Seminaires Mediterraneennes*, 42: 41–48.

Pachauri, R.K., Allen, M.R., Barros, V.R., Broome, J., Cramer, W., Christ, R., Church, J.A., Clarke, L., Dahe, Q., Dasgupta, P. and Dubash, N.K. (2014). *Climate Change 2014: Synthesis Report.* Contribution of Working Groups I, II and III to the fifth assessment report of the Intergovernmental Panel on Climate Change, IPCC, 151 pp.

Packer. (2014). California pomegranate crop growing to meet demand, *The Packer Newsletter*. California, USA.

Parvizi, H., Sepaskhah, A.R. and Ahmadi, S.H. (2016). Physiological and growth responses of pomegranate tree (*Punica granatum* (L.) cv. Rabab) under partial root zone drying and deficit irrigation regimes. *Agricultural Water Management*, 163: 146–158.

Porter, J. and Wetzstein, H.Y. (2014). *The Biology of Pomegranates: All about Flowers, Fruit and Arils.* Florida Pomegranate Association, University of Georgia, USA.

Pourghayoumi, M., Majid, R., Davood, B., Aalami, A., Akbar, A. and Haghighi, K. (2017). Responses of pomegranate cultivars to severe water stress and recovery: Changes on antioxidant enzyme activities, gene expression patterns and water stress responsive metabolites. *Physiology and Molecular Biology of Plants*, 23(2): 321–330.

Pourghayoumi, M., Majid, R., Davood, B., Aalami, A., Akbar, A. and Haghighi, K. (2017). Responses of pomegranate cultivars to severe water stress and recovery: changes on antioxidant enzyme activities, gene expression patterns and water stress responsive metabolites. *Physiology and Molecular Biology of Plants*, 23(2): 321–330.

Preece, J.E. (2017). What exactly is a pomegranate cultivar? A presentation. *National Clonal Germplasm Repository*, Davis, CA, USA.

Rahimi, H.R., Arastoob, M. and Ostad, S.N. (2012). A comprehensive review of *Punica granatum* (Pomegranate) properties in toxicological, pharmacological, cellular and molecular biology researches. *Iranian Journal of Pharmaceutical Research*, 11: 385–400.

Rai, P.K., Panda, L.L.S., Chutia, B.M. and Singh, M.M. (2013). Comparative assessment of air pollution tolerance index (APTI) in the industrial (Rourkela) and non-industrial area (Aizawl) of India: An eco-management approach. *African Journal of Environmental Science and Technology*, 7(10): 944–948.

Rath, S., Padhi, S.K., Kar, M.R. and Ghosh, P.K. (1994). Response of zinnia to sulphur dioxide exposure. *Indian Journal of Ornamental Horticulture*, 2: 42–45.

Raven, P. and Johnson, G. (1986). Biology, *Times Mirror*/Mosby College Publishing, St. Louis, USA.

Reddy, A.R., Chaitanya, K., Jutur, P. and Sumithra, K. (2004). Differential antioxidative responses to water stress among five mulberry (*Morus alba* L.) cultivars. *Environmental and Experimental Botany*, 52: 33–42.

Sage, R.F., Way, D.A. and Kubien, D.S. (2008). Rubisco, Rubisco activase, and global climate change. *Journal of Experimental Botany*, 59(7):1581–1595.

Samadia, D.K. and Pareek, O.P. (2006). Fruit quality improvement under hot arid environment. *Indian Journal of Horticulture*, 63: 126–132.

Samish, R.M. and Lavee, S. (1982). The chilling requirement of fruit trees. pp. 372–388. *Proceedings of XVI International Horticultural Congress*.

Seyyednejad, S.M., Niknejad, M. and Koochak, H. (2011). A review of some different effects of air pollution on plants. *Research Journal of Environmental Sciences*, 5(4): 302–309.

Shanmugasundaram, T. and Balakrishnamurthy, G. (2017). Exploitation of plant growth substances for improving the yield and quality of pomegranate under ultra-high-density planting. *International Journal of Current Microbiology and Applied Sciences*, 6(3): 102–109.

Silva, E.N., Ribeiro, R.V., Ferreira-Silva, S.L., Vieira, S.A., Ponte, L.F. and Silveira, J.A. (2012). Coordinate changes in photosynthesis, sugar accumulation and antioxidative enzymes improve the performance of *Jatropha curcas* plants under drought stress. *Biomass Bioenergy*, 45: 270–279.

Soloklui, A.A.G., Gharaghani, A., Oraguzie, N., Eshghi, S. and Vazifeshenas, M. (2017). Chilling and heat requirements of 20 Iranian pomegranate cultivars and their correlations with geographical and climatic parameters, as well as tree and fruit characteristics. *HortScience*, 52(4): 560–565.

Still, D.W. (2006). Pomegranate: A botanical perspective. pp. 199–209. *In*: Heber, D., Schulman R.N. and Seeram, N.P. (eds.). *Pomegranates: Ancient Roots to Modern Medicine*. CRC Press, Taylor & Francis, Florida, USA.

Stover, E.W. and Mercure, E.W. (2007). The pomegranate: a new look at the fruit of paradise. *Horticultural Science*, 42: 1088–1092.

Sturgeon, S.R. and Ronnenberg, A.G. (2010). Pomegranate and breast cancer: Possible mechanisms of prevention. *Nutrition Reviews*, 68: 122–128.

Sung, D.Y., Kaplan, F., Lee, K.L. and Guy, C.L. (2003). Acquired tolerance to temperature extremes. *Trends in Plant Science*, 8: 179–187.

Swain, S.C. and Padhi, S.K. (2015). Effect of sulphur dioxide on growth, chlorophyll and sulphur contents of pomegranate. *Tropical Agricultural Research and Extension*, 16(1): 21–24.

Terakami, S., Matsuta, N., Yamamoto, T., Sugaya, S., Gemma, H. and Soejima, J. (2007). Agrobacterium-mediated transformation of the dwarf pomegranate (*Punica granatum* L. var. nana). *Plant Cell Reports*, 26: 1243–1251.

Thongtham, C. (1986). Germplasm collection and conservation of pomegranate in Thailand. *International Board for Plant Genetic Resources Newsletter*, 10(8): 8–10.

Varshney, R.K., Terauchi, R. and McCouch, S.R. (2014). Harvesting the promising fruits of genomics: Applying genome sequencing technologies to crop breeding. *PLoS Biology*, 12: e1001883.

Venkatesha, H. and Yogish, S.N. (2016). Trends in area, production and productivity of pomegranate producing states of India. *International Journal of Multidisciplinary Research and Development*, 3(1): 356–359.

Verslues, P.E., Agarwal, M., Katiyar-Agarwal, S., Zhu, J. and Zhu, J.K. (2006). Methods and concepts in quantifying resistance to drought, salt and freezing, abiotic stresses that affect plant water status. *Plant Journal*, 45: 523–539.

Wang, S., Liang, D., Li, C., Hao, Y., Ma, F. and Shu, H. (2012). Influence of drought stress on the cellular ultrastructure and antioxidant system in leaves of drought-tolerant and drought-sensitive apple rootstocks. *Plant Physiology and Biochemistry*, 51: 81–89.

Wang, Y., Yin, X.L. and Yang, L.F. (2006). Breeding of Zaoxuan 018 and 027 pomegranate selections. *China Fruits*, 4: 6–8.

Westwood, M.N. (1999). *Temperate Zone Pomology: Physiology and Culture*. Timber Press, Portland, 482 pp.

Wetzstein, H.Y. (2011). A morphological and histological characterization of bisexual and male flower types in pomegranate. *Journal of the American Society for Horticultural Science*, 136(2): 83–92.

Winner, W.E. Mooney, H.A. and Goldstun, R.A. (1985). *Sulphur Dioxide and Vegetation: Physiology, Ecology and Policy Issues*. Stanford University Press, Stanford, CA, USA, 624 pp.

Xue, J.H., Zhuo, L.H. and Zhou, S.L. (2006). Genetic diversity and geographic pattern of wild lotus (*Nelumbo nucifera*) in Heilongjiang province. *Chinese Science Bulletin*, 51: 421–432.

Yezhov, V.N., Smykov, A.V., Smykov, V.K., Khokhlov, S.Y., Zaurov, D.E., Mehlenbacher, S.A., Molnar, T.J., Goffreda, J.C. and Funk, C.R. (2005). Genetic resources of temperate and subtropical fruit and nut species at the Nikita Botanical Gardens. *HortScience*, 40(1): 5–9.

Yuan, Z., Chen, X., He, T., Feng, J., Feng, T. and Zhang, C.H. (2007). Population genetic structure in apricot (*Prunus armeniaca* L.) cultivars revealed by fluorescent-AFLP markers in Southern Xinjiang, China. *Journal of Genetics and Genomics*, 34(11): 1037–1047.

Yuan, Z.H. and Zhao, X. (2019). Pomegranate genetic resources and their utilization in China. *Acta Horticulturae*, 1254: 49–56.

Zamani, Z., Sarkhosh, A., Fatahi, R. and Ebadi, A. (2007). Genetic relationships among pomegranate genotypes studied by fruit characteristics and RAPD markers. *Journal of Horticultural Science and Biotechnology*, 82: 11–18.

Zhao, G.R., Zhu, L.W., Zhang, S.M., Jia, B. and Li, S.W. (2007). A new soft-seeded pomegranate variety, Hongmanaozi. *Acta Horticulturae Sinica*, 34: 260–260.

Zhao, Y.L., Feng, Y.Z., Li, Z.H. and Cao, Q. (2006). Breeding of the new pomegranate cultivar 'Yushiliu 4'. *China Fruits*, 2: 8–10.

Part III

Cool Temperate Fruits

10
Behavior of Apricot (*Prunus armeniaca* L.) under Climate Change

Tomo Milošević[1],* and *Nebojša Milošević*[2]

1. Introduction

Climate change is a global phenomenon but is being felt more in the mountainous regions owing to the fragile nature of the ecosystem (Jangra and Sharma, 2013). This phenomenon is a change of climate over a comparable period of time and is attributed directly or indirectly to human activity that alters the composition of the global atmosphere (Rai et al., 2015). The effects of climate change are very numerous. However, one of the anticipated effects of this phenomenon is the possible increase in both frequency and intensity of extreme weather events, such as hurricanes, floods, droughts, heat waves, cold waves, tropical cyclones, tidal waves and severe storms (De et al., 2005). From this point of view, climate change, especially global warming is the greatest threat confronting the Earth. It is more pronounced in micro-climates and is being felt more in mountainous regions owing to the fragile nature of the ecosystem. Under these areas, weather fluctuations and variations are becoming evident in the form of untimely and irregular precipitations and extremes in temperatures and

[1] Department of Fruit Growing and Viticulture, Faculty of Agronomy, University of Kragujevac, Cara Dušana 34, 32000 Čačak, Republic of Serbia.
[2] Department of Pomology and Fruit Breeding, Fruit Research Institute, Čačak, Kralja Petra I/9, 32000 Čačak, Republic of Serbia.
Emails: nmilosevic@institut-cacak.org; mnebojsa@ftn.kg.ac.rs
* Corresponding author: tomomilosevic@kg.ac.rs

other weather events. According to (IPCC, 2013), changes in many extreme weather and climate events have been observed since about 1950.

The forecasts for the behavior of people, animals and plants in the near and far future are not too optimistic and urgent measures must be taken to find solutions to reduce the negative impact of global climate change.

Apricot has numerous positive fruit attributes, such as attractiveness, tasty flavor and ease of eating, as well as its multiple-use functionality and a non-surplus production (Zhebentyayeva et al., 2012). Besides its fresh consumption, all through the summer it is used in making marmalade, jam or jelly and also canned as slices or processed as fruit juice. Cultivars grown mainly in Turkey (Malatya), Iran, China, USA (California), Australia and South Africa are used in drying (Ercisli, 2009). On the other hand, apricot has many disadvantages that limit its cultivation, especially in the temperate and continental parts of the world, including Serbia. Dominant is the dying off of flower buds prior to flower opening during winter, flowers killed by spring frosts, sudden (premature) wilting—apoplexy, *Plum Pox Virus* (PPV) infection and absence of modern growing technologies (Milošević et al., 2010, 2013a,b). All these disadvantages cause irregularity of yield (Karakaş and Doğan, 2018) and irregular market supply (Zhebentyayeva et al., 2012). Recently, many studies indicate that apricot is very vulnerable to global warming (Viti and Monteleone, 1995; Viti et al., 2006; Bălan et al., 2009; Karakaş and Doğan, 2018; Bartolini et al., 2019), which is probably a greater problem than all the previous limiting factors.

The present chapter aims at analyzing the impacts of climate change on phenology, productivity and fruit quality in apricot orchards worldwide. The planning of suitable adaptation measures against these threats is critical for the future sustainability of apricot production sectors.

2. Impact of Climate Change on Agricultural Sector: An Overview

In most cases, climate change is often associated with extreme weather events (Stern, 2006; Karl et al., 2009). This phenomenon directly affects the agricultural sector and indirectly affects other sectors as well. In undeveloped and developing countries worldwide, such as, for example, Serbia, the agricultural sector is more sensitive to climate change than in developed countries (Mendelsohn, 2001; Maskrey et al., 2007) because it is largely dependent on climate factors. As a potential result, poor people living in agricultural communities are expected to be the most affected by climatic change (Maskrey et al., 2007) due to continuing negative impacts of climate factors on the yield of agricultural products (Chmielewski and Potts, 1995). However, although many studies have considered the

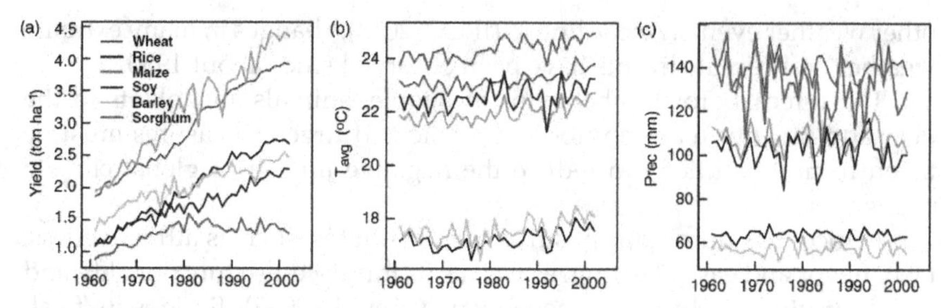

Fig. 1. Time series of (a) yields, (b) growing season average monthly temperature and (c) rainfall for six crops, 1961–2002 (*Source*: Lobell and Field, 2007).

influence of future climate changes on food production (Rosenzweig and Parry, 1994; Parry et al., 2005; Fischer et al., 2005; Edmonds and Rosenberg, 2005), the influences of these past changes on agriculture remain unclear. It is likely that, for example, warming has improved yields (food production per unit of land area) in some regions of the world, reduced them in others (Fig. 1) and had negligible impacts in still others (Lobell and Field, 2007).

Otherwise, about 30% of variations of global average yields for the world's most widely grown crops are the result of growing season precipitation and temperature variations (Lobell and Field, 2007). However, data related to the extremely harmful effects of global warming on agricultural sector, i.e., crop yield, are dominant. Thus, Temesgen (2007) reported the negative income effect of temperature increase and precipitation decrease in Ethiopia whereas Temesgen and Hassan (2009) revealed that climatic variables had a very important impact on the net crop revenue of farmers in this country. Similar tendency was found by Thapa and Joshi (2010) for Nepal. Crop productivity and climate change are highly inter-related: high negative correlation existed between sugar beet productivity and increasing temperature (Freckleton et al., 1999). On this line, Pidgeon et al. (2001) reported that drought losses of yield of sugar beet were greatest in east Ukraine and southern Russia, at over 40% of potential yield or 5 t ha^{-1}; losses were intermediate (15–30% or ≈ 2 t ha^{-1} in central Ukraine, west Poland, East Germany and England, and lowest in north-west Europe and west Ukraine. Later, Pidgeon et al. (2004) reported that climate change manifested by drought in future would reduce the sugar beet production in Europe by 50% on an average. Similar tendencies have been observed for other important crops, such as wheat, rice, maize, soybeans, barley and sorghum (Lobell and Field, 2007). Otherwise, about 30% of variations of global average yields for the world's most widely grown crops are the result of growing season precipitation and temperature variations (Lobell and Field, 2007; Majdancic et al., 2016).

Influence of climate change have been investigated for many cereals and oilseed crops, but relatively less is known about the potential impacts on fruit trees, including apricot (Černa et al., 2012; Rai et al., 2015).

Pedo-climatic conditions prevailing over the traditional area provide ideal conditions for the production of temperate fruits. Thanks to climate change, this is no longer the case. Reduced snowfall, delay in start of winter, threat of floods, hails, drought and wind storms are some of the major impacts of climate change. Global warming had caused loss of vigor and fruit-bearing ability, reduction in size of fruits, less juice content, low color, reduced shelf-life and increasing attack of pests and disease, resulting in the low production and poor quality of fruits (Jangra and Sharma, 2013). Based on pessimistic forecasts, maybe real in future, vulnerability, rarity and rapid extinction of many species of temperate fruits will be among the other consequences of climate change.

Climate change is expected to have an impact on future winter chill, which could potentially have a major influence on fruit species with chilling requirements (Luedeling et al., 2011). If the winter chill is not sufficient or fulfilled, dormancy is not broken (which occurs in extreme situations) and fruit yields and fruit quality can be severely affected (Denis, 2003). In addition, winter chilling refers to a physiological requirement for low temperatures to enable normal spring growth and failure to obtain sufficient winter chilling results in a marked decline in both yield and fruit quality. Some authors projected a 50% decrease in chill hours from 1950 to 2100, approaching the critical threshold for many fruit tree species in California (Baldocchi and Wong, 2006).

Regarding fruit trees from temperate and continental areas, such as, for example, apples, pears, plums, apricots, peaches and/or nectarines, they need to achieve a chilling requirement in order to break dormancy in winter and recommence growth in spring. Numerous new investigations focusing on dormancy-related traits and adaptation of genotypes to specific cultivation areas showed, in recent years, the frequent occurrence of warmer winters in Mediterranean regions (Bartolini et al., 2019) and also in Europe's moderate and continental areas (Černa et al., 2012; Sønsteby and Heide, 2014; Öncel and Tzanakis, 2018), causing problems with chilling requirement.

According to data from relevant literature, chilling requirement is defined as the number of effective chilling hours needed to restore bud-growth potential in spring (Richardson et al., 1974; Atkinson et al., 2013). It has been long known that the chilling requirement is measured in terms of numbers of hours, during which temperature remains at < 7.2°C or between 0–7.2°C during the winter season. The chilling requirements of different temperate fruit crops grown under Serbian conditions are presented in Table 1 (Milošević, 1997).

Table 1. Overview of chilling requirements of different temperate fruit crops grown under according to the Utah model proposed by Richardson et al. (1974) (*Source*: Milošević, 1997).

Fruit species	Chilling hours
Apple	800–1200
Pear	800–1000
Plum	800–1000 (European plum)
	700–1000 (Japanese plum)
Peach	200–1000
Apricot (European group)	300–1500
Cherries	1200–2500 (sweet cherry)
	1200–2000 (sour cherry)
Walnut	200–1000
Hazelnut	350–600 (male flower buds)
	600–800 (female flower buds)
Almond	600–800
Chestnut	200–1000
Strawberry	1200–1400
Red raspberry	800–1400
Blackberry	200–900
Currants	200–250

Generally, chilling hours vary year to year, depending on the presence of cold fronts each winter and also of fruit species and genotype (Richardson et al., 1974; Viti et al., 2006; Černa et al., 2012). Therefore, if present projections of future climate change (IPCC, 2007) are true, inadequate winter chill may become a limiting factor for successful apricot production in areas with mild winter climates, as has already been reported for other fruit crops (Chmielewski et al., 2004; Else and Atkinson, 2010; Bartolini et al., 2019). Some authors projected a 50% decrease in chill hours from 1950 to 2100, approaching the critical threshold for many fruit tree species in California (Baldocchi and Wong, 2006). As a consequence, several fruit species may show physiological and biological disturbances, producing detrimental effects on productivity.

Phenology, as a study of periodic biological events in the plant world and influenced by the environment (Cosmulescu and Bîrsanu Ionescu, 2018), are linked with chilling requirements. Inter-annual changes in spring plant phenology may be the most sensitive and observable indicators of the plant response to climate change (Beaubien and Freeland, 2000). Current climate change, represented by a substantial global warming, is impacting the biological dynamics of plants, showing phenological variations as a function of vulnerable environmental conditions. It is true that phenological phases are getting earlier and it is well known. Consequently, the timing of spring phenophases is considered to be under opposing pressures—earlier bud-burst increases the available growing season but later bud-burst decrease the risk of frost damage to actively growing parts (Černa et al., 2012). However, earlier blossom of fruit trees holds the danger of damage by last frosts.

These authors also reported that frosts before the beginning of blossom may cause masked injuries to flower buds and moreover the frost during the flowering period can harm the blossoms, so that total crop failures can occur. Similar observations were noted earlier by Szalay et al. (2006) on apricots in Hungary. They stated that one of the most spectacular effects of the warmer climate in Hungarian orchards is the faster bud development and earlier blooming of fruit trees, including apricot and which is extremely unfavorable because it increases the probability of frost damage. Many data from literature show that apricot most noticeably negatively reacts to climate change due to physiological and biological disturbances producing detrimental effects on productivity (Prudencio et al., 2018). Otherwise, apricot is a fruit species which is strongly influenced by weather-environmental factors which impact its floral biology patterns through the appearance of flower bud anomalies, drops and alterations of phenological processes (Legave et al., 2006, 2009). As a consequence of short or warm winter seasons, an incomplete resumption of flower bud dormancy occurs, resulting in scanty apricot blooming (Szalay et al., 2006; Bartolini et al., 2019).

Accordingly, as the flowering time of fruit crops is one of the most widely used indicators of climate change, the present study aims to assess, over a long-term period, the climatic trend and its influence on the flowering date of apricots.

3. Apricot Reaction to Climate Changes

3.1 *Origin of Apricot in Order to Better Understand the Requirements to Environment*

Apricot (*P. armeniaca* L.) is a member of the Rosaceae family and *Prunus* L. genus (stone fruits). The origin of the species is somewhat uncertain and disputed due to its early domestication. It was known in Armenia during ancient times (Neolithic era, also known as the New Stone Age) and has been cultivated there for so long that; it is often thought to have originated there. Seeds of the apricot have been discovered during archaeological excavations of the Garni Temple and Shengavit settlement in Armenia, having a history of 6,000 years (Arakelyan, 1968). However, Vavilov (1926, 1951, 1991) indicated that centers of origin of this species are multiple, i.e., (a) the Chinese center (central and western China), (b) the Central Asiatic center (Afghanistan, northwest India and Pakistan, Kashmir, Tajikistan, Uzbekistan, Xinjing province in China and western Tien-Shan), and (c) the Near-Eastern center (interior of Asia Minor), where the domestication of apricot would have taken place (Zhebentyayeva et al., 2012).

In addition, apricots have been cultivated in Persia since antiquity and dried ones were an important commodity on Persian trade routes. More recently, English settlers brought the apricot to the English colonies of the New World. Most of modern American production of apricots comes from the seedlings carried to the west coast by Spanish missionaries.

Today, apricot cultivation has spread to all parts of the globe with climates that support its growth needs. However, successful production is achieved in areas without explicit so-called transient seasons (without spring and autumn), such as parts of Central Asia, followed by warm Mediterranean areas and individual micro-areas that have moderate and/or continental climate regimes (Milošević and Milošević, 2019).

3.2 *Apricot Reaction to Climate Changes Worldwide*

Apricot is a stone fruit showing several problems related to its floral biology. The difficulty of several apricot cultivars to adapt to environments differing from their origin is well known (Carraut, 1968). This phenomenon can be particularly evident when cultivars originating from continental zones are introduced into coastal areas and vice versa (Viti et al., 2006; Bălan et al., 2009). For example, after the First and Second World Wars, hundreds of cultivars from southern Europe, the US and West Asia were introduced in Serbia, but only a small number was adapted (Milošević, 1997). According to the same author, apricots introduced in Serbia from countries and regions with similar environmental conditions, such as Hungary, were easily adapted. Today, 'Hungarian Best' with clones is the predominant cultivar in Serbia. However, similar to other European apricot production areas, great changes in apricot phenology in Serbia have been observed with this and other cultivars due to global warming (Milošević et al., 2018).

Deviation in flowering dates were observed between different genotypes under the same environmental conditions and which could be assigned to differences in temperature sensitivity (Vitasse et al., 2009). Previous studies have shown the influence of environmental factors on the budburst dates, the blooming period and duration of blooming in apricot cultivars (Cosmulescu and Gruia, 2016; Karakaş and Doğan, 2018; Bartolini et al., 2019). For example, one of the most spectacular effects of the warmer climate in Hungarian orchards is the faster bud development and earlier blooming of fruit trees, including apricots (Szalay et al., 2006).

This is extremely unfavorable, because it increases the probability of frost damage in moderate and continental areas worldwide (Karakaş and Doğan, 2018; Glišić et al., 2019). These tendencies are very pronounced in the Mediterranean area because most of the examined cultivars showed significant blooming changes in terms of date and intensity (Ighbareyeh and Carmona, 2017). Bartolini et al. (2019) reported that the early-cultivars showed a significant flowering delay—an occurrence that could enable them to face a minor risk of frost damage during anthesis in environments characterized by frequent spring cold which is not the case in continental areas (Milošević et al., 2018). Not only the frost, but also the high temperature—a prominent example from Egypt is a cultivar called 'Solitaire' with luxurious fruit characteristics which failed to bloom in the southern areas of Egypt though it succeeded in the northern areas near the Mediterranean Sea. This could be because of high temperatures in the southern parts of Egypt (personal communication). On this line, based on temperature relations for the 1927–2009 period in southern Moravia (Czech Republic), Černa et al. (2012) predicted that apricot phenophases in 2050 and 2100 will be 12–15 and 26 days earlier, respectively. These authors also reported that in these climatic conditions, the probability of damage by last frost days could be higher (63–73%). In Japan, global warming also affects the fruit trees behavior, especially Japanese apricot (*P. mume* Sieb. et Zucc.) phenology (Doi, 2007). Their experts divided the fruit species into two groups according to responses of fruit development to climate changes (Sugiura, 2010). One group was the earlier development type and the other was the prolonged development type. The first type included tree species in which both flowering and harvesting periods had accelerated. Examples are Japanese pears, peaches and apricots. Studies of several authors worldwide revealed that apricot yield in British Columbia (Canada) (Caprio and Quamme, 2006), Malatya province (Turkey) (Gunduz et al., 2011; Karakaş and Doğan, 2018) and Palestine (Ighbareyeh and Carmona, 2017) is negatively affected by global climate change.

As a consequence of global warming, many authors noted other emerging problems, such as diseases and pests. For example, many diseases have occurred in regions of Europe where they have not been previously observed, viz. very important pathogens (including quarantine) are spreading to the north and to higher altitudes originating in the southern, warmer areas. A typical example is some mycoplasmas and viruses, especially *Zuccini yellow mosaic virus* (ZYMV) and *Plum pox virus* (PPV). The first pathogen is fatal for some vegetable plants, while PPV is a global problem in the production of stone fruit, including apricot (Milošević et al., 2019). However, the spread of these pathogens is not only a consequence of global climate change, but also of human activity or a combination of both climate change and direct human activity. The occurrence of *Monilinia laxa* (Aderhol et Ruhl, Honey et Whetzel) has been intensified in many European apricot production areas (Bălan et al., 2009). A detailed overview and relevant data of this phenomenon are provided by Polak (2009), Rozák et al. (2019) and Milošević et al. (2019).

According to results of some studies, more than 10,000 insects, 600 weeds and 1,500 fungi, commonly named pests, adversely affect daily human life (Ibrahim, 2014). These organisms reduce the quality and quantity of food produced, by lowering production and destroying stored produce. Basically, global warming will increase pest populations, including weeds, invasive species, insects and insect-borne diseases, which will in all likelihood lead to large increases in the use of pesticides (Ladányi and Horváth, 2010). These authors also reported that changes in climate increase the chances of insect transport from regions to regions. In addition, increases in temperature have a number of implications for temperature-dependent insects, especially in the region of middle Europe. In the past few decades, thank to climate change, many invasive pests have been introduced in this cold arid region, i.e., codling moth in apple and apricot (*Cydia pomonella* L.), aphids (*Aphis* spp.) and several invasive species of moths (Gupta et al., 2015). For example, in recent years, severe infestation of defoliating caterpillar (*Euproctis* spp.) and aphids was observed in some important apricot growing areas in India, causing catastrophic damages through defoliation of apricot trees (Raghuvanshi et al., 2016). Otherwise, defoliator is the only native to Europe. Also, codling moth (*C. pomonella* L.) is a serious pest of apricot in fruit growing areas of Ladakh in India (Raghuvanshi et al., 2016).

A detailed overview of the impact of climate change on increasing weed, insect and fungal damage has been provided by Ladányi and Horváth (2010) and Ibrahim (2014).

3.3 *Some Experience of Apricot Reaction to Climate Changes under Moderate or Continental Climate. Overview through Serbian Climatic Conditions*

The Republic of Serbia is a small country in the Balkan peninsula. It is located between 40.13° and 46.15° N latitude and 18.9° and 22.9° E longitude (Fig. 2). The altitude varies from 28 m (the mouth of River Timok in the Danube) to 2,656 m (top of mountain Djeravica).

The climate of Serbia is predominantly moderate-continental with steppe characteristics in the north in Vojvodina and east in the Negotin region. On the rugged parts of the Old Vlach in south-western Serbia and on the highest mountains, the climate has mountain characteristics with short coldish summers and long and cold winters. In general, the climate of Serbia belongs to the Cfb type, according to the Köppen-Geiger climate classification described by Kottek et al. (2006).

Many studies show that global climate change is registered in Serbia. Changes in main climate conditions since 1961 and until now show significant increase in temperature and change in precipitation patterns (Vuković et al., 2018). These authors also reported that temperature increase averaged 1.2°C for the period 1996–2015 and with respect to the period 1961–1980, highest increase of maximum daily temperature was seen during the summer season at 2.2°C. On the basis of these data and other observations, it can be said that spatial distribution of temperature increase, intensification of high precipitation events and decrease of summer precipitation show intrusion of subtropical climate over Serbia and increase of high temperature and high precipitation risks. For these reasons, agriculture in Serbia has suffered significant losses due to adverse weather conditions and distinctive climatic anomalies. The most important damages arose from the effects of drought, spring frosts, hailstorms and floods. Since 2000, Serbia went through several episodes of severe drought, resulting in significant losses. Extreme droughts were recorded in the years 2000, 2003, 2007, 2011, 2012 and partial ones in 2019 and 2021 (personal data).

Čačak region is located in western part of Serbia and produces 10–15% of total Serbian apricot production on an average. In the last few decades, it is obvious that there have been important changes in the climate, i.e., weather conditions, in this region. On this occasion an analytical review of the average annual temperature and precipitation is provided for the period 1965–2018 (Fig. 3).

Data in Fig. 4 revealed that the mean annual temperatures had a tendency to increase over the period 1965–2018; of course, with

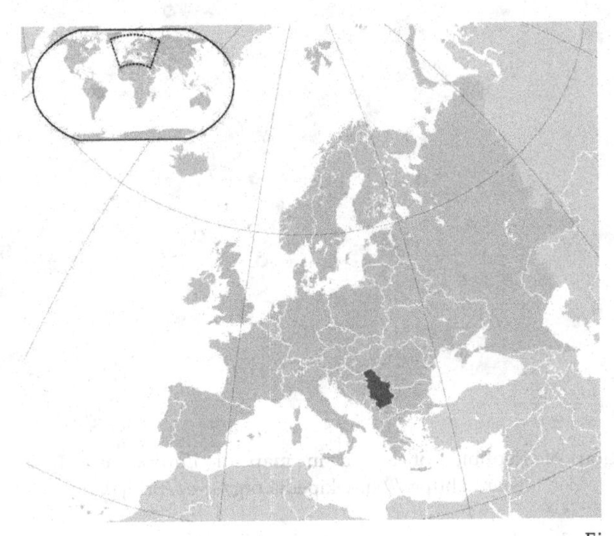

Figure 2 contd. ...

...*Figure 2 contd.*

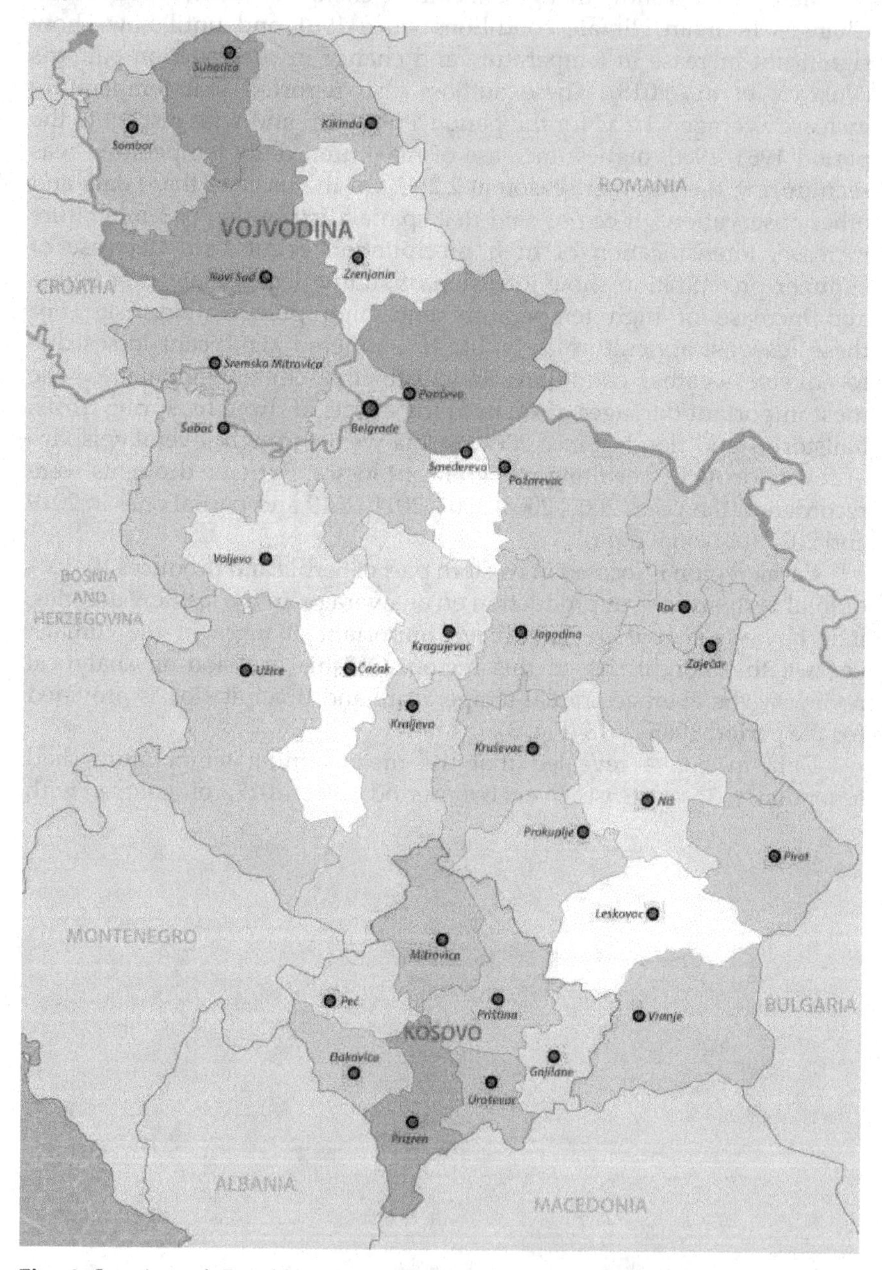

Fig. 2. Location of Republic of Serbia in map of Europe and Balkan peninsula (*Source*: https://sr.wikipedia.org/sr-ec/Srbija).

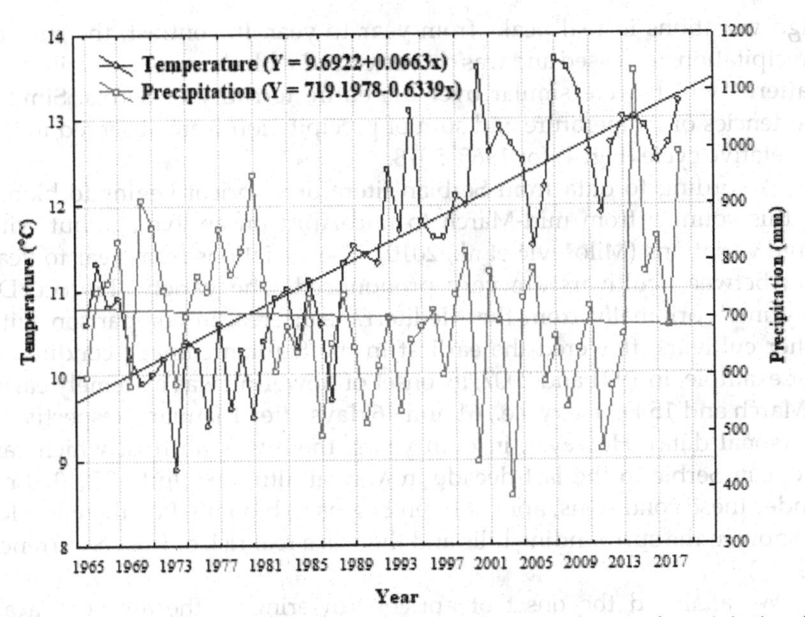

Fig. 3. Trend of average annual air temperatures and annual sum of precipitation for Čačak region from 1965 to 2018 (*Source*: Republic Hydrometeorological Service of Serbia, Belgrade, 2020).

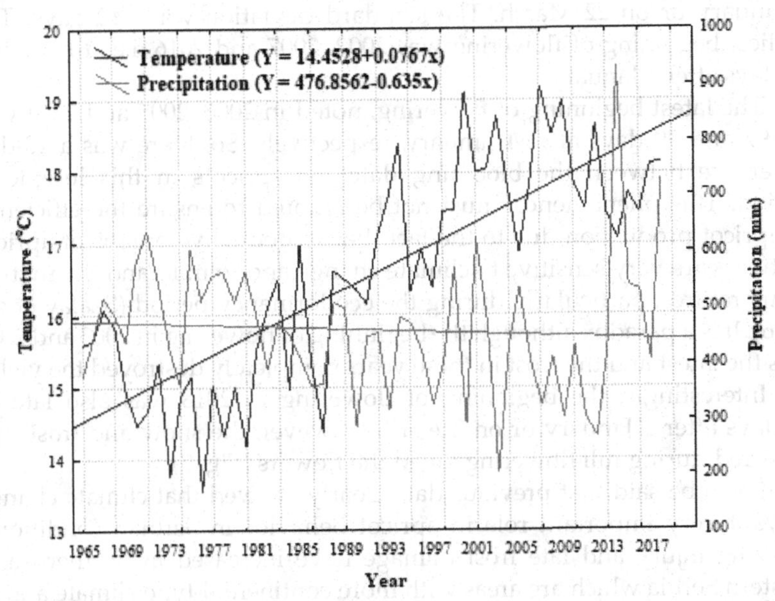

Fig. 4. Trend of average air temperatures and sum of precipitation for vegetative cycle for Čačak region from 1965 to 2018 (*Source*: Republic Hydrometeorological Service of Serbia, Belgrade, 2020).

high variations in real peaks from year to year. In contrast, the sum of precipitation decreased and was characterized with changes in distribution patterns which were similar over the entire territory of Serbia. Similar tendencies of temperature and sum of precipitation were observed in the vegetative cycle (Fig. 4) for 1965–2018.

According to data from Serbian literature, apricot begins to bloom in this country from mid-March to mid-April on an average but with large variations (Milošević et al., 2010). The variations from year to year and between cultivars are very pronounced. The variety 'Precoce De Thyrinte', originally from the Mediterranean area, in comparison with other cultivars, flowered the earliest in the Serbian climatic conditions. For example, in 1994 and 2007, its onset of flowering was extremely early, 2 March and 15 February, i.e., 61 and 46 days after 1 January, respectively (personal data). However, in both years, there was no frost which can occur in Serbia in the last decade in May at altitudes up to 250–300 m. Under these conditions, apricot often begins to bloom when there is a lot of snow in the surrounding hills and there is a real risk of frost occurrence (Fig. 5).

We analyzed the onset of apricot flowering in the area of Čačak for the period 1994–2019 (Fig. 6). During this long-term period, frost was registered six times, i.e., 1995, 2001, 2002, 2003, 2005 and 2018. The average date of beginning of flowering for this period was 81 days, from 1 January or on 22 March. The standard deviation was ±12 days. The earliest beginning of flowering was 2002, 2007 and 2016, i.e., 61, 65 and 59 days after 1 January.

The latest beginning of flowering, noted in 2003, 2005 and 2014 was 99, 97 and 99 days after 1 January, respectively. So, there was a 40-day difference between the blooming dates of apricots in this long-term period. This phenomenon must not be ignored to ensure the efficiency of apricot production due to the fact that reproductive organs of apricot cultivars are very sensitive to climate change and to mild and fluctuating temperatures, particularly during the ecodormancy period (Szalay et al., 2006). It is a paradox although the beginning of flowering in 2003 and 2005 was the latest and the frost in these years completely destroyed the yield.

Interestingly, the beginning of flowering in 2018 was also late or 92 days after 1 January or on 2 April. However, the snow and frost that occurred during full flowering killed the flowers (Fig. 7).

It can be said that previous data clearly showed that climate change plays a very important role in apricot behavior in Serbian conditions. Freezing injury and late frost damage have increased in northern and western Serbia which are areas with more continental-type climate and at an altitude below 250 m (Radičević et al., 2011; Glišić et al., 2019).

Fig. 5. Apricot orchard at the beginning of flowering in Prislonica village near Čačak city (*Photo*: Tomo Milošević).

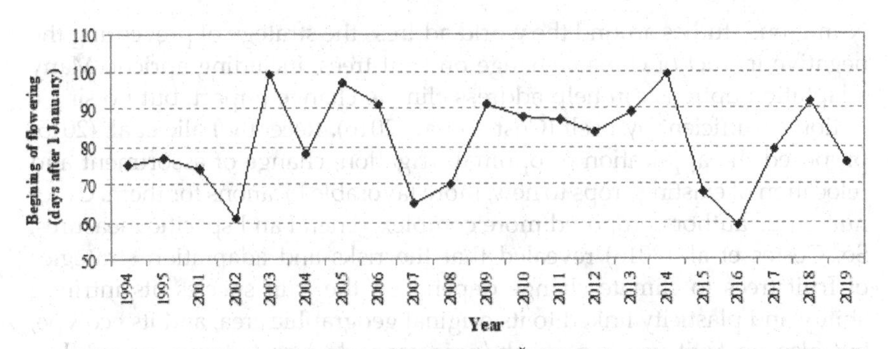

Fig. 6. Date of beginning of flowering of apricot in Čačak region (*Source*: personal data).

Fig. 7. Frost followed by snow and wind killed apricot flowers in 2018 in the Čačak region (*Photo*: Tomo Milošević).

Climate change in this country appears to have caused the following occurrences on deciduous fruit trees, including apricot (Lalić et al., 2015): (a) change in the start of the growing season, (b) the warmer vegetation period leads to early ripening and loss of fruit quality, (c) increased damage from drought, hail and spring frost, and (d) increased incidence of diseases and pests. Hence, it is clear that apricot production may be particularly vulnerable to climate change in Serbia (Milošević et al., 2010, 2018) and other similar production areas worldwide (Viti et al., 2006; Legave et al., 2006; Sugiura, 2010; Bartolini et al., 2019). Lalić et al. (2015) predict the following negative occurrences in future: (a) decrease in yield and fruit quality due to increased temperature and droughts, (b) increased risk of spring frost, and (c) increase in the risk of disease and pests and of new ones.

4. Strategies for Damage Reduction Caused by Global Climate Changes

Numerous studies around the world address the strategy of preventing the negative impact of climate change on fruit trees, including apricot. Many adaptation options can help address climate change impact, but no single option is sufficient by itself (Costes et al., 2016). Recently Lalić et al. (2015) proposed the application of optimal irrigation, change of assortment and relocation of existing crops to new, more favorable locations for them. Other numerous authors proposed more complex general and specific measures. So, Costes et al. (2016) revealed that the risks and adaptation strategies of fruit trees to climate change depend on the fruit species, its intrinsic ability and plasticity linked to its original geographic area, and its ecotype, but also on fruit use (e.g., fresh/processed/dry) and storage possibility. These authors also recommended combining several approaches, notably breeding for new material and innovative cropping systems. On this line, intensive investigations on the impact of changing climate on apricot's biological processes should be carried out in order to assess the performance of genotypes in specific environmental areas (Bartolini et al., 2019). In order to reduce harmful effect of climate change on fruit crops, including apricot, Cosmulescu and Gruia (2016) recommended measures to reduce the effects of drought, to prevent and combat desertification and land degradation, to create new cultivars tolerant to drought and high temperatures, delimitation of regions convenient for apricot growing and promoting tolerant and high-quality cultivars as well as to adapt technology which includes pruning, irrigation, fertilizing and disease and pest control to specific conditions. So, choosing suitable cultivar and rootstock, as well as protecting the apricot orchards by using irrigation systems and/or anti-hail nets in order to reduce adverse effects of fluctuations on climatic variables, play an important role (Lalić et al., 2015).

Regarding cultivars, Tresson et al. (2020) reported that under the combined effects of shifts of blooming period and changing climatic conditions, late apricot cultivars, such as 'Bergarouge' might see a reduction in the risk of blossom blight (down to 31%) because of warmer but dryer blooming periods in south France. According to the same authors, a traditional French cultivar called 'Bergeron' could experience an increase in this risk by up to 27% with a shift in the blooming period towards rainier conditions at the highest altitudes in this area. Other authors, such as Krška and Bauer (2020), revealed that choice of late and market-aluable cultivars are much more important than before in order to reduce the risk of frost damage to flowers and young fruits under climate change. For Lednice (southern Moravia part of the Czech Republic) and other regions with similar climatic conditions, these authors recommended several apricot cultivars, such as 'Bergeval', 'Sefora', 'Bigred', 'Vynoslivyi', 'Tomcot', 'Bronzovyi', 'Kostinskyi', 'Harglow', 'NS-4' and 'Strepet'. We found that apricot varieties in western part of Serbia with moderate climate—such as 'Fardao', 'Čačansko Zlato', 'Spring Blush', 'Orange Red', especially 'Roxana' and 'Novosadska Kasnocvetna', bloom several days later as compared to some traditional ones (Milošević et al., 2021).

When it comes to apricot rootstocks, the question is very complicated and open. However, according to several studies, the rootstocks can reduce the harmful effects of climate change on apricots through their impacts on flowering phenology and their consequences on production (spring frost, poor pollination, extended fruit maturity, etc.) (Milošević et al., 2021). There is no universal rootstock for apricot, like M.9 for apple or quince clones for pear. Each country or region has a specific rootstock(s) for this fruit type. An additional problem is compatibility with cultivars. For the temperate-continental area affected by climate change, rootstocks originate from myrobalan (*Prunus cerasifera* Ehrh.), apricots, almonds or peaches or their interspecific hybrids (propagated vegetatively or by seed) are not a good solution, due to an earlier beginning of vegetative cycle, i.e., earlier flowering. Rootstocks of European plum (*P. domestica* L.) origin are a better choice for apricots in moderate climatic conditions in comparison with myrobalan (Milošević et al., 2015; Milošević and Milošević, 2019).

On the other hand, drought, as a direct result of global and regional climate change, seems to be the most devastating stress affecting overall apricot production, with no exception. On this point of view, there are controversial results of scientists regarding tolerance or resistance of apricot rootstocks to drought. Obviously, rootstocks that reduce the negative impact of drought in the Mediterranean area are not suitable for use in the temperate-continental area and vice versa. For example, Rezq et al. (2020) reported that under Palestine's conditions, rootstocks for apricot, such as bitter almond, local apricot (*P. armeniaca* L.), Mahaleb and myrobalan-GF81 are resistant or tolerant to drought stress. However, these rootstocks, as previously said, are unsuitable for apricots grown in moderate and/or continental climate, i.e., in countries on Balkan peninsula or central Europe (Krška and Bauer, 2020).

Gunduz et al. (2011) reported that using special heater in apricot orchards, mixing the air in apricot orchards by using wind machine, using irrigation sprinklers to spray water on the apricot trees, burning the waste into the orchards may reduce the adverse effects of freeze on flowers when the expected lows are just below freezing. Also, modern irrigation systems, such as drips, sprinklers, etc. should be used to counter the overflowing temperature (Gunduz et al., 2011). A reduction in evapotranspiration can be achieved by a reduction in wind speed (using hedgerows and windbreaks), an increase in soil conductivity (mulching) and a reduction in available energy (shading) which can be achieved by anti-hail nets (Lalić and Mihailović, 2011).

However, most of previously described measures cannot be implemented without a new approach to apricot-growing technology. Primarily, training and pruning must be quite different from traditional ones. It has been long known that apricot and its cultivars have a vigor tree, which, in traditional planting, reaches a height of more than 5–6 m. In these plantations, it is not possible to install anti-hail nets. Anti-hail nets also protect against excessive sunburn that can cause burns on the fruit. Under these conditions, trees receive 10–15% less solar radiation and have a lower evaporative demand, since extreme temperatures and high vapor-pressure deficits are avoided. Therefore, trees should not be higher than 4 m. It was found in our previous research that it is possible to grow apricots in semi-high (600 tree ha^{-1}) and high dense planting system (> 1,000 tree ha^{-1}) on vigorous rootstock with the tree height less than 4 m (Milošević et al., 2013a,b). In this case careful canopy management (e.g., summer pruning) is required. It is also proved that pruning before paradormancy, i.e., between 15 June and 15 July causes the appearance of premature (sylleptic) shoots on which flower buds form (Fig. 8).

Flowering on these shoots is a few days late. Potentially, spring frost damage can be avoided in this way. So, it can be said that ongoing global warming threatens fruit tree industry all around the world, but also opens new challenges and opportunities for renewing apricot material (cultivars and rootstocks) and growing technology (Milošević et al., 2013a,b; Milošević and Milošević, 2019).

Fig. 8. Formation of premature (sylleptic) shoots on apricot crown after summer pruning between 15 June and 15 July under Čačak conditions (western Serbia) (*Photo*: Tomo Milošević).

5. Conclusion and Prospects

Apricot (*P. armeniaca* L.) originates from China and Central Asia, i.e., from an area which has two seasons—cold and warm (without transitional seasons of spring and autumn). In general, as a plant, it is difficult to adapt to different environmental conditions which significantly limit its commercial cultivation. Compared with other fruit types, apricot cultivation and production is strongly restricted by climatic conditions and also by soil conditions and technology used for its cultivation. In the last few decades, climate change has been an additional problem in its cultivation. Given this, new strategies and management of apricot plantations are an imperative. Many studies around the world have tried and are trying to provide answers to the challenges posed by climate change. Some of them are pessimistic and indicate that it is relatively late to react. However, based on relevant scientific research and our own experience, we believe that it is possible to reduce the negative impact of climate change on apricots through the following short-term and long-term measures: (1) securing water reservoirs; (2) regulation of water rights; (3) implementing water-saving measures, such as drip irrigation, deficit irrigation, partial root zone drying or no irrigation; (4) use of suitable and/or breeding new cultivars and rootstocks with efficient water usage and tolerant to drought, pests and disease and other biotic and abiotic stresses; (5) extensive low plant/tree density; (6) reduction in evapotranspiration, e.g., by shading with nets and mulching with shredded tree-pruning residues; (7) zero tillage; (8) grassy alleyways or cover crops;

(9) and apricot translocation towards to north and/or upwards into colder hillsides. Finally, will apricot cultivation become a climate protector?

Acknowledgement

This research was conducted under the support of Ministry of Education, Science and Technological Development of the Republic of Serbia through contracts No. 451-03-9/2021-14/200215 and 451-03-9/2021-14/200088.

References

Arakelyan, B. (1968). Excavations at Garni, 1949–50. *In*: Field, H. (ed.). *Contributions to the Archaeology of Armenia*. Cambridge, 29 p.

Atkinson, C.J., Brennan, R.M. and Jones, H.G. (2013). Declining chilling and its impact on temperate perennial crops. *Environmental and Experimental Botany*, 91: 48–62. Doi: 10.1016/j.envexpbot.2013.02.004.

Bălan, V., Tudor, V., Topor, E., Chireceanu, C., Dobrin, I. and Iacom, B. (2009). Apricot adaptability under the Romanian climatic conditions. *Scientific Papers USAMV Bucharest, Series A, Agronomy*, 52: 445–450.

Baldocchi, D. and Wong, S. (2006). An Assessment of the Impacts of Future CO_2 and Climate on Californian Agriculture. *California Climate Change Center Report*, University of California, Berkeley.

Bartolini, S., Massai, R., Iacona, C., Guerriero, R. and Viti, R. (2019). Forty-year investigations on apricot blooming: Evidences of climate change effects. *Scientia Horticulturae*, 244: 399–405. Doi: 10.1016/j.scienta.2018.09.070.

Beaubien, E.G. and Freeland, H.J. (2000). Spring phenology trends in Alberta, Canada: Links to ocean temperature. *International Journal of Biometeorology*, 44(2): 53–59. Doi: 10.1007/s004840000050.

Caprio, J.M. and Quamme, H.A. (2006). Influence of weather on apricot, peach and sweet cherry production in the Okanagan Valley of British Columbia. *Canadian Journal of Plant Science*, 86(1): 259–267. Doi: 10.4141/P05-032.

Carraut, A. (1968). *Contribution a l'étude de la levêè de dormance des bourgeons afleur de l'abricotier. Acta Horticulturae*, 11: 479–484. Doi: 10.17660/ActaHortic.1968.11.48.

Černa, H., Bartošova, L., Trnka, M., Bauer, Z., Štěpanek, P., Možny, M., Dubrovsky, M. and Žalud, Z. (2012). The analysis of long-term phenological data of apricot tree (*Prunus armeniaca* L.) in southern Moravia during 1927–2009. *Acta Universitatis Agriculturae et Silviculturae Mendelianae Brunensis*, 60(3): 9–18. Doi: 10.11118/actaun201260030009.

Chmielewski, F.M. and Potts, J.M. (1995). The relationship between crop yields from an experiment in southern England and long-term climatic variation. *Agricultural and Forest Meteorology*, 73(1-2): 43–66. Doi: 10.1016/0168-1923(94)02174-I.

Chmielewski, F.M., Muller, A. and Bruns, E. (2004). Climate changes and trends in phenology of fruit trees and field crops in Germany, 1961–2000. *Agricultural and Forest Meteorology*, 121: 69–78. Doi: 10.1016/S0168-1923(03)00161-8.

Cosmulescu, S. and Gruia, M. (2016). Climatic variability in Craiova (Romania) and its impacts on fruit orchards. *South-western Journal of Horticulture, Biology and Environment*, 7(1): 15–26.

Cosmulescu, S. and Bîrsanu Ionescu, M. (2018). Phenological calendar in some walnut genotypes grown in Romania and its correlations with air temperature. *International Journal of Biometeorology*, 62(11): 2007–2013. Doi: 10.1007/s00484-018-1606-3.

Costes, E. Khadari, B., Zaher, H., Moukhli, A., Morillon, R., Legave, J-M. and Regnard, J-L. (2016). Adaptation of Mediterranean fruit tree cultivation to climate change. pp. 503–510.

In: Moatti, J-P. and Thiébault, S. (eds.). *The Mediterranean Region under Climate Change: A Scientific Update*. IRD Éditions, Marseille, France. Doi: 10.4000/books.irdeditions.23838.

De, U.S., Dube, R.K. and Prakash Rao, G.S. (2005). Extreme weather events over India in the last 100 years. *Indian Journal of Geophysics*, 9(3): 173–187.

Denis, F.G. (2003). Problems in standardizing methods for evaluating the chilling requirements for the breaking of dormancy in bud woody plants. *HortScience*, 38(3): 347–350. Doi: 10.21273/HORTSCI.38.3.347.

Doi, H. (2007). Winter flowering phenology of Japanese apricot *Prunus mume* reflects climate change across Japan. *Climate Research*, 34: 99–104.

Edmonds, J.A. and Rosenberg, N.J. (2005). Climate change impacts for the Conterminous USA: An integrated assessment summary. *Climatic Change*, 69(1): 151–162. Doi: 10.1007/s10584-005-3613-8.

Else, A.M. and Atkinson, J.C. (2010). Climate change impacts on UK top and soft fruit production. *Outlook on Agriculture*, 39(4): 257–262. Doi: 10.5367/oa.2010.0014.

Ercisli, S. (2009). Apricot culture in Turkey. *Scientific Research and Essay*, 4(8): 715–719.

Fischer, G., Shah, M., Tubiello, F.N. and van Velhuizen, H. (2005). Socio-economic and climate change impacts on agriculture: An integrated assessment, 1990–2080. *Philosophical Transactions of the Royal Society-B, Biological Sciences*, 360(1463): 2067–2083. Doi: 10.1098/rstb.2005.1744.

Freckleton, R.P., Watkinson, A.R., Webb, D.J. and Thomas, T.H. (1999). Yield of sugar beet in relation to weather and nutrients. *Agricultural and Forest Meteorology*, 93(1): 39–51. Doi: 10.1016/S0168-1923(98)00106-3.

Glišić, I., Milošević, T., Ilić, R. and Paunović, G. (2019). Freezing of flower buds of apricot (*Prunus armeniaca* L.) during winter dormancy. pp. 524–530. *The Proceedings of the XXIV Conference of Biotechnology*. 15–16 March 2019, Faculty of Agronomy, Čačak, Serbia.

Gunduz, O., Ceyhan, V. and Bayramoglu, Z. (2011). Influence of climatic factors on apricot (*Prunus armeniaca* L.) yield in the Malatya province of Turkey. *Asian Journal of Agricultural Sciences*, 3(2): 150–155.

Gupta, V., Namgyal, D., Raghuvanshi, M.S. and Stanzin, L. (2015). Major pests of cold arid region of Leh and their integrated management. pp. 104–107. *In: Proceedings of National Symposium on Sustaining Agricultural Productivity in Arid Ecosystems: Challenges and Opportunities (SAPECO-2015)*, held at RRS, CAZRI, 19–22 August 2015, Leh, India.

Ibrahim, Z.I. (2014). Climate change impacts on pests and pesticide use. *ARCA Working Paper, Alexandria Research Center for Adaptation to Climate Change (ARCA)*. Alexandria, Egypt, 33 p.

Ighbareyeh, J.M.H. and Carmona, C.E. (2017). Impact of climate and bioclimate factors on apricot (*Prunus armeniaca* L.) yield to increase economy and achieve maintaining food security of Palestine. *Open Access Library Journal*, 4: e4119, 13 pages. Doi: 10.4236/oalib.1104119.

IPCC. (2007). Climate Change (2007). The physical science basis. *In: Contribution of Working Group I to the Fourth Assessment Report of the Intergovernmental Panel on Climate Change*. Cambridge University Press, Cambridge, UK and New York, USA, 996 pp.

IPCC. (2013). Summary for Policymakers. *In*: Stocker, T.F., Qin, D., Plattner, G.-K., Tignor, M., Allen, S.K., Boschung, J., Nauels, A., Xia, Y., Bex, V. and Midgley, P.M. (Eds.). Climate Change 2013: The Physical Science Basis, *Contribution of Working Group I to the Fifth Assessment Report of the Intergovernmental Panel on Climate Change*. Cambridge University Press, Cambridge, United Kingdom and New York, NY, USA.

Jangra, M.S. and Sharma, J.P. (2013). Climate resilient apple production in Kullu valley of Himachal Pradesh. *International Journal of Farm Sciences*, 3(1): 91–98.

Karakaş, G. and Doğan, G.H. (2018). The effect of climate change on apricot yield: a case of Malatya Province. *In*: Tanritanir, C.B. and Özer, S. (eds.). *Academic Research in Social, Human and Administrative Sciences – I*. Gece Kitaplığ, İskitler, Ankara.

Karl, T., Melillo, J., Peterson, T. and Hassol, S. (2009). *Global Climate Change: Impacts in the United States*. Cambridge University Press, New York, NY.

Kottek, M.J., Grieser, C., Beck, B., Rudolf, B. and Rubel, F. (2006). World map of the Köppen-Geiger climate classification updated. *Meteorologische Zeitschrift*, 15(3): 259–263. Doi: 10.1127/0941-2948/2006/0130.

Krška, B. and Bauer, Z. (2020). The impact of climate change on the selection of well adapted apricots. *Acta Horticulturae*, 1290: 217–220. Doi: 10.17660/ActaHortic.2020.1290.38.

Ladányi. M. and Horváth. L. (2010). A review of the potential climate change impact on insect populations—general and agricultural aspects. *Applied Ecology and Environmental Research*, 8(2): 143–152. Doi: 10.15666/aeer/0802_143151.

Lalić, B. and Mihailović, D.T. (2011). Impact of climate change on food production in northern Serbia (Vojvodina). pp. 67–85. *In:* Ivanyi, Z. (ed.). *The Impact of Climate Change on Food Production in the Western Balkan Region*. Regional Environmental Center for Central and Eastern Europe (REC). Available at: http://documents.rec.org/topic-areas/Impacts-climage-change-food-production.pdf.

Lalić, B., Janković, D., Jančić, M., Ejcinger, J. and Firanj, A. (2015). Heating of crops—How to respond? Effects of climate change on Serbian agriculture. *'Preparation of the Second Report of the Republic of Serbia under the United Nations Framework Convention on Climate Change'*. United Nations Development Program (UNDP) in Serbia, the Global Environment Facility (GEF). Available at: http://www.klimatskepromene.rs/wp-content.

Legave, J.M., Richard, J.C. and Fournier, D. (2006). Characterization and influence of floral abortion in French apricot crop area. *Acta Horticulturae*, 701: 63–68. Doi: 10.17660/ActaHortic.2006.701.6.

Legave, J.M., Christen, D., Giovannini, D. and Oger, R. (2009). Global warming in Europe and its impact on floral bud phenology in fruit species. *Acta Horticulturae*, 838: 21–26. Doi: 10.17660/ActaHortic.2009.838.1.

Lobell, D. and Field, C. (2007). Global scale climate-crop yield relationships and the impacts of recent warming. *Environmental Research Letters*, 2: 014002, 6 p. Doi: 10.1088/1748-9326/2/1/014002.

Luedeling, E., Girvetz, E.H., Semenov, M.A. and Brown, P.H. (2011). Climate change affects winter chill for temperate fruit and nut trees. *PLoS ONE*, 6: e20155, 13 p. Doi: 10.1371/journal.pone.0020155.

Majdancic, M., Basic, M., Salkic, B., Kovacevic, V., Rastija, M. and Jovic, J. (2016). Weather conditions and yield of wheat in bosnia and herzegovina with emphasis on climatic change and Tuzla Canton. *Journal of Agriculture and Ecology Research International*, 7(2): 1–9.

Maskrey, A., Buescher, G., Peduzzi, P. and Schaerpf, C. (2007). *Disaster Risk Reduction: 2007 Global Review*. Consultation Edition, Prepared for the Global Platform for Disaster Risk Reduction First Session, Geneva, Switzerland, pp. 5–7.

Mendelsohn, R. (2001). Global warming and the american economy: a regional assessment of climate change impacts. *In*: Oates, E.W. and Folmer, H. (eds.). *New Horizons in Environmental Economics*. Edward Elgar Publishing, Cheltenham, GB.

Milošević, T. (1997). Special topics in fruit growing. *Faculty of Agronomy and Fruit and Vegetable Community*. Čačak-Belgrade, pp. 181–213 (in Serbian).

Milošević, T., Milošević, N., Glišić, I. and Krška, B. (2010). Characteristics of promising apricot (*Prunus armeniaca* L.) genetic resources in Central Serbia based on blossoming period and fruit quality. *Horticultural Science*, 37(2): 46–55. Doi: 10.17221/67/2009-HORTSCI.

Milošević, T., Milošević, N. and Glišić, I. (2013a). Tree growth, yield, fruit quality attributes and leaf nutrient content of 'Roxana' apricot as influenced by natural zeolite, organic and inorganic fertilizers. *Scientia Horticulturae*, 156(6): 131–139. Doi: 10.1016/j.scienta.2013.04.002.

Milošević, T., Milošević, N., Glišić, I., Bošković-Rakočević, L. and Milivojević, J. (2013b). Fertilization effect on trees and fruits characteristics and leaf nutrient status of apricots which are grown at Cacak region (Serbia). *Scientia Horticulturae*, 164(16): 112–123. Doi: 10.1016/j.scienta.2013.09.028.

Milošević, T., Milošević, N. and Glišić, I. (2015). Apricot vegetative growth, tree mortality, productivity, fruit quality and leaf nutrient composition as affected by Myrobalan rootstock and Blackthorn inter-stem. *Erwerbs-Obstbau*, 57(2): 77–91. Doi: 10.1007/s10341-014-0229-z.

Milošević, T., Milošević, N. and Glišić, I. (2018). Yield and fruit quality of newly-breed domestic and foreign apricot cultivars (*Prunus armeniaca* L.). pp. 162–171. *The Proceedings of the XXIII Conference of Biotechnology*. 9–10 March 2018, Faculty of Agronomy, Čačak, Serbia.

Milošević, T., Milošević, N., Mladenović, J. and Jevremović, D. (2019). Impact of Sharka disease on tree growth, productivity and fruit quality of apricot (*Prunus armeniaca* L.). *Scientia Horticulturae*, 244(2): 270–276. Doi: 10.1016/j.scienta.2018.09.055.

Milošević, T. and Milošević, N. (2019). Behavior of some cultivars of apricot (*Prunus armeniaca* L.) on diferent rootstocks. *Mitteilungen Klosterneuburg*, 69(1): 1–12.

Milošević, T., Milošević, N. and Glišić, I. (2021). Early tree performances, precocity and fruit quality attributes of newly introduced apricot cultivars grown under western Serbian conditions. *Turkish Journal of Agriculture and Forestry*. Doi: 10.3906/tar-2010-39 (in press).

Öncel, A.G. and Tzanakis, T. (2018). Legal and statistical framework of climate change from the EU and international point of view. *Athens Journal of Sciences*, 5(4): 307–328. Doi: 10.30958/ajs.5-4-1.

Parry, M., Rosenzweig, C. and Livermore, M. (2005). Climate change, global food supply and risk of hunger. *Philosophical Transactions of the Royal Society B, Biological Sciences*, 360(1463): 2125–2138. Doi: 10.1098/rstb.2005.1751.

Pidgeon, J.D., Werker, A.R., Jaggard, K.W., Richter, G.M., Lister, D.H. and Jones, P.D. (2001). Climatic impact on the productivity of sugar beet in Europe, 1961–1995. *Agricultural and Forest Meteorology*, 109(1): 27–37. Doi: 10.1016/S0168-1923(01)00254-4.

Pidgeon, J.D., Jaggard, K.W., Lister, D.H., Richter, G.M. and Jones, P.D. (2004). Climatic impact on the productivity of sugar beet in Europe. *Zuckerindustrie (Sugar Industry)*, 129(1): 20–25.

Polák, J. (2009). Influence of climate changes in the czech republic on the distribution of plant viruses and phytoplasmas originally from the mediterranean subtropical region. *Plant Protection Science*, 45: S20–S26. Doi: 10.17221/2806-PPS.

Prudencio, A.S., Martínez-Gómez, P. and Dicenta, F. (2018). Evaluation of breaking dormancy, flowering and productivity of extra-late and ultra-late flowering almond cultivars during cold and warm seasons in south-east of Spain. *Scientia Horticulturae*, 235: 39–46. Doi: 10.1016/j.scienta.2018.02.073.

Radičević, Z., Radenković, T., Mlakara, S. and Bojović, J. (2011). The risk of severe winter and late spring frost on apricot production in Serbia. pp. 365–370. *The Proceedings of the XVI Conference of Biotechnology*, 4–5 March 2011, Faculty of Agronomy, Čačak, Serbia, pp. 524–530.

Rai, R., Joshi, S., Roy, S., Singh, O., Samir, M. and Chandra, A. (2015). Implications of changing climate on productivity of temperate fruit crops with special reference to apple. *Journal of Horticulture*, 2(2): 1000135, 6 p. Doi: 10.4172/2376-0354.1000135.

Raghuvanshi, M.S., Gupta, V., Stanzin, J., Dorjey, N. and Chauhan, S.K. (2016). Introduction of new insect-pests on apricot and its preliminary management options in cold arid region of Ladakh. *Indian Journal of Ecology*, 43(2): 590–592.

Rezq, B.-S., Fatina, H., Abdul-Jalil, H. and Mohamad, Al-S. (2020). Breeding rootstocks for fruit trees in Palestine: Status and prospective toward future climate change scenarios. *Hebron University Research Journal (A)*, 9: 11–28.

Richardson, E.A., Seeley, S.D. and Walker, D.R. (1974). A model for estimating the completion of rest for Redhaven and Elberta peach trees. *HortScience*, 9(4): 331–332.

Rosenzweig, C. and Parry, M.L. (1994). Potential impact of climate-change on world food-supply. *Nature*, 367: 133–138. Doi: 10.1038/367133a0.

Rozák, J., Gálová, Z. and Glasa, M. (2019). *Molekulárna a biologická diagnostika vybraných vírusových fytopatogénov ovocných drevín* (Molecular and biological diagnostics of selected viral phytopathogens of fruit trees). SUA Nitra, Slovakia (in Slovak).

Sønsteby, A. and Heide, O.M. (2014). Cold tolerance and chilling requirements for breaking of bud dormancy in plants and severed shoots of raspberry (*Rubus idaeus* L.). *Journal of Horticultural Science & Biotechnology*, 89(6): 631–638. Doi: 10.1080/14620316.2014.11513131.

Stern, N. (2006). What is the Economics of Climate Change? *World Economics*, 7(2): 1–10.

Sugiura, T. (2010). Characteristics of responses of fruit trees to climate changes in Japan. *Acta Horticulturae*, 872: 85–88. Doi: 10.17660/ActaHortic.2010.872.8.

Szalay, L., Pedryc, A., Szabo, Z. and Papp, J. (2006). Influence of the changing climate on flower bud development of apricot varieties. *Acta Horticulturae*, 717: 75–78. Doi: 10.17660/ActaHortic.2006.717.12.

Tresson, P., Brun, L., de Cortázar-Atauri, I.G., Audergon, J.-M., Buléon, S., Chenevotot, H., Combe, F., Dam, D., Jacquot, M., Labeyrie, B., Mercier, V., Parveaud, C.-E. and Launay, M. (2020). Future development of apricot blossom blight under climate change in southern France. *European Journal of Agronomy*, 112: 125960. Doi: 10.1016/j.eja.2019.125960.

Temesgen, T.D. (2007). Measuring the economic impact of climate change on Ethiopian agriculture: A Ricardian approach. *World Bank Policy Research Working Paper No. 4342.* Washington, DC. Available at SSRN: https://ssrn.com/abstract=1012474.

Temesgen, T.D. and Hassan, M.R. (2009). Economic impact of climate change on crop production in Ethiopia: Evidence from cross-section measures. *Journal of African Economies*, 18(4): 529–554. Doi: 10.1093/jae/ejp002.

Thapa, S. and Joshi, R.G. (2010). *A Ricardian Analysis of Climate Change Impact on Nepalese Agriculture.* Munich Personal RePEc Archive (MPRA), Ministry of Environment, Kathmandu, Nepal.

Vavilov, N.I. (1926). Centres of origin of cultivated plants. *Bulletin of Applied Botany, Genetic and Plant Breeding*, 16: 1–248.

Vavilov, N.I. (1991). Near the Pamir (Darvaz, Roshan, Shugna): Agricultural Essay. *Bulletin of Applied Botany, Genetic and Plant Breeding*, 140: 1–12.

Vitasse, Y., Delzon, S., Bresson, C.C., Michalet, R. and Kremer, A. (2009). Altitudinal differentiation in growth and phenology among populations of temperate-zone tree species growing in a common garden. *Canadian Journal of Forest Research*, 39(7): 1259–1269. Doi: 10.1139/X09-054.

Viti, R. and Monteleone, P. (1995). High temperature influence on the presence of flower bud anomalies in two apricot varieties characterized by different productivity. *Acta Horticulturae*, 384: 283–289. Doi: 10.17660/ActaHortic.1995.384.43.

Viti, R., Bartolini, S. and Guerriero, R. (2006). Apricot floral biology: The evolution of dormancy and the appearance of bud anomalies in several Italian genotypes. *Advances in Horticultural Science*, 20(4): 267–274.

Vuković, J.A., Vujadinović, P.M., Rendulić, M.S., Djurdjević, S.V., Ruml, M.M., Babić, P.V. and Popović, P.D. (2018). Global warming impact on climate change in Serbia for the period 1961–2100. *Thermal Science*, 22(6A): 2267–2280.

Zhebentyayeva, T., Ledbetter, C., Burgos, L. and Llácer, G. (2012). Apricot. pp. 415–458. *In:* Badenes, M.L. and Byrne, D.H. (eds.). *Handbook of Plant Breeding*, vol. 8, *Fruit Breeding*. Springer Science + Business Media. Doi: 10.1007/978-1-4419-0763-9_12.

11

Impact of Climate Change on Plum (*Prunus domestica* L.)

Nebojša Milošević[1],* and *Tomo Milošević*[2]

1. Introduction

After almost 10,000 years of relatively stable climate on the Earth, since the 1900s more rapid changes have taken place and have been mostly caused by the human factor. Global mean temperatures increased by 0.74°C during the last 100 years and best estimates predict that to increase global annual mean temperatures in the range of 1.8–4°C during the year 2100, resulted in increase variability of rainfall and enhanced frequency of extreme weather events, such as heat waves, cold waves, droughts and floods (IPCC, Climate Change, 2007). Climate changes and their biologic effects have shown that they impact species physiology, species distribution, organisms' phenology and biocenosis composition and dynamics (Parmesan and Yohe, 2003). Increasing temperatures, modified rainfalls and climate variability lead to significant changes in productivity of different crops and their distribution in familiar areas in the future period with large concern for their subsistence in some regions. All these negative consequences call into question the safe production of food in some regions of the world. High quality and greater yield of all crops can only be achieved in optimum climatic, soil and water requirements. The

[1] Department of Pomology and Fruit Breeding, Fruit Research Institute, Čačak, Kralja Petra I/9, 32000 Čačak, Republic of Serbia.
[2] Department of Fruit Growing and Viticulture, Faculty of Agronomy, University of Kragujevac, Cara Dušana 34, 32000 Čačak, Republic of Serbia.
Email: tomomilosevic@kg.ac.rs
* Corresponding author: nmilosevic@institut-cacak.org; mnebojsa@ftn.kg.ac.rs

changed climatic parameters affect the crop physiology, biochemistry, floral biology, biotic stresses, like disease-pest incidence, etc. and ultimately result in the reduction of yield and quality of fruit crops (Rajatiya et al., 2018). These changes also have very strong economic impacts on agriculture, causing price fluctuations, trade and profitability of producers and companies.

Almost all European regions are expected to be negatively affected by future climate changes, which will cause changes in many economic spheres. In many European regions, the most prevalent abiotic stress factors caused by climate change are rainfall deficiency or surplus, very low temperatures during winter and early spring, summer heat, hailstorms as well as unsatisfying soil fertility. To these abiotic stress factors must be added the biotic factors, such as diseases, pests and competing plants whose impact increases.

Temperature has different-types of impact on biological properties of different fruit species. It has direct influence as mean temperature, heat waves, hot days, etc. and indirect effect through water regime, and storm and hail occurrence. Due to climate change, extreme temperature effects very often occur; so different types of adaptions have been required.

Discrepancies with water availability had very large effect on fruit production among climate factors. Climate impacts on water resources varied in different regions of the world and led to a high possibility of precipitation decrease. It is assumed that the frequency of droughts and floods will increase under future climate conditions. Efficient water use and integrated management will be increasingly important for reducing water scarcity and droughts. Climate change also has significant effects on the global water cycle and could have far-reaching impacts on humans and natural ecosystems. Changes include variations in the distribution, timing and intensity of precipitation events (rainfall, snow, hail and cloud mist) and changes in the timing of seasonal waterflows (Rehman et al., 2015). This phenomenon has significant impact on most agricultural crops including fruit species. Access to water, both in terms of quality and quantity and possibility of irrigation, has a major influence on the structure and quality of agricultural production. Although many water-management approaches have been adapted to mitigate climate impacts, there is still a need to determine local solutions (Kang et al., 2009).

Climate change could also have a significant impact on winter chill, which could potentially have a major impact on fruit species with chilling requirements, such as apricots, apples, pears, nectarines and plums. So they need to achieve a chilling requirement in order to break dormancy in winter and recommence growth in spring (Luedeling et al., 2011; Ogundeyi and Jordaan, 2017). If the winter chill is not sufficient or fulfilled, dormancy is not broken (which occurs in extreme situations) and fruit yield and quality can be severely affected (Dennis, 2003). In this regard, it is up to scientists and producers around the world to find ways to overcome climate change problems and ensure safe food production.

2. Impact of Climate Change on Fruits

Global food security threatened by climate change is one of the most important challenges in the 21st century so as to be able to supply sufficient food for the increasing population while sustaining the already stressed environment (Lal, 2005). Climate change has already caused significant impacts on water resources, food security, hydropower, human health in the entire world (Kang et al., 2009). On the other hand, in some regions, such as northern areas in Europe, increasing temperatures could improve crop production. This phenomenon also affects water availability and represents one of the greatest pressures on the hydrological cycle along with population growth, pollution, land use changes and other factors (Aerts and Droogers, 2004). If climate change trends continue in this direction, production of fruits could be significantly altered.

Influence of climate change on fruit crops is miscellaneous and multiplex. Temperate zone fruit species, including plum, undergo essential physiological and morphological changes which are progressing and will do so only at the low temperature to which evolution of plants has adapted. Such imposed temperature regimes are very precise: there are sharp differences in temperature requirements, not only among genera, but even between individual cultivars (Van Tuyl, 1983). Optimal temperature is necessary for pollination and fruit set in all fruit species. It can affect different stages of the reproductive process, such as pistil receptivity (Hedhly et al., 2004), embryo sac vitality (Cerović and Ružić, 1992; Cerović et al., 2000), pollen germination and pollen tubes growth (Delph et al., 1997). Low temperatures in flowering time decrease pollen germination and pollen tubes growth in pistil, while high temperature on the one hand decrease pistil receptivity and pollen germination and on the other, significantly accelerate pollen tubes' growth in the pistil (Hedhly et al., 2004; Milošević, 2013). Very high temperatures can even limit fruit setting of citrus fruits and not only temperate-zone fruits (Rehman et al., 2015). Winter chill is one of the most important determinants of good yields in many fruit trees. Rehman et al. (2015) stated that a species' chilling requirement is not necessarily fulfilled by a certain number of hours below a given temperature threshold. Warm temperatures might instead compensate for earlier chilling hours, or the breaking of dormancy might happen in two phases. The easiest and most common approach to approximating winter chill conditions, however, is to sum up the seasonal hours between 0–7.2°C (Bennett, 1994). Among temperate-zone fruits, the minimum requirements for chilling period are as follows: for peach

(*Prunus persica* L.) – 200 h and apricot (*Prunus armeniaca* L.) – 300 h; apple (*Malus × domestica* Borkh.), pear (*Pyrus communis* L.) and plum (*Prunus domestica* L.) need at least 800 h, while sour cherry and sweet cherry demand 1,200 h (Milošević, 1997).

According to Rötzer and Chmielewski (2001), phenology is one of the most suitable indicators accepted and used to monitor climate change and it is an ideal way to demonstrate the effects of global warming on the living world (Sparks and Menzel, 2002). Phenology presents the study of life cycle appearances in plants and animals (Schwartz, 2003). These phenomena, known as phenophases (which include occurrence of the first leaves and the flowering time) are dependent on environmental conditions, such as temperature and length of day (Pudas et al., 2008). Generally, different fruit phenophases, their onset and duration are highly dependent on local climatic conditions and fluctuate from year to year (Montagnon, 2007). Among the factors that directly influence flowering is the number of hours it takes for a tree to spend at temperatures below 7°C (chilling period) during winter dormancy (Alburquerque et al., 2007). Also, phenophases in fruit trees depend on air temperature (Rodrigo and Herrero, 2002), wind (Dennis, 1979), frost (Rodrigo, 2000), rain and humidity (Gradziel and Weinbaum, 1999). Climate change, primarily related to global warming, affects biological dynamics of plants showing phenological variations as a function of vulnerable environmental conditions. As a result, phenological phases started earlier and lasted shortly. In some regions, this phenomenon increased the risk of frost damage to actively growing parts. Also, appearance of late spring frosts before the start of flowering may cause covert damages in flower buds and moreover, the frost during the blossoming time can destroy the flower elements, so that the yield may be completely absent.

3. Plum Importance and Reaction to Climate Change

Plum production in the world rose from 8.1 million tons in 1999 to over 10.3 million tons in 2008 and to 11.8 million tons in 2017 (FAOSTAT, 2019). The most important producers of domestic plum (*Prunus domestica* L.) are countries situated in the regions where moderate continental climate is dominant, such as Romania, Serbia, USA, Germany, Bosnia and Herzegovina, Bulgaria, etc. In these areas, plum has been cultivated for centuries. Roman emperors, Probus (232–282) and Diocletian (243–316), planted plums in the territories of present-day Bosnia, Serbia and Croatia. According to Ramming and Cocciu (1991) via Milatović, 2019 cultivar 'De Bosnia' was grown in this region and as it is supposed, presents progenitor of cultivar 'Požegača'. This was the most important plum cultivar in Europe until the Sharka virus occurred in the middle of the 20th century. The second most important cultivar was 'French Prune' which was brought to France from Persia or Turkey by Crusaders. European colonists introduced this cultivar to America and its large-scale growing started at the end of the 19th century. Today it is the most important cultivar in the USA (DeJong et al., 2008).

A large number of autochthonous or local plum cultivars (accessions) exist on the Balkan peninsula and some of them are highly presented in orchards in Serbia, Bosnia and Herzegovina, North Macedonia, Montenegro, Croatia, Romania and Bulgaria (Milošević et al., 2012). These plum accessions are well adapted to agro-ecological conditions. The first selection from a diverse gene pool was conducted by local growers in order to obtain certain desirable traits and hardiness to ecological conditions. Thanks to its long-standing presence in these areas, autochthonous genotypes have acquired some level of resistance to most unfavorable factors, including those caused by climate change. However, due to certain shortcomings, their share in commercial production is diminishing.

Significantly more important are cultivars that have been developed in different breeding programs, such as 'Stanley', 'Čačanska Lepotica', 'Čačanska Najbolja' and 'Čačanska Rodna'. These cultivars show a high level of adaptability to different climatic conditions and achieve good productive results in many parts of Europe and the world. Among the most important plum breeding objectives is tolerance to stressful environmental factors (low temperatures, frost occurrence, high temperature and lack of water). Most plum cultivars tolerate low temperatures till –25°C and some cultivars can withstand temperatures even as low as –30°C, but variable temperatures during the winter months, which have been particularly pronounced in recent years as a consequence of climate change, are an even bigger problem. Bark and flower buds can be damaged in this period. Breeding to winter hardiness was mostly carried out in Russia and cultivars 'Vengerka Moskovskaya', 'Zuysinskaya' and 'Reine Claude Reform' were used as donors of this trait (Eremin and Zaremuk, 2014; Simonov and Kulemekov, 2017); in Sweden cultivars Opal, Herman, Violetta, etc. were created (Okie and Ramming, 1999) as well as in Latvia, where cultivars Agra Dzeltena, Minjona, Zemgale, Minjona, Ance, Sonora, Adelyn and Laine were developed (Ikase, 2015; Kaufmane et al., 2012). Among the commercially important cultivars, there are differences in tolerance to winter frosts. Cultivars 'Weingenheims', 'Anna Spath', 'President', 'Čačanska Najbolja' and 'Požegača' are tolerant to winter frosts, while cultivars 'Stanley', 'Čačanska Lepotica', 'Blufree' and 'Ruth Gerstetter' are susceptible (Grzyb and Rozpara, 2000; Szabó, 2003). Spring frosts can also harm flower buds and the degree of damage depends on the developing stage of the flower buds and frost intensity. Early flowering cultivars are more exposed to spring frost while old autochthonous cultivars are mostly less susceptible (Paunović, 1988).

Most of domestic plum cultivars have intermediate to high chilling requirements, which could have a positive impact on frost tolerance but could cause problems in areas with low chilling possibilities (Neumuller, 2011). Insufficient chilling can lead to inadequate flowering and poor yield. Generally, in the regions where moderate continental climate is dominant, there was no serious problem with chilling of domestic plum cultivars.

Many physiological processes in fruit trees are directly affected by temperature. Among them, the most important are developmental processes, such as flower bud formation and induction and release of winter dormancy (Ruiz et al., 2007; Horvath, 2009). Flower bud formation in plum occurs in late summer and early autumn in the year before flowering and fruiting. Some members of Rosaceae family, including plum, unlike other temperate fruit species, are not sensitive to short photoperiods for induction of growth cessation and winter dormancy (Heide and Prestrud, 2005; Heide, 2008). Positive correlation between temperatures in the period July–September and fruit yield in next year was found in plums (Døving, 2009; Wozniacki et al., 2019), as well as in pear (Atkinson and Lucas, 1996) and small fruits belonging to Rubus and Ribes family (Heide and Sønsteby, 2011; Sønsteby and Heide, 2011). The correlation with August and September temperatures together in flowering abundance was stronger than those for the separate months (Wozniacki et al., 2019). Warm temperatures during this period are convenient to flower bud formation in plum trees and are highly correlated with flowering and yield in the following season. This phenomenon is probably also dependent on cultivars, which was confirmed by Wozniacki et al. (2019), who determined that cultivar 'Victoria' evinces weak correlation between warm temperatures during August–September and flowering abundance. In the last few years in Serbia, due to the very warm and dry weather during July–September, slightly lower yields of some plum cultivars have been observed in the following season (personal data).

Among the many factors that affect the flowering time of plum, temperature is one of the most important. Flower induction is deeply influenced by temperature, especially low temperature. However, strong interaction between genotype, photoperiod and temperature interactively control flowering (Rai et al., 2015). The change in timing of different physiological activities, i.e., phenology, is one of the most pronounced effects of climate change (Cleland et al., 2007). The fact is that all phenological phases, including flowering, occur earlier from year to year. This phenomenon can increase the risk of damage caused by late spring frost. An earlier date of full bloom of up to 10 days was observed in apple cultivars 'Boskoop', 'Cox's Orange Pippin' and 'Golden Delicious' when comparing the last 25 years with the previous 30 years, which is less than the 14 days and reported generally for Germany (Kunz and Blanke, 2016).

Wolfe et al. (2005) reported that three woody perennials (lilac, apple and grapevine) started phenological phases two to eight days earlier during the period 1965–2001 in north-eastern USA. Regarding the domestic plum, numerous authors reported differences in flowering time depending on years and regions (Cosmulescu et al., 2010; Milošević et al., 2012; Milošević et al., 2016). Gitea et al. (2019) found that in 2016 as compared to 2009, the blooming period recorded a difference of eight days for the 'Centenar' and 'Anna Spath' cultivars and of nine days for the 'Minerva' and 'Stanley' cultivars in temperate continental climate of Romania. In cooler Nordic climate, increasing spring temperature was associated with an earlier mean date of full bloom for the plum cultivars: 'Ive', 'Mallard' and 'Victoria', resulting in approximately 10 days of advancement in full blooming across the years 1985–2016 (Wozniacki et al., 2019). According to Menzel et al. (2007), the most prominent temperature-driven changes in plant phenology are an earlier start of spring in the last three to five decades with an average of 2.5 days/decade mainly observed in mid latitudes and higher latitudes of the northern hemisphere.

Temperature is also a vital factor affecting pollination and fruit set in plums. Even if plum trees have bloomed satisfactorily, temperature can be a determinant of whether a good yield will occur or not. Pollination is normally done by honeybees and if the temperature is too low or too high, the bees quit flying and that can mean a very poor yield indeed. When the pollen is spread, it must germinate and the pollen tube needs to grow down to the ovule. High temperatures have a twofold effect on the fruit set. On the one hand, they negatively affect pollen germination, slowing down this process and reducing the number of germinated pollen grains, while on the other hand, they accelerate the growth of pollen tubes. Temperature significantly influences stigmatic receptivity (Hedhly et al., 2003), pollen germination and pollen tube growth (Pirlak, 2002; Hedhly et al., 2004), ovule longevity (Cerovic and Ruzic, 1992; Hedhly et al., 2007), and fruit set (Hedhly et al., 2007). Incidences of a season characterized by high temperatures during the flowering period reduced fertility of cultivated plants, due to sensitivity of different phases of the reproductive process to higher temperatures (Hedhly et al., 2008). This phenomenon is bound to result in gradual changes in distribution of cultivars, by favoring those better adapted to higher temperatures (Radičević et al., 2016). According to DeCeault and Polito (2010), temperatures around 22–24°C were optimal for fruit set in cultivars 'Improved French' and 'Muir Beauty', while temperatures at 20–26°C were optimal for cultivars 'Katinka', 'Hanita' and 'Jojo' according to Milošević (2013). Too high temperatures are more likely to affect fruit set of citruses than of temperate-zone fruits. Trouble is more apt to come from a combination of high temperature and high humidity, affecting pollen tube growth and ovule longevity. And even when pollination has been successful, growth of individual fruits can be restricted by both too high and too low temperatures.

During the period of fruit growth and ripening time, very high temperatures can result in damage to fruits; viz. temperatures higher than 35°C can cause occurrence of heat scalds usually presented on Japanese plums or in weak-colored cultivars, such as 'Ersinger' and 'Jalomita' (Neumuller, 2011) as well as 'Mildora' and 'Zlatka' (personal data). Unusually high temperatures during the flower-bud formation very often cause occurrence of twin fruits. Some cultivars, such as 'Stanley' and 'Jojo', tend to form a larger number of double fruits as compared to other cultivars. Neumuller (2011) reported that this phenomenon is probably highly correlated with large cropping potential. However, these fruits are useless.

Climate change has a strong impact on water availability and water productivity. By warming the planet, the atmosphere will retain more water and the water cycle will intensify. For every 1°C rise in Earth's mean surface temperature, global precipitation is expected to increase by approximately 1% (Rehman et al., 2015). Increasing global temperatures will also result in more precipitation falling as rain, rather than snow, and more intense precipitation events. Variable rainfall and temperatures could lead to a consequent higher incidence of droughts and floods, sometimes in the same places. Higher temperatures influence enhancement of evaporation rates, which can have significant impacts on the availability of soil moisture at different times of the year. Global average precipitation (rain or snowfall) and evaporation are projected by climate models to increase by about 1–9% by 2100, depending on the scenario and climate model used (Rehman et al., 2015). All these changes are expected to result in increased frequency and intensity of both droughts and floods. Water productivity is a concept to express the value or benefit derived from the use of water and includes essential aspects of water management, such as production in arid and semi-arid regions (Singh et al., 2006). Increasing water productivity means either production of the same yield with less water resources or higher crop yield with the same water resources (Zwart and Bastiaanssen, 2004). Climate change impacts on crop water productivity are affected by many uncertain factors, such as soil characteristics, soil water storage, long-term condition in soil fertility, climate variables and enhanced atmospheric CO_2 levels.

Domestic plum has large demands for available water, among temperate fruit species, immediately after apple and quince (Stanković and Jovanović, 1990). It shows best results in areas with an annual rainfall of 700–1000 mm, a vegetation rainfall of 350–600 mm and a relative humidity of 75–85% (Mišić, 2006). The lower limit of the annual rainfall for plum growing without irrigation is 500 mm. However, at least 700 mm of precipitation is necessary for intensive production and high cropping. Therewith, the timing and intensity of rainfall is also very important. It is preferable that most of the precipitation occurs in the vegetation period. Plum requirements for water depend on many factors, such as cultivars (late ripening cultivars need more water), rootstock (generative rootstocks are more tolerant to lack of water compared to vegetative), age of trees (younger trees need more water), cropping potential (for higher yield, more water is necessary), phenophase (the highest demands are in phases of largest branch growth, hardening of stone, fruit growth and flower bud formation), soil characteristics and air and soil temperature. In severe droughts during the summer months, the leaves draw water from the branches and even fruits. As a result, the fruits remain small, ripen prematurely and fall off before harvest. Lack of water also leads to poorer flower-bud formation which causes a decrease in yield the following year. Further, water scarcity can lead to bud and branches reserve nutrients deficit which results in frost sensitivity during hard winters. On the other hand, water excess can also be disadvantageous. Plum root doesn't tolerate water surplus in the soil as it can lead to root rot and dying of the tree. A long-term rainfall at the flowering time disables the flight of bees, as well as germination of pollen to the pistil stigma. Also, high amount of rainfall in the harvest time can cause cracking of the fruits which become useless. Climate change over the last twenty years has made commercial plum cultivation, with high quality fruits and large yields, practically unsustainable without irrigation in countries of the temperate climate zone which are traditionally large plum producers, such as Serbia, Romania, Bulgaria and Bosnia and Herzegovina. The exception to this rule is if orchards are planted at higher altitudes or at northern latitudes.

Alterations of the climatic factors, such as temperature and moisture, can cause changes in disease, pest and weed occurrence in different ways. These organisms adversely affect fruit quality, yield and the overall condition of the fruit trees. Temperature has potential impacts on plant disease through both the host fruit tree and the pathogen (Rehman et al., 2015). Basically, global warming will increase pest population, including that of weeds, invasive species, insects and diseases, which may lead to large increases in the use of pesticides. Enhancement of temperature, often followed by drought, can cause the plant to become more susceptible to diseases and pests due to exhaustion. Temperature increase can lead to occurrence of more disease cycles or generations of insects throughout the year. Enhanced temperatures can potentially affect insect survival, development, geographic range and population size. Temperature can affect insect physiology and development, directly or indirectly, through the physiology or existence of the host. Also, thanks to climate change, many invasive pests including aphids (*Aphis* spp.) have been introduced in colder regions, significantly affecting the spread of different viruses, especially with Sharka virus being the most important. Adaptation to higher temperatures may favor the epidemiological impact of Sharka virus isolate in the presence of aphids and sensitive hosts during the warm period of the growing season. High temperatures, often combined with lack of water, indirectly influence the development of weeds that are better adapted to adverse environmental factors than cultivated plants. Similar to temperature, moisture can affect both plum as a host plant and pathogen organisms in various ways. Generally, humidity favors the development of most pathogens, especially in combination with temperature and susceptible cultivars. For example, popular cultivars, such as 'Stanley' and 'Jojo' are very susceptible to *Monilinia* spp. and epidemics of the disease are possible in case of a wet spring. Soil moisture favors the development of particular root pathogens, such as *Armillaria mellea* (Vahl ex Fr.) and *Verticillium* spp., which can lead to tree dying.

4. Climate Change Impact on Plum in Serbia as One of the Largest World Producers

Republic of Serbia is located in the Balkan peninsula, in south-eastern Europe. It is located between 40.13° and 46.15° N latitude and 18.9° and 22.9° E longitude. Predominant climate is moderate-continental whereas mountainous climate occurs in the high mountains of the south-western Serbia. According to Köppen-Geiger climate classification, climate of Serbia belongs to Cfb type (Beck et al., 2018). However, the climate of Serbia is influenced by the Alps, the Mediterranean Sea, the Genoa Gulf, the Pannonian basin, the Moravian valley, the Carpathian and Rhodope mountains, the hilly-mountainous part with ravines and the highland plains, as well as the deep southward penetration of polar air masses, which lead to high spatial variability (Radinović, 1979). Winters are variable from year to year in both length and sharpness, while summers are very warm constantly; in recent years, even dry. Autumn and spring, as transitional seasons, are characterized by high temperature fluctuations during one day and between days. The beginning of autumn is always warmer than spring. The second half of summer is warmer than the first, and the second half of winter is generally colder than the first. In this region, temperature extremes associated with temperature maxima are characteristic. Thus, this extremity occurs in summer and often the highest temperature amplitude of air is over 30°C. In some years, absolute minimums also occur when the temperature drops below –27°C.

Considering that the temperature amplitude for plum varieties originating from *Prunus domestica* L. in the conditions of Serbia, as well as in other production areas of Europe, varies from –25°C to 35°C, the general view is that temperature movements are not a limiting factor for intensive cultivation of plums (Milosevic, 2002). According to Vuković et al. (2018), changes in climate conditions since 1961 until now influenced significant temperature increase and change in precipitation patterns. These authors indicated that temperature increase average was 1.2°C for the period 1996–2015 and with respect to the period 1961–1980, the highest increase of maximum daily temperature during the summer period was 2.2°C. It can be said that annual temperature increase, intensification of high precipitation events and decrease of summer precipitation have occurred in the last few decades. Also, hailstorms and floods have become more frequent than in the past. Čačak region, situated in the western part of Serbia, is one of the largest and the most important plum production regions in Serbia. This area, as well as the largest part of Serbia, is affected by changes in climate, which is especially noticeable through frequent occurrence of drought periods and hailstorms and changes in precipitation distribution patterns. Overview of annual temperature and precipitation, as well as temperature and sum of precipitation for the vegetative cycle in the period 1965–2018 are presented in Fig. 1 and Fig. 2.

The data provided in Fig. 1 shows enhancement of annual temperature during the period 1965–2018, with deviations from year to year. The trend of changes in mean annual temperatures modifies significantly in the positive direction because it is strongly correlated with the observation years. Based on this mathematical model, it can be concluded that it is

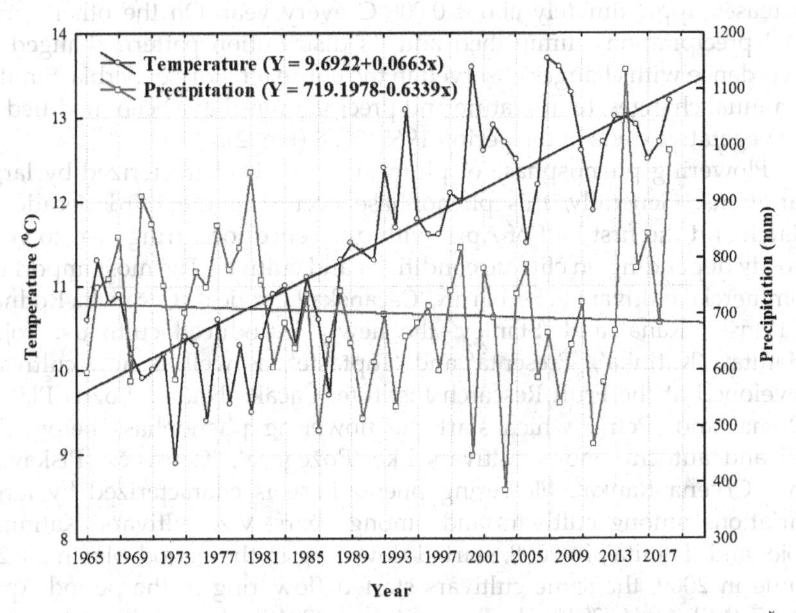

Fig. 1. Trend of average annual air temperatures and annual sum of precipitation for Čačak during 1965–2018 (*Source*: Republic Hydrometeorological Service of Serbia, Belgrade, 2020).

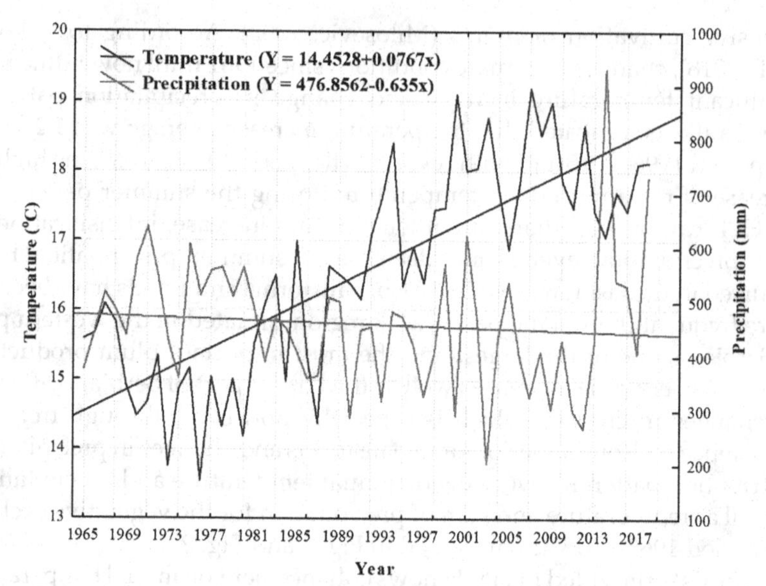

Fig. 2. Trend of average air temperatures and sum of precipitation for vegetative cycle for Čačak region during 1965–2018 (*Source*: Republic Hydrometeorological Service of Serbia, Belgrade, 2020).

getting warmer in Čačak area and that the average annual temperature increases approximately about 0.008°C every year. On the other hand, total precipitations diminished and its distribution pattern changed in accordance with changes observed in the entire territory of Serbia. Similar to annual changes, temperature and precipitations have been modified in the vegetative cycle in the period 1965–2018 (Fig. 2).

Flowering phenosphase of plum in Serbia is characterized by large variations. Generally, this phenophase occurs in the third decade of March and the first half of April with differences occurring year to year, mostly depending on climate conditions and cultivar. The most important commercial cultivars in Serbia are 'Čačanska Lepotica', 'Čačanska Rodna', 'Čačanska Rana' and 'Stanley' the newly introduced cultivars: 'Jojo', 'Hanita', 'Katinka', 'Presenta' and 'Toptaste', as well as new cultivars developed at the Fruit Research Institute, Čačak: 'Nada', 'Pozna Plava', 'Divna' and 'Petra' which start the flowering phenophase before the old and autochthonous cultivars like 'Požegača', 'Cerovački Piskavac' and 'Crvena Ranka'. Flowering phenophase is characterized by large variations among cultivars and among years, viz. cultivars 'Katinka', 'Jojo' and 'Hanita', in 2008, started flowering in the period March 23–25, while in 2009, the same cultivars started flowering in the period April 14–17 (Milošević, 2013). In the next year (2010), these cultivars started flowering in April 2–4. Commonly, flowering onset in the most grown cultivars in Serbia ('Stanley', 'Čačanska Lepotica' and 'Čačanska Rodna') as well as new cultivars ('Mildora', 'Nada', 'Divna' and 'Petra') occurs in the first half of April in Čačak conditions (Ogašanović and Ranković, 1996; Glišić et al., 2012; Milošević et al., 2016; Milošević et al., 2017). In this period, the late frosts are very rare and damage to flowers is avoided. In addition, autochthonous cultivars, which are mostly grown on higher altitudes, start flowering a few days after commercial cultivars (Glišić and Milošević, 2015).

In the last few years, plum flowering has been characterized with explosive and short-term flowering due to very high temperatures during the flowering time (personal data). Because of that, in cultivars with short effective pollination period, such as 'Čačanska Rana' and 'Pozna Plava', a high yield is frequently jeopardized. On the other hand, occurrence of late frosts in this period is unusual, except on higher altitudes and river valleys. So there is no high risk of damage. Very high temperatures during summer months and frequent drought periods have led to the commercial production of plum in Serbia becoming dependent on irrigation, which was not the case 10–20 years ago. In this country, during decades, plum have been grown with minimal application of pomo- and agrotechnical measures, including irrigation. New circumstances have led to a change in growing technology approach. Establishment of new plum orchards implies usage of irrigation systems and more often, installation of anti-hail nets. Despite that, climate change still affects plum production in many ways. Warmer vegetative period expedites ripening of some cultivars which causes reduced fruit quality and smaller yields; disturbed precipitation distribution pattern can lead to frequent occurrence of floods, increasing risk of disease and pests damage and occurrence of new ones.

5. How to Overcome Negative Climate Change Impact on Plum Production?

Nowadays different strategies to mitigate the negative effects of climate change on the production of different fruit species are adopted worldwide. Many approaches entail single options to diminish unfavorable impact of individual negative factors caused by climate change, such as late frosts, drought, floods or hailstorms. Some of these single options showed positive results but only the complex effect of different factors can significantly mitigate the negative consequences of this phenomenon.

To achieve high plum yields, Gitea et al. (2019) suggested that the following combination of factors is required: modern technological equipment, hybrids and cultivars with high productive potential, modern cropping technologies based on tillage according to the fruit tree type, modern methods of irrigation, usage of both organic and mineral fertilizers and integrated management of diseases, pests and weeds, all of which imply the preservation of biodiversity and recourse to a skilled workforce. According to Lalić et al. (2015), assorted modifications according to new conditions, application of optimal irrigation and growing of different crop species at more suitable locations are the most important options to overcome negative effects of climate change.

Generally, the first step in adapting to climate change in plum production is breeding of new cultivars which are more tolerant or resistant to unfavorable biotic and abiotic factors, including frosts, heat, drought and diseases. Development of cultivars tolerant or resistant to Sharka virus (Figs. 3 and 4) is the most important breeding objective due to destructive impact of this disease om plum production worldwide (Neumüller, 2011; Milošević and Milošević, 2018; Milošević et al., 2019). It is well known that trees infected with this virus, among other things, are more susceptible to other negative factors, such as drought, freeze and other diseases; hence they are prone to extinction. Further, development of cultivars with later blooming period or tolerant to late frosts is also very important, especially in some breeding programs of northern countries, such as Russia (Eremin

Fig. 3. Destructive impact of Sharka virus on plum fruits (*Source*: Dr. Darko Jevremović).

Fig. 4. Sharka-infected plum fruits fall before harvest (*Source*: Dr. Darko Jevremović).

and Zaremuk, 2014; Simonov and Kulemekov, 2017), Belarus (Matveev and Vasileva, 2015) and Latvia (Ikase, 2015; Kaufmane et al., 2012).

Some of the widespread cultivars, such as 'Prune d'Agen', 'Italian Prune', 'Stanley', 'Čačanska Rodna' and 'Čačanska Najbolja' show a high adaptability to different climate conditions and could be grown in very large areas (Neumüller, 2011; Lukić et al., 2016). In addition, cultivars such as 'Opal', 'Valor', 'Boranka' and 'Mildora' evince high level of tolerance to frosts and could be grown in northern latitudes or on higher altitudes (Ogašanović et al., 2007; Milatović, 2019). So, proper selection of suitable cultivars is the basis for successfully overcoming the negative effects of climate change.

Availability of appropriate rootstock is also an important factor to suit the changing climatic conditions. For example, rootstocks which had almond as one of the parents showed higher tolerance to drought, while waterlogging at the satisfactory level had been inherited by rootstocks which had myrobalan as one of parents (Xiloyannis et al., 2007). On the other hand, new increasingly-used plum rootstocks like 'Wavit', 'Weiva', 'Docera 6' and 'Dospina 235' require almost optimal conditions for growing and should be used in the usual plum-growing regions (personal data).

Modern growing technology and orchard management are among the key factors for advanced plum production. Various management practices suitable to changing climate have to be adopted. Winter pruning is necessary in all plum orchards while in modern high-density orchards, both winter and summer pruning are required. Summer pruning has the positive effect on reducing transpiration due to partial leaf removal as well as on flower-bud formation. Reduction of evapotranspiration can be brought about by reducing wind power (using hedgerows and windbreaks), increasing soil permeability (applying adequate tillage and mulching) and mitigating the negative impact of solar energy (using anti-hail nets for providing shade and protection from hail) (Lalić and Mihailović, 2011).

Also, improvement of irrigation and water-use efficiency by optimizing irrigation techniques and methods is necessary to overcome problems caused by high temperatures and lack of rainfall. Spraying irrigation systems can be also used to reduce the harmful effects of late frosts. Adverse impact of this negative occurrence can be diminished to a certain extent by utilization of special heaters, sprinklers, wind machines, surface irrigation, burning of waste in orchards and some combinations of these options (Snyder and Melo-Abrey, 2005).

To mitigate the damaging effects caused by diseases and pests, we should collect information on specific climatic conditions and interaction with diseases and pest development conditions (Subedi, 2019). Adoption of weather-based monitoring strategies for rapid and effective diagnosis of diseases and pests is required. Growing of resistant/tolerant cultivars to most important diseases is of great importance in overcoming this problem. Also, cultivation of these cultivars enables significantly less use of chemicals in plant protection, thus contributing to production of healthy food as well as reduction of air, water and soil pollution.

Finally, climate change makes it possible to successfully grow plum on northern latitudes in unusual countries, such as Norway (Meland et al., 2019; Woznicki et al., 2019), Sweden (Hjalmarsson, 2019) and Latvia (Kaufmane et al., 2019); at higher altitudes (Mathur et al., 2012) as well as in Asian counties with temperate monsoon climate (Ohata et al., 2017).

6. Conclusion and Prospects

The climate changes through alterations in temperature and water regime which further affect many other factors, have a very strong impact on fruit production worldwide. Domestic plum (*Prunus domestica* L.), as one of the most important temperate-zone fruit species, is also affected with this phenomenon. An increase in temperature brings about changes in the phenological phases that can begin earlier and last shorter, increasing the risk of damage caused by late frosts. Also, higher temperatures may trigger certain abnormalities in pollination and fruit set of plum which could reduce yield, especially in cultivars with short effective pollination period. During the period of fruit growth and ripening time, very high temperatures can result in damage to fruits. Developmental processes, such as flower-bud formation and induction and release of winter dormancy, could be also affected due to high temperatures. For intensive production and high cropping, plum requires at least 700 mm of precipitation with optimal timing and intensity of rainfall, mostly during the vegetation period. Climate change influences different water regimes which result in increased frequency and intensity of both droughts and floods. Lack of water induces lower fruit quality and yield as well as poorer flower-bud formation which causes decrease in yield in the next year. On the other hand, water excess can lead to root rot and dying of the tree. Long-duration rainfall at the flowering time disables the flight of bees, as well as germination of pollen on the pistil stigma. Also, high amounts of rainfall at harvest time can cause cracking of the fruit which becomes useless. Alterations of climatic factors, such as temperature and moisture, influence deviations in diseases, pests and weeds occurrence in different ways. These pathogens cause reduced fruit quality, yield and the overall condition of the fruit trees. To overcome problems in plum productions caused by climate change it is necessary to combine different factors. Among them, breeding of new tolerant/resistant cultivars and rootstocks, modern orchard management, different types of irrigation techniques, anti-hail nets and adequate disease and pest protection are the most important. Additionally, climate change makes it possible to successfully grow plum on northern latitudes in unusual countries, such as Norway, Sweden and Latvia as well as at higher altitudes in plum-growing countries.

Acknowledgement

This research was conducted under the support of Ministry of Education, Science and Technological Development of the Republic of Serbia through contracts No. 451-03-68/2022-14/200215 and 451-03-9/2021-14/2000088.

References

Aerts, J. and Droogers, P. (2004). *Climate Change in Contrasting River Basins: Adaptation Strategies for Water, Food and Environment*. The Netherlands: CABI Publishing, pp. 1–264.

Alburquerque, N., Carrillo, A. and García-Montiel, F. (2007). *Estimación de las necesidades de frío para florecer en variedades de cerezo. Fruticultura Profesional*, 164: 5–12.

Atkinson, C.J. and Lucas, A.S. (1996). The response of flowering time and cropping of *Pyrus communis* cv. conference to autumn warming. *Journal of Horticultural Science and Biotechnology*, 71: 427–434. Doi: org/10.1080/14620316.1996.11515423.

Beck, H.E., Zimmermann, N.E., McVicar, T.R., Vergopolan, N., Berg, A. and Wood, E.F. (2018). Present and future Köppen-Geiger climate classification maps at 1-km resolution. *Scientific Data*, 5: 180214. Doi: 10.1038/sdata.2018.214.

Bennett, J.P. (1994). Temperature and bud rest period. *California Agriculture*, 3: 9–12.

Cerovic, R. and Ruzic, D. (1992). Senescence of ovules at different temperatures and their effect on the behavior of pollen tubes in sour cherry. *Scientia Horticulturae*, 51: 321–327.

Cerović, R., Ružić, D. and Mićić, N. (2000). Viability of plum ovules at different temperatures. *Annals of Applied Biology*, 137: 53–59.

Cleland, E.E., Chuine, I., Menzel, A., Moonie, H.A. and Schwartz, M.D. (2007). Shifting plant phenology in response to global change. *Ecology and Evolution*, 22: 357–365.

Cosmulescu, S., Baciu, A., Cichi, M. and Gruia, M. (2010). The effect of climate changes on phenological phases in plum tree (*Prunus domestica* L.) in south-western Romania. *South Western Journal of Horticulture, Biology and Environment*, 1(1): 9–20.

DeCeault, M.T. and Polito, V.S. (2010). High temperatures during bloom can inhibit pollen germination and tube growth, and adversely affect fruit set in the *Prunus domestica* cultivars 'Improved French' and 'Muir Beauty'. *Acta Horticulturae*, 874: 163–168.

DeJong, T.M., DeBuse, C.J. and Bradley, S.J. (2008). Dried plum cultivar development and evaluation, California Dried Plum Board. *Research Reports*, 1–19.

Delph, L.F., Johannsson, H.M. and Stephenson, G.A. (1997). How environmental factors affect pollen performance: Ecological and evolutionary perspectives. *Ecology*, 78: 1632–1639.

Dennis, F.G. (1979). Factors affecting yield in apple with emphasis on 'Delicious'. *Horticultural Review*, 1: 395–422.

Dennis, F.G. (2003). Problems in standardizing methods for evaluating the chilling requirements for the breaking of dormancy in buds of woody plants. *HortScience*, 38: 347–350.

Døving, A. (2009). Modelling plum (*Prunus domestica*) yield in Norway. *European Journal of Horticultural Science*, 74: 254–259.

Eremin, T. and Zaremuk, R. (2014). Advance of assortment of plums in the south of Russia. *Russian Journal of Horticulture*, 1(1): 1–6.

FAOSTAT. (2019). Available at http://www.faostat.fao.

Gitea, M.A., Gitea, D., Tit, D.M., Purza, L., Samuel, A.D., Bungau, S., Badea, G.E. and Aleya, L. (2019). Orchard management under the effects of climate change: Implications for apple, plum, and almond growing. *Environmental Science and Pollution Research*, 26(10): 9908–9915. Doi: org/10.1007/s11356-019-04214-1.

Glišić, I. and Milošević, N. (2015). Evaluation of some autochthonous plum cultivars grown in Čačak region. *Journal of Mountain Agriculture on the Balkans*, 18(1): 148–161.

Glišić, S.I., Cerović, R., Milošević, N., Đorđević, M. and Radičević, S. (2012). Initial and final fruit set in some plum (*Prunus domestica* L.) hybrids under different pollination types. *Genetika*, 44(3): 583–593.

Gradziel, T.M. and Weinbaum, S.A. (1999). High relative humidity reduces anther dehiscence in apricot, peach and almond. *Hortscience*, 34: 322–325.

Grzyb, Z.S. and Rozpara, E. (2000). *Nowoczesna uprava śliw, Hortpress Sp. z.o.o.* Warszava, Poland (in Polish).

Hedhly, A., Hormaza, J.I. and Herrero, M. (2003). The effect of temperature on stigmatic receptivity in sweet cherry (*Prunus avium* L.). *Plant Cell & Environment*, 26: 1673–1680.

Hedhly, A., Hormaza, J.I. and Herrero, M. (2004). Effect of temperature on pollen tube kinetics and dynamics in sweet cherry, *Prunus avium* (Rosaceae). *American Journal of Botany*, 91: 558–564.

Hedhly, A., Hormaza, J.I. and Herrero, M. (2007). Warm temperatures at bloom reduce fruit set in sweet cherry. *Journal of Applied Botany and Food Quality*, 81: 158–164.

Hedhly, A., Hormaza, J.I. and Herrero, M. (2008). Global warming and sexual plant reproduction. *Trends in Plant Science*, 14: 30–36.

Heide, O.M. and Prestrud, A.K. (2005). Low temperature, but not photoperiod, controls growth cessation and dormancy induction in apple and pear. *Tree Physiology*, 25: 109–114. Doi: org/10.1093/treephys/25.1.09.

Heide, O.M. (2008). Interaction of photoperiod and temperature in the control of growth and dormancy of Prunus species. *Scientia Horticulturae*, 115: 309–314. Doi: org/10.1016/j.scienta.2007.10.005.

Heide, O.M. and Sønsteby, A. (2011). Physiology of flowering and dormancy regulation in annual- and biennial-fruiting red raspberry (*Rubus idaeus* L.)—a review. *The Journal of Horticultural Science and Biotechnology*, (86): 433–442. https://doi.org/10.1080/14620316.2011.11512785.

Hjalmarsson, I. (2019). Plum cultivars in Sweden: history and conservation for future use. Proceedings of the Latvian Academy of Sciences, Section B, *Natural, Exact, and Applied Sciences*, 73(3): 207–213. Doi: 73. 207-213. 10.2478/prolas-2019-0033.

Horvath, D. (2009). Common mechanisms regulate flowering and dormancy. *Plant Science*, 177: 523–531. Doi: 10.1016/j.plantsci.2009.09.002.

Ikase, L. (2015). Results of fruit breeding in Baltic and Nordic states. pp. 31–37. *In*: Zeverte-Rivza, S. (ed.). Proceedings of 25th Congress of the Nordic Association of Agricultural Scientists (NJF), *Nordic View to Sustainable Rural Development*, 16–18 June, 2015, Riga, Latvia.

IPCC, Climate Change. (2007). Mitigation of Climate Change, *Fourth Assessment Synthesis Report of Intergovernmental Panel on Climate Change*.

Kang, Y., Khan, S. and Xiaoyi, M. (2009). Climate change impacts on crop yield, crop water productivity and food security—A review. *Progress in Natural Science*, 19: 1665–1674.

Kaufmane, E., Gravite, I. and Trajkovski, V. (2012). Results of Latvian plum breeding programme. *Acta Horticulturae*, 968: 55–60.

Kaufmane, E., Grāvīte, I. and Ikase, L. (2019). Plum Research and Growing in Latvia, Proceedings of the Latvian Academy of Sciences, Section B, *Natural, Exact, and Applied Sciences*, 73(3): 195–206. Doi: 73. 195-206. 10.2478/prolas-2019-0032.

Kunz, A. and Blanke, M.M. (2016). Effects of climate change on fruit tree physiology, based on 55 years of meteorological and phenological data at Klein-Altendorf. *Acta Horticultura*, 1130: 49–54. Doi: 10.17660/ActaHortic.2016.1130.7.

Lal, R. (2005). Climate change, soil carbon dynamics and global food security. pp. 113–143. *In*: Lal, R., Stewart B., Uphoff, N. et al. (eds.). *Climate Change and Global Food Security*. CRC Press, Boca Raton (FL).

Lalić, B. and Mihailović, D.T. (2011). Impact of climate change on food production in northern Serbia (Vojvodina). pp. 67–85. *In*: Ivanyi, Z. (ed.). *The Impact of Climate Change on Food Production in the Western Balkan Region*. Regional Environmental Center for Central and Eastern Europe (REC). Available at: http://documents.rec.org/topic-areas/Impacts-climage-change-food-production.pdf.

Lalić, B., Janković, D., Jančić, M., Ejcinger, J. and Firanj, A. (2015). Heating of crops—How to respond? Effects of climate change on Serbian agriculture. Project *'Preparation of the Second Report of the Republic of Serbia under the United Nations Framework Convention on Climate Change'*. United Nations Development Program (UNDP) in Serbia, the Global Environment Facility (GEF). Available at: http://www.klimatskepromene.rs/wp-content.

Luedeling, E., Girvetz, E.H., Semenov, M.A. and Brown, P.H. (2011). Climate change affects winter chill for temperate fruit and nut trees. *PLoS ONE*, 6(5): e20155. Doi: 10.1371/journal.pone.0020155.

Lukić, M., Pešaković, M., Marić, S., Glišić, I., Milošević, N., Radičević, S., Leposavić, A., Đorđević, M., Miletić, R., Karaklajić-Stajić, Ž., Tomić, J., Paunović, S.M., Milinković, M., Ružić, Đ., Vujović, T., Jevremović, D., Paunović, S.A., Popović, B., Mitrović, O. and Kandić, M. (2016). *Fruit Cultivars Developed at the Fruit Research Institute*. Čačak, Чачак (1946–2016), Fruit Research Institute, Čačak, Republic of Serbia, 1–182.

Mathur, P.N., Ramirez-Villegas, J. and Jarvis, A. (2012). The impacts of climate change on tropical and subtropical horticultural production. pp. 27–44. *In:* Sthapit, B.R., Ramanatha Rao, V. and Sthapit, S.R. (eds.). *Tropical Fruit Tree Species and Climate Change*. New Delhi, India: Biodiversity International.

Matveev, V.A. and Vasileva, M.N. (2015). Breeding characteristics of plum seedlings of hybrid family Narach × Oda on the complex of economic characters. *Plodovodstvo*, 27: 87–92 (in Belarusian with English summary).

Meland, M., Frøynes, O. and Maas, F. (2019). Performance of dwarfing and semi-dwarfing plum rootstocks on three European plum scion cultivars in a Nordic climate. *Acta Horticulturae*, 1260: 181–186. Doi: org/10.17660/ActaHortic.2019.1260.28.

Menzel, A., Estrella, N. and Schleip, C. (2007). Impacts of climate variability, trends and NAO on 20th century european plant phenology. pp. 221–233. *In*: Brönnimann, S. et al. (eds.). *Advances in Global Change Research, Climate Variability and Extremes during the Past 100 Years*. Springer, The Netherlands.

Milatović, D. (2019). *Šljiva, Naučno voćarsko društvo Srbije*, 1–531 (in Serbian).

Milošević, N., Mratinić, E., Glišić, I. and Milošević, T. (2012). Precocity, yield and postharvest physical and chemical properties of plums resistant to Sharka grown in Serbian conditions. *Acta Scientiarum, Polonorum, Hortorum Cultus*, 11(6): 23–33.

Milošević, N. (2013). *Degree of Fertilization and Biological Traits of New Plum Cultivars (Prunus domestica* L.). doctoral dissertation, pp. 1–169 (in Serbian).

Milošević, N., Glišić, I., Lukić, M., Đorđević, M. and Karaklajić-Stajić, Ž. (2016). Properties of some late season plum hybrids from Fruit Research Institute, Čačak. *Agriculturae Conspectus Scientificus*, 81(2): 65–70.

Milošević, N., Glišić, I., Popović, B. and Mitrović, O. (2017). Productive traits of new cultivar 'Nada' grown on three localities in Serbia. *Journal of Mountain Agriculture on the Balkans*, 20(5): 197–207.

Milošević, N., Glišić, I., Lukić, M., Popović, B. and Đorđević, M. (2019). Plum breeding in the Fruit Research Institute, Čačak, Serbia - Results in the last fifteen years. *Acta Horticulturae*, 1260: 29–34. Doi: 10.17660/ActaHortic.2019.1260.6.

Milošević, T. (1997). *Specijalno voćarstvo. Agronomski fakultet Čačak* (in Serbian).

Milošević, T. (2002). *Šljiva – Tehnologija gajenja, Agronomski fakultet Čačak* (in Serbian).

Milošević, T. and Milošević, N. (2012). Phenotypic diversity of autochthonous European (*Prunus domestica* L.) and Damson (*Prunus insititia* L.) plum accessions based on multivariate analysis. *Horticultural Science*, 39(1): 8–20.

Milošević, T. and Milošević, N. (2018). Plum (*Prunus* spp.) Breeding. pp. 165–215. *In*: Al-Khayri, J.M., Jain, M.S. and Johnson, D.V. (eds.). *Advances in Plant Breeding Strategies: Fruits*. vol. 3, Springer International Publishing AG, part of Springer Nature 2018. Doi: 10.1007/978-3-319-91944-7_5.

Mišić, P. (2006). *Šljiva. Partenon, Institut za istraživanja u poljoprivredi 'Srbija'*, Beograd (in Serbian).

Montagnon, J.M. (2007). *Las ciruelas japonesas. Elección de las variedades polinizadoras. Fruticultura Profesional*, 164: 25–32.

Neumüller, M. (2011). Fundamental and applied aspects of plum (*Prunus domestica*) breeding. *Fruit, Vegetable and Cereal Science and Biotechnology*, 5(special issue 1): 139–156.

Ogašanović, D. and Ranković, M. (1996). *Važne karakteristike nekih hibrida šljive otpornih na šarku šljive. Jugoslovensko voćarstvo*, 30(113-114): 117–122 (in Serbian).

Ogašanović, D., Plazinić, R., Rankovic, M., Stamenković, S. and Milinković, V. (2007). Pomological characteristics of new plum cultivars developed in Čačak. *Acta Horticulturae*, 734: 183–186. Doi: 10.17660/ActaHortic.2007.734.22.

Ogundeyi, A.A. and Jordaan, H.A. (2017). Simulation study on the effect of climate change on crop water use and chill unit accumulation. *South African Journal of Science*, 113(7/8): Art. #2016-0119. Doi: 10.17159/sajs.2017/20160119.

Ohata, K., Togano, Y., Matsumoto, T., Uchida, Y., Kurahashi, T. and Itamura, H. (2017). Selection of Prune (*Prunus domestica* L.) cultivars suitable for the East Asian temperate monsoon climate: Ripening characteristics and fruit qualities of certain prunes in a warm southwest region of Japan. *The Horticulture Journal*, 86. Doi: 10.2503/hortj.OKD-044.

Okie, W.R. and Ramming, D.W. (1999). Plum breeding worldwide. *HortTechnology*, 9(2): 162–176.

Parmesan, C. and Yohe, G. (2003). A globally coherent fingerprint of climate change impacts across natural systems. *Nature*, 421: 37–42. Doi: 10.1038/nature01286.

Paunovic, S. (1988). Plum genotypes and their improvement in Yugoslavia. *Fruit Varieties Journal*, 42: 143–151.

Pirlak, L. (2002). The effects of temperature on pollen germination and pollen tube growth of apricot and sweet cherry. *Gartenbauwissenschaf*, 67: 61–64.

Pudas, E., Tolvanen, A., Poikolainen, J., Sukuvaara, T. and Kubin, E. (2008). Timing of plant phenophases in Finnish Lapland in 1997–2006. *Boreal Environment Research*, 13: 31–43.

Radičević, S., Cerović, R., Nikolić, D. and Đorđević, M. (2016). The effect of genotype and temperature on pollen tube growth and fertilization in sweet cherry (*Prunus avium* L.). *Euphytica*, 209: 121–136.

Radinović, Đ. (1979). Weather and Climate of Yugoslavia. *Civil Engineering Book*, Belgrade, pp. 283.

Rai, R., Joshi, S., Roy, S., Singh, O., Samir, M. and Chandra, A. (2015). Implications of changing climate on productivity of temperate fruit crops with special reference to apple. *Journal of Horticulture*, 2: 135. Doi:10.4172/2376-0354.1000135.

Rajatiya, J., Varu, D.K., Gohil, P., Solanki, M., Halepotara, F., Gohil, M., Mishra, P. and Solanki, R. (2018). Climate change: impact, mitigation and adaptation in fruit crops. *International Journal of Pure and Applied Bioscience*, 6(1): 1161–1169. Doi: 10.18782/2320-7051.6161.

Ramming, D.W. and Cociu, V. (1991). Plums (*Prunus*). *Acta Horticulturae*, 290: 235–290.

Rehman, M.U., Rather, G.H., Gull, Y., Mir, M.R., Mir, M.M., Waida, U.I. and Hakeem K.R. (2015). Effect of climate change on horticultural crops. pp. 211–239. *In*: *Crop Production and Global Environmental Issues*. Springer International Publishing, Switzerland. Doi: 10.1007/978-3-319-23162-4_9.

Rodrigo, J. (2000). Review: spring frost in deciduous fruit trees-morphological damage and flower hardiness. *Scientia Horticulturae*, 83: 155–173.

Rodrigo, J. and Herrero, M. (2002). Effects of pre-blossom temperatures on flower development and fruit set in apricot. *Scientia Horticulturae*, 31: 125–135.

Rötzer, T. and Chmielewski, F.M. (2001). Phenological maps of Europe. *Climate Research*, 18(3): 249–257.

Ruiz, D., Campoy, J.A. and Egea, J. (2007). Chilling and heat requirements of apricot cultivars for flowering. *Environmental and Experimental Botany*, 61: 254–263. Doi: 10.1016/j. envexpbot. 2007.06.008.

Schwartz, M.D. (2003). *Phenology: An Integrative Environmental Science*. Kluwer Academic Publishers, The Netherlands.

Simonov, V.S. and Kulemekov, S.N. (2017). Modern assortment of the plum and ways to increase its adaptivity in central region. *Proceedings of Science Works GNBS*, 144(1): 143–150 (in Russian).

Singh, R., van Dam, J.C. and Feddes, R.A. (2006). Water productivity analysis of irrigated crops in Sirsa district, India. *Agricultural Water Management*, 82: 253–278.

Snyder, R.L. and Melo-Abreu, J.P. (2005). Frost Protection: Fundamentals, practice and economics. *FAO Environment and Natural Resources Service Series*, No. 10, FAO, Rome, 2005, vol. 1. Available at http://www.fao.org/3/a-y7223e.pdf.

Sønsteby, A. and Heide, O.M. (2011). Elevated autumn temperature promotes growth cessation and flower formation in black currant cultivars (*Ribes nigrum* L.). *The Journal of Horticultural Science and Biotechnology*, (86): 120–127. https://doi.org/10.1080/14620316.2011.11512736.

Sparks, T.H. and Menzel, A. (2002). Observed changes in seasons: An overview. *International Journal of Climatology*, 22(14): 1715–1726.

Stanković, D. and Jovanović, M. (1990). *Opšte voćarstvo, Naučna knjiga*, Beograd (in Serbian).

Subedi, S. (2019). Climate change effects of Nepalese fruit production. *Advances in Plants and Agriculture Research*, 9(1): 141–145. Doi: 10.15406/apar.2019.09.00426.

Szabó, Z. (2003). Frost injuries of the reproductive organs in fruit species. pp. 59–74. *In*: Kozma, P., Nyéki, J., Soltész, M. and Szabó, T. (eds.). *Floral Biology, Pollination and Fertilization in Temperate Zone Fruit Species and Grapes, Akadémiai Kiadó*. Budapest, Hungary.

Van Tuyl, J. (1983). Effect of temperature treatments on the scale propagation of Lilium longiflorum 'White Europe' and Lilium x 'Enchantment'. *HortScience*, 18: 754–756.

Vuković, J.A., Vujadinović, P.M., Rendulić, M.S., Djurdjević, S.V., Ruml, M.M., Babić, P.V. and Popović, P.D. (2018). Global warming impact on climate change in Serbia for the period 1961–2100. *Thermal Science*, 22(6A): 2267–2280.

Wolfe, D., Schwartz, M., Lakso, A., Otsuki, Y., Pool, R. and Shaulis, N. (2005). Climate change and shifts in spring phenology of three horticultural woody perennials in northeastern USA. *International Journal of Biometeorology*, 49(5): 303–309. Doi: 10.1007/s00484-004-0248-9.

Woznicki, T.L., Heide, O.M., Sønsteby, A., Mage, F. and Remberg, S.F. (2019). Climate warming enhances flower formation, earliness of blooming and fruit size in plum

(*Prunus domestica* L.) in the cool Nordic environment. *Scientia Horticulturae*, 257: 108750. Doi: org/10.1016/j.scienta.2019.108750.

Xiloyannis, C., Dichio, B., Tuzio, A.C., Kleinhentz, M., Salesses, G., Gomez-Aparisi, J., Rubio-Cabetas, M.J. and Esmenjaud, D. (2007). Characterization and selection of prunus rootstocks resistant to abiotic stresses: Waterlogging, drought and iron chlorosis. *Acta Horticulturae*, 732: 247–251. Doi: 10.17660/ActaHortic.2007.732.35.

Zwart, S.J. and Bastiaanssen, W.G.M. (2004). Review of measured crop water productivity values for irrigated wheat, rice, cotton and maize. *Agricultural Water Management*, 69: 115–133.

Index

D

E

F

G

Printed in the United States
by Baker & Taylor Publisher Services